Lecture Notes in Physics

Edited by H. Araki, Kyoto, J. Ehlers, München, K. Hepp, Zürich
R. Kippenhahn, München, D. Ruelle, Bures-sur-Yvette
H. A. Weidenmüller, Heidelberg, J. Wess, Karlsruhe and J. Zittartz, Köln
Managing Editor: W. Beiglböck

319

Luis Garrido (Ed.)

Far from Equilibrium Phase Transitions

Proceedings of the Xth Sitges Conference
on Statistical Mechanics,
Sitges, Barcelona, Spain, June 6–10, 1988

Springer-Verlag
Berlin Heidelberg GmbH

Editor

Luis Garrido
Facultad de Física, Departamento de Física Fundamental
Universidad de Barcelona
Diagonal 647, E-08028 Barcelona, Spain

ISBN 978-3-662-13690-4 ISBN 978-3-540-46060-2 (eBook)
DOI 10.1007/978-3-540-46060-2

Originally published by Springer-Verlag Berlin Heidelberg New York in 1988
Softcover reprint of the hardcover 1st edition 1988

2158/3140-543210 – Printed on acid-free paper

ACKNOWLEDGEMENTS

I would like to take this opportunity to express my sincere thanks to all those who collaborated in the organization of this conference.

Also, I wish to extend my warmest thanks to the Generalitat de Catalunya , to the Consejo Superior de Investigaciones Científicas and the Ministry of Education for their economic support. To the city of Sitges I express my gratitude for allowing us again to use the Museum "Maricel" as a lecture hall.

Finally, I wish to thank my wife for her unremitting cooperation.

L. Garrido

CONTENTS

A REVIEW OF CURRENT ISSUES IN THE QUANTUM THEORY OF ENVELOPE SOLITONS

Katja Lindenberg
Department of Chemistry, B-040 and Institute for Nonlinear Science, R-002
University of California at San Diego, La Jolla, CA 92093 U.S.A.

David Brown
Institute for Nonlinear Science, R-002
University of California at San Diego, La Jolla, CA 92093 U.S.A.

Xidi Wang
Department of Physics, B-019
University of California at San Diego, La Jolla, CA 93093 U.S.A.

1. Introduction

The concept of the molecular soliton as a mechanism for the transport of energy in biological systems has received a great deal of attention in recent years. Such a mechanism is attractive because it offers a transport channel that minimizes both energy delocalization through dispersion and energy loss through dissipation. The idea of a molecular soliton was first introduced in the mid 1970's by Davydov and Kisluka [1] to describe a single electronic or vibronic excitation (which we shall call an "exciton") propagating along a deformable molecular chain [2-7]. The deformability of the molecular chain affects the dynamics of the mobile excitation through the dependence of the excitation energies on the configuration of the chain.

The theory of molecular solitons depends on the nonlinear interaction of sets of excitations, and on the condition that the resulting nonlinear equations of motion possess solitary-wave solutions for meaningful values of the model parameters. The two sets of excitations considered in this context are often the acoustic vibrations of the chain and an intramolecular vibration or electronic excitation. For example, a physical system often considered in bioenergetics is the polypeptide backbone of the α helix. In this system, the amide-I vibration ($C=O$ stretch) and the acoustic vibrations of the polypeptide chain constitute two sets of modes whose excitations interact to induce nonlinear behavior [8,10].

The theory of Davydov is developed for systems described by the Fröhlich Hamiltonian familiar from the theory of polarons in condensed matter [11,12]:

$$H = \sum_n E a_n^+ a_n - \sum_{m,n} J_{mn}\, a_m^+ a_n + \sum_q \hbar\omega_q b_q^+ b_q + \sum_{q,n} \chi_n^q \hbar\omega_q (b_q^+ + b_{-q}) a_n^+ a_n \quad . \tag{1.1}$$

Here a_n^+ and a_n are, respectively, the creation and annihilation operators of the exciton on the nth molecule; b_q^+ and b_q respectively create and annihilate a phonon in normal mode q. Both the exciton and phonon operators are taken to satisfy Bose commutation relations. The coupling coefficients χ_n^q in a translationally invariant system are of the form $\chi_n^q = \chi^q \exp(-iqR_n)$ where R_n is the equilibrium position of the nth molecule. The dependence of χ^q and of the phonon frequencies ω_q on the phonon mode index q varies from system to system. The parameter J_{mn} is responsible for the motion of the exciton among the molecules and may arise from dipole, exchange, or purely mechanical interactions.

The Fröhlich Hamiltonian can be diagonalized exactly only in the limit $J_{mn} = 0$, where there is no intermolecular transport of excitations. The canonical transformation that diagonalizes the problem in this limit gives precise meaning to the concept of the polaron, which has been central to one avenue of treatment of the Fröhlich Hamiltonian even when J_{mn} is nonzero [12-14]. The transportless problem is by no means trivial since it contains the dynamics of exciton-phonon interactions and can be used as a benchmark against which to assess the transportless limit of approximate theories developed for $J_{mn} > 0$ [15,16].

Since the Hamiltonian (1.1) can not in general be exactly diagonalized, a number of theories have emerged that attempt an approximate solution. The ones of interest to us are those that lead to possible solitonic solutions for the amplitude of excitation. Most of these latter theories have been developed for zero temperature [1-10,17-19]; some theoretical work as well as numerical simulations exist for finite temperatures [17,20,21]. The approximate nature of these treatments has elicited a certain amount of discussion as to the merits of the assumptions that must be made; herein we deal with some of these issues.

If one accepts one or the other of the approaches that have been suggested to deal with the Hamiltonian (1.1) then one must study its consequences. This is the second portion of our task: we consider one of the proposed (approximate) solutions to the problem and explore the physical predictions that it leads to. This analysis tells us a great deal about parameter regimes for solitary wave behavior. Most of this analysis is carried out for zero temperature, but we make a few comments about ways to extend the analysis to finite temperatures.

The Fröhlich Hamiltonian (1.1) is seen to conserve the number of excitons since each of its exciton-dependent terms is of the form a^+a. The conservation of number is reasonable if one is dealing with electronic excitations, but it is inappropriate if the excitations are vibrational. In the latter case the Fröhlich Hamiltonian must be viewed as a rotating wave approximation of a more general form that does not conserve excitation number (but that has other appropriate physical characteristics). We complete this chapter with a short discussion of the Takeno Hamiltonian and briefly indicate some of its important physical consequences [22].

We end this introductory section by providing a simple qualitative picture of the spontaneous localization of vibrational energy [23]. Consider, for example, a translationally invariant aggregate (polymer or crystal) having several optical phonon bands and the usual acoustic phonons; let us concentrate on the lowest optical phonon band. Our picture requires that for this branch $d^2\omega/dk^2 > 0$ near $k=0$, *i.e.* that the frequency of long-wavelength phonons is lower than that of short-wavelength phonons. We also assume that this optical mode shares with the simple mechanical pendulum the property that oscillation frequencies decrease with increasing amplitude (due to nonlinear processes at the moment undetermined). Suppose that one can excite long-wavelength, finite-amplitude oscillations belonging to this optical band in a small region of the solid. The dispersion curve in the excited region would then be depressed in frequency. If the depressed dispersion curve lies within the optical-acoustic band gap, then we have created an excitation whose central frequency lies below the lowest optical propagating frequency and above the highest acoustic propagating frequency in the surrounding unperturbed regions of the solid. If the harmonics of the frequency of the nonlinear excitation also do not overlap with higher optical bands, then the only decay channels for this excitation are nonpropagating evanescent waves. This scenario then leads to a localized vibrational state, perhaps a new *nonlinear* elementary excitation of the solid. Note that optical modes other than the lowest one may also lead to localization of vibrational energy, but higher modes have more decay paths and hence the likelihood of a long-lived local excitation decreases.

The essential element in this qualitative argument is the nonlinear process which is assumed to result in a red-shifting of vibrational frequencies. As mentioned above, the nonlinear effects can arise either from anharmonicities of individual bonds or from the nonlinear coupling of different vibrational fields. Thus, for example, in liquid benzene the apparent localization of vibrational energy in a single C–H bond can be adequately explained on the basis of the anharmonicity of the C–H stretch oscillation: the energy of the molecule can be lowered by concentrating vibrational energy in one bond instead of having it shared by all six bonds [24-26]. On the other hand, spectral shifts in acetanilide have been ascribed to soliton mechanisms arising from the anharmonic coupling of vibrational fields (acoustic and optical modes) that are relatively weakly excited [9,10,27].

2. The Davydov Ansätze

The Hamiltonian (1.1), as well as being the point of departure for the construction of solitary wave solutions, has an even older history as the basis of most studies of polaron transport. Polaron transport theory makes extensive use of the canonical transformation [11,12,14,28]

$$U = \exp\left[-\sum_{q,m} (\chi_m^q b_q^+ - \chi_m^{q*} b_q) a_m^+ a_m \right] \tag{2.1}$$

which diagonalizes (1.1) in the limit $J_{mn} = 0$. The diagonalized Hamiltonian

$$H = \left[E - \sum_q |\chi^q|^2 \hbar\omega_q \right] \sum_m A_m^+ A_m + \sum_q \hbar\omega_q B_q^+ B_q \tag{2.2}$$

involves the "dressed" operators

$$A_m = U a_m U^+ = a_m \exp\left[\sum_q (\chi_m^q b_q^+ - \chi_m^{q*} b_q) \right] , \tag{2.3a}$$

$$B_q = U b_q U^+ = b_q + \sum_m \chi_m^q a_m^+ a_m \tag{2.3b}$$

that represent the new excitations. In writing (2.2) we have omitted a two-body term as irrelevant when only one excitation is present.

We use the term "polaron" to indicate the particle created by A_m^+. A single-polaron state has the form

$$|\psi(t)\rangle = \sum_m \psi_m(t) A_m^+ |0\rangle$$

$$= \sum_m \psi_m(t) a_m^+ \exp\left[-\sum_q (\chi_m^q b_q^+ - \chi_m^{q*} b_q) \right] |0\rangle \tag{2.4}$$

where $|0\rangle$ is the vacuum for both phonons and excitons. The polaron thus consists of an exciton accompanied by a "phonon cloud" organized around the exciton in a specified way determined by the coupling constants. The state (2.4) is an exact solution of the Schrödinger equation only when $J_{mn} = 0$. Otherwise, polaron states are used as a basis set for a perturbative treatment of the "residual" J-dependent terms that appear added to (2.2).

Davydov and co-workers constructed a different kind of state for the Hamiltonian (1.1): one in which the "phonon cloud" surrounding the exciton can itself respond dynamically to the evolution of the excitation [1-7]. Davydov actually proposed two such Ansatz states, which we [29] have called $|D_1(t)\rangle$ and $|D_2(t)\rangle$, given by

$$|D_1(t)\rangle = \sum_m \alpha_m(t) a_m^+ \exp\left[-\sum_q (\beta_{qm}(t)\, b_q^+ - \beta_{qm}^*(t) b_q) \right] |0\rangle \tag{2.5}$$

$$|D_2(t)\rangle = \sum_m \alpha_m(t) a_m^+ \exp\left[-\sum_q (\beta_q(t) b_q^+ - \beta_q^*(t) b_q) \right] |0\rangle \; . \tag{2.6}$$

In these superpositions of products of single exciton states and coherent phonon states the dynamic response of the medium is contained in the $\beta(t)$'s whose time dependence must be determined and is in general expected to depend on that of the $\alpha(t)$'s (and vice versa). In Davydov's original work the $\alpha(t)$'s and $\beta(t)$'s are treated as generalized coordinates with corresponding generalized momenta $i\hbar\alpha^*(t)$ and $i\hbar\beta^*(t)$. The equations of motion for these generalized dynamical variables are taken to be the classical Hamilton equations in which the expectation value of the quantum Hamiltonian appears as the Hamilton function. For instance, for the state $|D_2(t)\rangle$ following this procedure and setting $\langle H\rangle \equiv \langle D_2(t)|H|D_2(t)\rangle$ one obtains

$$i\hbar\dot{\alpha}_n(t) = \frac{\partial}{\partial\alpha_n^*(t)} \langle H\rangle = E\,\alpha_n(t) - J\,[\alpha_{n+1}(t) + \alpha_{n-1}(t)]$$

$$+ \sum_q \chi_n^q \omega_q\, [\beta_q^*(t) + \beta_{-q}(t)]\alpha_n(t) \tag{2.7a}$$

and

$$i\hbar\dot{\beta}_q(t) = \frac{\partial}{\partial\beta_q^*(t)} \langle H\rangle = \hbar\omega_q \beta_q(t) + \hbar\omega_q \sum_m \chi_m^q\, |\alpha_m(t)|^2 \; , \tag{2.7b}$$

where we have assumed only nearest neighbor transport, *i.e.* $J_{mn} = J\,\delta_{m,n\pm1}$. Equations (2.7a) and (2.7b) constitute the "Davydov system of equations". Elimination of β_q and β_q^* from Eq. (2.7a) by explicit use of (2.7b) together with subsequent approximations [including a continuum limit in which $\alpha_n(t)$ is replaced with an amplitude $\alpha(x,t)$] finally leads to the nonlinear Schrödinger (NLS) equation

$$i\hbar \frac{\partial}{\partial t} \alpha(x,t) = -\frac{\hbar^2}{2m} \frac{\partial^2}{\partial x^2} \alpha(x,t) + E(0)\alpha(x,t) - G_\upsilon |\alpha(x,t)|^2 \alpha(x,t) \; . \tag{2.8}$$

Here $E(0) = E - 2J$ is the bottom of the bare exciton band with an associated effective mass m. In writing (2.8) from (2.7) a solution of the D'Alembert form depending on x and t only through the combination $(x - \upsilon t)$ has been assumed, and the constant G_υ depends on the speed υ and the sound speed υ_a (the latter having entered via ω_q). In writing (2.8) the lattice has been assumed to be a linear chain with acoustic phonons of frequency ω_q; in this case the coupling function χ_n^q is given by

$$\chi_n^q = -\frac{2i\,\chi \sin(ql)}{(2NM\hbar\omega_q{}^3)^{1/2}}\, e^{-iqR_n} \tag{2.9}$$

where l is the lattice constant, M is the molecular mass and R_n is the equilibrium position of the nth molecule. In terms of these various parameters and the lattice stiffness coefficient w the nonlinearity parameter G_υ in (2.8) is

$$G_\upsilon = \frac{4\chi^2 l}{w}\,\frac{\upsilon_a^2}{\upsilon_a^2 - \upsilon^2}\ . \tag{2.10}$$

The NLS equation (2.8) has, among others, the well-known soliton solution

$$\alpha(x,t) = \left(\frac{\kappa}{2}\right)^{1/2} \frac{e^{-i(kx - \omega t)}\, e^{-i[E(0)/\hbar]t}}{\cosh[\kappa(x - \upsilon t)]} \tag{2.11}$$

where

$$\hbar k \equiv -m\upsilon \tag{2.12a}$$

$$\hbar\kappa \equiv \frac{mG_\upsilon}{2\hbar} \tag{2.12b}$$

$$\hbar\omega \equiv \frac{mG_\upsilon^2}{8\hbar^2} - \tfrac{1}{2} m\upsilon^2 \tag{2.12c}$$

and (2.11) is normalized to unity.

The conclusion drawn from this analysis is that the Fröhlich Hamiltonian can support soliton transport, at least at T=0. [Note the absence of T from (2.8) but its potential presence in the system (2.7) via phonon initial conditions.] Numerical simulations of the full system (2.7) have confirmed soliton-like structures at T=0 whose stability with increasing T is still a point of argument in the literature [17,20,21].

One criticism leveled at the derivation of the Davydov system of equations (2.7) is the use of Hamilton's equations. Kerr and Lomdahl have presented an alternative derivation of (2.7) starting with an Ansatz state of the form $|D_2(t)\rangle$ and using strictly quantum-mechanical manipulations [18]. The two methods result in Ansatz state vectors which differ by a time-dependent global phase and thus lead to the same expectation values for physical observables but to different time correlation functions.

A similar analysis starting with the Ansatz state $|D_1(t)\rangle$ is also possible but considerably more complicated [30]. For this reason, most of the literature has concentrated on the state $|D_2(t)\rangle$.

At this point one should stop and formulate some questions that one might ask in the above analysis. These questions occur at various levels. First, one should consider whether the Ansatz states are in fact solutions of the Schrödinger equation for the Fröhlich Hamiltonian under any circumstances.

If they are, then one has solved the problem exactly; if they are not, then one should consider how well they approximate the actual dynamics. Second, if one accepts the Ansatz states as a starting point, then one can explore its consequences more extensively and carefully than has been done in going from Eq. (2.7) to (2.8). In particular, there are interesting questions concerning initial preparation of the system, the effects of the discreteness of the lattice, and temperature that one can address at the level of the Davydov system of equations. We address these issues and others in this and subsequent sections.

The first question that we ask concerns the Ansatz states themselves: is one or the other a solution of the Schrödinger equation for the Fröhlich Hamiltonian (1.1)? The relevance of this question has not always been appreciated: one encounters the misconception that equations of motion such as (2.7) that arise from the Ansatz state exactly represent the evolution of the system provided one arrives at them via correct quantum mechanical manipulations [31]. This is of course not correct [32]: to represent this evolution correctly $|D(t)\rangle$ must first of all be a solution of the Schrödinger equation

$$i\hbar \frac{\partial}{\partial t} |D(t)\rangle = H\,|D(t)\rangle \ . \tag{2.13}$$

Since we only know the exact solution of this equation with the Hamiltonian (1.1) when $J_{mn} = 0$, it is only in this limit that we can arrive at precise conclusions [15,16,29]. As we shall see, even this restricted limit provides valuable insight as to the validity of the Ansatz states.

Let us first consider the Ansatz state $|D_1(t)\rangle$. The equations of motion that have previously been derived from this state using *Hamilton's equations* are more complicated than (2.7) and have been analyzed by us in detail in two ways, both of which show that the state $|D_1(t)\rangle$ constructed from these equations of motion does not solve equation (2.13). First, we have shown that $|D_1(t)\rangle \neq \exp(-iHt)|D_1(0)\rangle$, *i.e.* the Hamiltonian (1.1) does not evolve $|D_1(0)\rangle$ into $|D_1(t)\rangle$ [33]. This conclusion can be shown to be true whether or not J_{mn} vanishes. When $J_{mn} = 0$ we know the explicit form of $\exp(-iHt)|D_1(0)\rangle$ and can therefore carry out an explicit comparison. For this purpose we have calculated a number of explicit diagnostic tests at zero temperature [29]:

a. **Stationary States.** The stationary states obtained from the D_1-Hamilton equations are precisely of the one-polaron eigenstate form (2.4), *i.e.* the D_1-Hamilton dynamics yields the same stationary states as does the exact dynamics.

b. **Lattice deformalibility.** The lattice deformation around an initially bare excitation evolves towards the correct value at long times but approaches this state on an incorrect time scale (too slow) unless the excitation is localized on a single site.

c. **Transfer of energy to the phonon bath.** The expectation value of the energy of the lattice starting from a bare excitation initial condition evolves towards the correct value at long times, but again does so on a time scale that is too slow unless the excitation is localized on a single site.

d. **Overlap of exact and approximate states.** The overlap of the D_1-Hamilton state and the exact state both starting from a bare exciton initial condition (initial unit overlap) decays to a value that deviates from unity for even the smallest deviations from complete localization of the excitation.

e. **Spectra.** The D_1-Hamilton dynamics leads to time-dependent changes in global phases which lead to spectral errors not correctable or predictable with this scheme. Spectra can therefore not be predicted within this scheme.

One must conclude that the dynamics obtained from the $|D_1(t)\rangle$ Ansatz state with parameters obtained via the Hamilton equation approach do not lead to the correct evolution of the excitation when $J_{mn} = 0$. In the extreme case that the excitation is completely localized on one site, the D_1-Hamilton dynamics predicts certain properties correctly but still leads to incorrect spectra. It is expected that these deviations from the correct behavior will persist when $J_{mn} \neq 0$.

Let us now turn to the dynamics for the $|D_2(t)\rangle$ Ansatz state (2.6) together with Hamilton's equations as reflected in the equations of motion (2.7). As with $|D_1\rangle$, one can show that $|D_2(t)\rangle \neq \exp(-iHt)|D_2(0)\rangle$, *i.e.* the Hamiltonian (1.1) does not evolve $|D_2(0)\rangle$ into $|D_2(t)\rangle$ for any value of the J_{mn} [33]. As for the explicit diagnostics when $J_{mn} = 0$ [29]:

a. **Stationary states.** The stationary states obtained from (2.7) lead to $|D_2(t)\rangle$*'s* that are of the correct one-polaron from only when the excitation is localized on a single site.

b. **Lattice deformation.** The lattice deformation around an initially bare excitation is correctly predicted by the D_2-dynamics at all times.

c. **Transfer of energy to the phonon bath.** The expectation value of the energy of the lattice starting from a bare excitation initial condition evolves on the correct time scale but towards an incorrect value. The final amount of energy in the lattice is only reproduced correctly when the excitation is localized on a single site.

d. **Overlap of exact and approximate states.** As with D_1, the D_2-state overlap with the exact state starting from a bare initial condition decays to a value that is less than unity unless the excitation is localized on one site.

e. **Spectra.** D_2-spectra differ from the correct one in ways that are neither predictable nor correctable within this scheme.

One must therefore conclude that $|D_2(t)\rangle$ with parameters determined via (2.7) does not describe the correct evolution of the system when $J_{mn} = 0$ and is likely to contain errors when $J_{mn} \neq 0$ as well.

In view of these negative conclusions about the D_1- and D_2-dynamics, should one conclude that the Ansatz approach should be thrown out altogether? The answer is that it would be premature to do so for a number of reasons. First, it is possible that, although imperfect, predictions provided by the Ansätze are nevertheless extremely useful: after all one does not discard good approximate solutions simply because they are not exact. Second, one must question whether the problems lie in the Ansätze themselves or in the subsequent application of Hamilton's equations to them (as opposed to purely quantum mechanical manipulations). We have stated earlier that Eqs. (2.7) *can* in fact be obtained via purely quantum mechanical operations once one invokes the $|D_2\rangle$ state [18]; from this one must conclude that problems with the D_2-dynamics are inherent in the Ansatz itself and must arise at least in part from the omission of any exciton-phonon phase mixing, as implied by the product form of the state. The situation with the D_1-dynamics is somewhat different: here the problems exhibited above are indeed partly a consequence of the application of Hamilton's equations. It turns out that the $|D_1>$ Ansatz state with parameters determined by entirely quantum mechanical manipulations yields the *correct* solution of the Schrödinger equation for $J_{mn}=0$ (but *not* for $J_{mn}\neq 0$) [34-36].

Recently Zhang *et al.* [35] and Škrinjar *et al.* [36] applied the variational principle of time-dependent quantum mechanics,

$$\delta\int_{t_1}^{t_2} dt \left\{ <D_1(t)|i\hbar\frac{d}{dt}-H|D_1(t)> - \lambda <D_1(t)|D_1(t)> \right\} = 0 , \tag{2.14}$$

to the $|D_1(t)>$ Ansatz state and they obtained equations of motion that are indeed able to reproduce the *exact* $J_{mn}=0$ evolution of the system and hence all known exact results. The resulting equations of evolution (not shown here) are considerably more complicated than those obtained from Hamilton's equations. To gain some insight into the more general $J_{mn}\neq 0$ case, we proceed somewhat differently [34]. Let us begin by substituting $|D_1>$ into both sides of the Schrödinger equation (2.13) and let us obtain the necessary conditions for their equality:

$$i\hbar\frac{\partial}{\partial t}|D_1> = H|D_1> . \tag{2.15}$$

To establish these conditions we expand each side in a complete set of orthogonal basis states (for which we choose the number states) and equate appropriate coefficients. The coherent phonon states

$$|\beta_m> \equiv \exp\left[\sum_q(\beta_{qm}b_q^+ - \beta_{qm}^* b_q)\right]|0> \tag{2.16}$$

appearing in (2.5) have the expansion

$$|\beta_m> = \prod_q e^{-\frac{1}{2}|\beta_{qm}|^2}\sum_{\nu_q=0}^{\infty}\frac{(\beta_{qm})^{\nu_q}}{\sqrt{\nu_q!}}|\nu_q> \tag{2.17}$$

where $|\nu_q>$ is the phonon state consisting of ν_q phonons of wavevector q. The right hand side of (2.15) involves the calculation of the action of b^+'s, b's, and b^+b's on (2.17) while the left side brings out $\dot\beta_{qm}(t)$'s and $\dot\alpha_m(t)$'s. The operations are tedious but straightforward. Focusing on each basis state

$$|m,\{\nu_q\}> \equiv a_m^+\prod_q\frac{b_q^{+\nu_q}}{\sqrt{\nu_q!}}|0> \tag{2.18}$$

which involves an arbitrary site m and a specific but arbitrary set of phonon occupation numbers $\{\nu_q\}$ and equating the coefficients of this state on both sides of (2.15) finally yields

$$i\hbar\dot\alpha_m - \frac{i}{2}\hbar\,\alpha_m\sum_q(\dot\beta_{qm}^*\beta_{qm}+\beta_{qm}^*\dot\beta_{qm}) - \alpha_m\sum_q\chi_m^{-q}\,\hbar\omega_q\,\beta_{qm}$$

$$= \sum_n J_{mn}\,\alpha_n\prod_q e^{\frac{1}{2}[|\beta_{qm}|^2-|\beta_{qn}|^2]}\left[\frac{\beta_{qn}}{\beta_{qm}}\right]^{\nu_q}$$

$$- i\hbar\alpha_m\sum_q\nu_q\frac{\dot\beta_{qm}+i\omega_q\,\beta_{qm}+i\omega_q\chi_m^q}{\beta_{qm}} . \tag{2.19}$$

Since the α_m and β_{qm} must be independent of the phonon occupation numbers ν_q, the coefficients of *each* ν_q must independently sum to zero. When $J_{mn}=0$ Eq. (2.19) yields the balance equation

$$\dot{\beta}_{qm} = -i\,\omega_q\,\beta_{qm} - i\,\omega_q\,\chi_m^q \tag{2.20}$$

which integrates to

$$\beta_{qm}(t) = e^{-i\omega_q t}\,\beta_{qm}(0) + (e^{-i\omega_q t}-1)\chi_m^q \quad . \tag{2.21}$$

One can easily verify that (2.21) with any initial condition leads to a family of $|D_1\rangle$ states that are solutions of the Schrödinger equation (2.15). Therefore these states correctly describe the evolution of the system when $J_{mn}=0$.

A glance at (2.19) shows that when $J_{mn}\neq 0$ the dependence on the ν_q is nonlinear and hence can not be balanced for every distribution $\{\nu_q\}$ by choosing ν_q-independent coefficients. Therefore the $|D_1\rangle$ states do *not* solve the Schrödinger equation when there is excitation transport, *i.e.* when $J_{mn}\neq 0$.

Although we need not repeat this analysis for the $|D_2\rangle$ states because we already understand that they do not exactly solve the Schrödinger equation, it is nevertheless instructive (and simple) to do so. Instead of Eq. (2.19) one here obtains [34]

$$i\hbar\dot{\alpha}_m - \frac{i}{2}\hbar\alpha_m\sum_q(\dot{\beta}_q^*\beta_q + \beta_q^*\dot{\beta}_q) - \alpha_m\sum_q\chi_m^{-q}\hbar\omega_q\beta_q$$
$$= \sum_n J_{mn}\alpha_n - i\hbar\,\alpha_m\sum_q\nu_q\,\frac{\dot{\beta}_q + i\,\omega_q\beta_q + i\,\omega_q\chi_m^q}{\beta_q} \quad . \tag{2.22}$$

The balance equation for each ν_q and for arbitrary J_{mn} that one obtains from (2.22) is clearly

$$\dot{\beta}_q = -i\,\omega_q\beta_q - i\,\omega_q\chi_m^q \quad . \tag{2.23}$$

However, the coefficients $\beta_q(t)$ are independent of the site index m and hence this condition can only be satisfied when $\chi_m^q=0$, *i.e.* when there is no coupling between excitons and phonons. In this trivial case the $|D_2\rangle$ states solve the Schrödinger equation exactly.

Let us pause to take stock of the situation and provide a context for the material to follow. The $|D_1\rangle$ Ansatz state (2.5) provides an exact solution for the Schrödinger equation with the Fröhlich Hamiltonian (1.1) when $J_{mn}=0$ but not otherwise. The $|D_2\rangle$ Ansatz state (2.6) is a solution in the trivial case $\chi_m^q=0$ but not otherwise. Although we have some measures of the deviation of the predictions of the D_2-dynamics from the known behavior when $J_{mn}=0$, we really do not know how good (or bad) an approximation either $|D_1\rangle$ or $|D_2\rangle$ is to the actual evolution in general. The continued activity on the analysis of these states indicates the optimism that they do provide useful approximations and therefore important information. Although the $|D_1\rangle$ states, having more flexibility than the $|D_2\rangle$ states through the possible site dependence of the $\beta(t)$'s, are expected to provide a more accurate representation of the dynamics even when $J_{mn}\neq 0$, they are considerably more complicated and have not been used extensively. In the next section we carry out a detailed analysis of the D_2-states and their physical consequences. We follow this presentation with the discussion of an alternate approach that we believe overcomes the limitations of either of these Ansatz states and that can be extended to finite temperatures in a natural way. This new approach leads to an interesting *nonlinear* equation for the reduced density matrix of the system and reduces to the correct known result when $J_{mn} = 0$. Even when $J_{mn} \neq 0$ the equation incorporates all the constraints imposed by the known (exact) fluctuation-dissipation relation and commutation relations of the exciton and phonon operators.

3. Analysis of Davydov's $|D_2\rangle$ Ansatz State [37]

The Davydov Ansatz state $|D_2\rangle$ leads to the equations of motion (2.7), and this is the starting point of our analysis. For an acoustic chain the dimensionless coupling function χ_n^q is given in (2.9), and the acoustic frequency ω_q is

$$\omega_q = 2\left(\frac{w}{M}\right)^{1/2} \sin \frac{|ql|}{2} \quad . \tag{3.1}$$

Integration of (2.7b) yields the relation

$$\beta_q(t) = e^{-i\omega_q t}\beta_q(0) - i\int_0^t d\tau\, e^{-i\omega_q(t-\tau)} \sum_m \chi_m^q \omega_q |\alpha_m(\tau)|^2 \quad . \tag{3.2}$$

Insertion of (3.2) into (2.7a) followed by an integration by parts leads to

$$\begin{aligned} i\hbar\dot{\alpha}_n(t) &= E\,\alpha_n(t) - J[\alpha_{n+1}(t)+\alpha_{n-1}(t)] + f_n(t)\alpha_n(t) \\ &\quad - \frac{\chi^2}{w}[|\alpha_{n+1}(t)|^2 + 2|\alpha_n(t)|^2 + |\alpha_{n-1}(t)|^2]\alpha_n(t) \\ &\quad + \sum_m [K_{nm}(t)|\alpha_m(0)|^2]\alpha_n(t) + \left[\int_0^t d\tau \sum_m K_{nm}(t-\tau)\frac{\partial}{\partial\tau}|\alpha_m(\tau)|^2\right]\alpha_n(t) \quad . \end{aligned} \tag{3.3}$$

Here

$$f_n(t) \equiv \sum_q \chi_n^q \hbar\omega_q \left[e^{i\omega_q t}\beta_q^*(0) + e^{-i\omega_q t}\beta_{-q}(0)\right] \tag{3.4}$$

and

$$K_{mn}(t) = 2\sum_q \chi_m^q \chi_n^{-q}\hbar\omega_q \cos\omega_q t \quad . \tag{3.5}$$

In writing (3.3) we have taken the lattice to be an infinite linear monotomic chain, for which

$$K_{mn}(t) = \frac{\chi^2}{w}\left[J_{2(m-n+1)}(2\omega_B t) + 2J_{2(m-n)}(2\omega_B t) + J_{2(m-n-1)}(2\omega_B t)\right] \tag{3.6}$$

where $J_n(t)$ is the Bessel function of the first kind and $\omega_B = 2(w/M)^{1/2}$ is the maximum acoustic branch frequency. Note that (3.6) is entirely equivalent to (2.7) for the infinite linear chain.

The initial condition-dependent term $f_n(t)$ contains information about the preparation of the phonon state of the system. The initial values $\beta_q(0)$ may contain two contributions: a "systematic" portion related to the way in which the system is excited, and a "random" portion that represents the thermally induced variations at a finite temperature (in this latter context we can think of our system as a member of an ensemble of systems at a given temperature). The thermal portion of $f_n(t)$ is related to the kernel

$K_{mn}(t)$ via an appropriate fluctuation-dissipation relation. Except for a few subsequent remarks, we do not deal in this section with the finite-temperature problem and hence we shall only attend to the preparation-specific systematic contributions to $f_n(t)$. We consider two extreme initial conditions: the "bare exciton" and the "preformed soliton". The bare exciton is created without concurrent initial excitation of the lattice, *i.e.* it corresponds to the choice $\beta_q(0) = f_n(0) = f_n(t) = 0$ [the lattice *does* become subsequently excited through its interaction with the exciton, as seen in Eq. (3.2)]. The preformed soliton initial condition involves a nonzero choice for $f_n(0)$ that can only be stated when we have specified the way in which (3.3) may lead to soliton solutions. This is most easily (and customarily) done in the continuum limit, to which we now proceed (later in this section we consider important corrections to this limit).

In going to the continuum limit, certain combinations of parameters must be held constant as the lattice constant l vanishes:

$$Jl^2 = \frac{\hbar^2}{2m}\,, \quad \chi l = \varepsilon\,, \quad wl = \zeta\,, \quad \frac{M}{l} = \eta\,, \quad \upsilon_a \equiv (\zeta/\eta)^{1/2}\,. \tag{3.7}$$

Here m is the effective mass of the free exciton, ε is the energy change resulting from a molecular displacement of one lattice constant in the linearized exciton field, ζ is the tension, η is the mass density, and υ_a is the velocity of sound. The energy $E(0)$ (bottom of the free-exciton band) diverges in the strict continuum limit (divergent bandwidth); its value is irrelevant to transport but must be maintained at its proper microscopic value for spectral calculations since it does play a role in the position of spectral features. In the continuum limit we have

$$R_n \rightarrow x \tag{3.8a}$$

$$\sum_n |\alpha_n(t)|^2 = 1 \rightarrow \int dx\, |\alpha(x,t)|^2 = 1 \tag{3.8b}$$

$$\frac{O_{n+1} - O_n}{l} \rightarrow \frac{\partial}{\partial x} O(x) \tag{3.8c}$$

$$K_{mn}(t) \rightarrow K(x,y,t) = \frac{2\varepsilon^2}{\zeta}\, [\delta(x-y+\upsilon_a t) + \delta(x-y-\upsilon_a t)] \tag{3.8d}$$

and in this limit (3.3) is transcribed to

$$\begin{aligned} i\hbar\dot{\alpha}(x,t) = &-\frac{\hbar^2}{2m}\frac{\partial^2}{\partial x^2}\alpha(x,t) + E(0)\alpha(x,t) - \frac{4\varepsilon^2}{\zeta}|\alpha(x,t)|^2\alpha(x,t) \\ &+ f(x,t)\alpha(x,t) + \frac{2\varepsilon^2}{\zeta}[|\alpha(x+\upsilon_a t,0)|^2 + |\alpha(x-\upsilon_a t,0)|^2]\alpha(x,t) \\ &+ \frac{2\varepsilon^2}{\zeta}\int_0^t d\tau \left[\frac{\partial}{\partial \tau}|\alpha(y,\tau)|^2 \Big|_{y=x+\upsilon_a(t-\tau)} + \frac{\partial}{\partial \tau}|\alpha(y,\tau)|^2 \Big|_{y=x-\upsilon_a(t-\tau)} \right] \alpha(x,t)\,. \end{aligned} \tag{3.9}$$

Note that (3.9) permits an arbitrary initial phonon distribution via $f(x,t)$ (including one appropriate to finite temperature) and also fully includes the dissipative effects implicit in the kernel $K_{mn}(t)$.

Equation (3.9) does not appear significantly simpler than the discrete version (3.3). Meaningful simplification can be achieved if we restrict our search to nondissipative solutions by assuming that

$|\alpha(x,t)|^2 = \rho(x-\upsilon t)$ *i.e.* by assuming that the amplitude of the solution has the D'Alembert form. The velocity υ is a new constant that parametrizes a family of solutions. With this condition the integrations in (3.9) can be carried out explicitly and one finds the simpler form

$$i\hbar\dot{\alpha}(x,t) = -\frac{\hbar^2}{2m}\frac{\partial^2}{\partial x^2}\alpha(x,t) + E(0)\alpha(x,t) + f(x,t)\alpha(x,t)$$
$$+\frac{2\varepsilon^2}{\zeta}\left[\frac{\upsilon_a}{\upsilon_a+\upsilon}\rho(x+\upsilon_a t) - \frac{2\upsilon_a^2}{\upsilon_a^2-\upsilon^2}\rho(x-\upsilon t) + \frac{\upsilon_a}{\upsilon_a-\upsilon}\rho(x-\upsilon_a t)\right]\alpha(x,t) . \quad (3.10)$$

After one has solved (3.10), one must check that the amplitude of the solution in fact has the assumed D'Alembert form.

To proceed further we must specify the initial condition of the medium. Let us first consider the "preformed soliton," wherein the exciton wave packet is prepared together with an appropriate distortion of the medium designed to yield a coherent structure that can survive for an indefinite period (whether this can actually be done experimentally is another question). A straightforward normal-mode transformation relating $\{\beta_q\}$ to lattice momentum and position amplitudes $\{P_n, Q_n\}$ yields in the continuum limit the expression equivalent to (3.4)

$$f(x,t) = \varepsilon\left[\frac{\partial Q(y,0)}{\partial y} + \frac{P(y,0)}{\eta\upsilon_a}\right]_{y=x+\upsilon_a t} + \varepsilon\left[\frac{\partial Q(y,0)}{\partial y} - \frac{P(y,0)}{\eta\upsilon_a}\right]_{y=x-\upsilon_a t} . \quad (3.11)$$

The initial choices

$$Q(y,0) = -\frac{G_\upsilon}{2\varepsilon}\int^{y} dy'\rho(y') , \quad (3.12a)$$

$$P(y,0) = \eta\upsilon\frac{G_\upsilon}{2\varepsilon}\rho(y) , \quad (3.12b)$$

where G_υ is the velocity-dependent parameter (2.10), leads to the following identity (at all times) of terms occurring in (3.10):

$$f(x,t) = -\frac{2\varepsilon^2}{\zeta}\left[\frac{\upsilon_a}{\upsilon_a+\upsilon}\rho(x+\upsilon_a t) + \frac{\upsilon_a}{\upsilon_a-\upsilon}\rho(x-\upsilon_a t)\right] . \quad (3.13)$$

The medium initial conditions have thus been chosen to exactly cancel the sound pulses that would otherwise interfere with soliton propagation. With this initial condition, (3.10) reduces to the nonlinear Schrödinger equation (2.8) whose soliton solution (2.11) has the well-known sech-amplitude of the D'Alembert form and hence is also a solution of (3.9). The appropriate initial conditions (3.12) for this solution have the explicit form

$$Q(y,0) = -\frac{G_\upsilon}{\kappa\varepsilon}\tanh(\kappa y) , \quad (3.14a)$$

$$P(y,0) = \eta\upsilon\frac{G_\upsilon}{4\varepsilon}\kappa\,\mathrm{sech}^2(\kappa y) . \quad (3.14b)$$

Many aspects of the behavior of the soliton solution of the nonlinear Schrödinger equation have been studied extensively. It propagates undistorted at velocity υ. It is instructive to note that the nonlinear Schrödinger equation implies a length and a time scale over which coherence must exist for the moving soliton to be stable. This space-time region contributes to the speed-dependent nonlinear potential experienced by the soliton, and coherence over this region is required for soliton stability. The appropriate spatial scale of coherence is

$$\Lambda = \frac{1}{\kappa}\left[\frac{\upsilon}{\upsilon_a}(1+\frac{\upsilon}{\upsilon_a})+1\right] . \tag{3.15}$$

The temporal scale over which coherence must persist is

$$\tau_D \equiv \Lambda/\upsilon_a . \tag{3.16}$$

The continuum limit that leads to the nonlinear Schrödinger equation as a limit of the Davydov system (2.7) (for appropriate initial conditions) is perhaps appropriate for extended excitations but becomes inappropriate when the lattice microstructure influences the dynamics of the excitation. Inspection of (2.11) shows that the soliton solution indeed collapses to dimensions smaller than a lattice constant for soliton velocities approaching the speed of sound ($\upsilon \to \upsilon_a$) and also in the limit $J \to 0$ (which implies $\upsilon \to 0$). One can include some effects of the microstructure by retaining low-order corrections in a gradient expansion of certain quantities in the theory. This procedure allows us to build in some discreteness corrections while retaining the advantages of a continuum formulation. In particular, we have shown that the kernel $K(x,y,t)$ can be written as an infinite series in increasing derivatives of delta functions of which (3.8d) is the leading term. Retention of the next term in the series replaces (3.8d) with

$$K_{mn}(t) \to K(x,y,t) \approx \frac{2\chi^2 l}{w}\left[1+\frac{l^2}{4}\frac{\partial^2}{\partial x^2}\right][\delta(x-y+\upsilon_a t)+\delta(x-y-\upsilon_a t)] . \tag{3.17}$$

We also retain the first correction in the replacement

$$|\alpha_{n+1}(t)|^2+2|\alpha_n(t)|^2+|\alpha_{n-1}(t)|^2 \to 4\left[1+\frac{l^2}{4}\frac{\partial^2}{\partial x^2}\right]|\alpha(x,t)|^2 . \tag{3.18}$$

Choosing "preformed soliton" initial conditions according to the recipe (3.11)-(3.13) leads us from (3.3) to a *modified* nonlinear Schrödinger equation

$$i\hbar\dot{\alpha}(x,t) = -\frac{\hbar^2}{2m}\frac{\partial^2}{\partial x^2}\alpha(x,t)+E(0)\alpha(x,t)-G_\upsilon\alpha(x,t)\left[1+\frac{l^2}{4}\frac{\partial^2}{\partial x^2}\right]|\alpha(x,t)|^2 . \tag{3.19}$$

The derivatives in the last term cause a softening of the nonlinear potential in regions of high curvature. The steps that lead to (3.19) include the assumption (as with the nonlinear Schrödinger equation) of the existence of nondissipative solutions, whence we again seek solutions of the form

$$\alpha(x,t) = \phi(x-\upsilon t)e^{-i(kx-\omega t)}e^{-iE(0)t/\hbar} . \tag{3.20}$$

Substitution of (3.20) into (3.19) leads to the relation $\hbar k = -m\upsilon$ as in (2.12a) and allows (3.19) to be transformed into a nonlinear equation for the envelope function,

$$\frac{\partial^2\phi}{\partial x^2} - A\phi + B\phi^3 + C\phi\frac{\partial^2\phi^2}{\partial x^2} = 0 \tag{3.21}$$

where the C-term is the new discreteness contribution. The coefficients are

$$A = \frac{Jl^2k^2 + \hbar\omega}{Jl^2} = \frac{2m}{\hbar^2}(\hbar\omega + \tfrac{1}{2}mv^2) \tag{3.22a}$$

$$B = \frac{4\chi^2}{wJl}\frac{v_a^2}{v_a^2 - v^2} = \frac{2m}{\hbar^2}G_v \tag{3.22b}$$

$$C = \frac{\chi^2 l}{wJ}\frac{v_a^2}{v_a^2 - v^2} = \frac{2m}{\hbar^2}G_v\frac{l^2}{4} \tag{3.22c}$$

The behavior of the solution of (3.21) can be assessed if one rewrites it as follows. Note that

$$\frac{\partial^2\phi^2}{\partial x^2} = 2\left(\frac{\partial\phi}{\partial x}\right)^2 + 2\phi\frac{\partial^2\phi}{\partial x^2}\ . \tag{3.23}$$

The change of variables $\partial\phi/\partial x \rightarrow y^{1/2}(\phi)$ and use of (3.23) in (3.21) leads to

$$\frac{\partial}{\partial\phi}[(1 + 2C\phi^2)y] = -(-2A\phi + 2B\phi^3) \tag{3.24}$$

and integration with respect to ϕ gives

$$\frac{const.}{1+2C\phi^2} = \frac{1}{2}\left(\frac{\partial\phi}{\partial x}\right)^2 + \frac{1}{2}\left[\frac{-A\phi^2 + \frac{1}{2}B\phi^4}{1+2C\phi^2}\right]. \tag{3.25}$$

The soliton solution we seek will have appropriate localization properties if the envelope satisfies the boundary conditions

$$\phi(\infty) = \phi(-\infty) = 0,\quad \frac{\partial\phi}{\partial x}(\infty) = \frac{\partial\phi}{\partial x}(-\infty) = 0 \tag{3.26}$$

and it follows that the constant in (3.25) must vanish for these solutions. We have therefore re-expressed $\partial\phi/\partial x$ in terms of ϕ; insertion into (3.23) and thence into (3.21) then permits us to rewrite the latter in the form of a "Newton's Law",

$$\frac{\partial^2\phi}{\partial x^2} = -\frac{\partial V(\phi)}{\partial\phi}\ , \tag{3.27}$$

with the "potential"

$$V(\phi) = \frac{1}{2}\left[\frac{-A\phi^2 + \frac{1}{2}B\phi^4}{1+2C\phi^2}\right]. \tag{3.28}$$

The quartic form obtained when $C=0$ is appropriate to the nonlinear Schrödinger equation. When $C \neq 0$, the potential is softened by the presence of the denominator. The envelope $\phi(x)$ of the modified nonlinear Schrödinger equation soliton can be viewed as the trajectory of a "mass" starting from the top of the central peak of $V(\phi)$ with an infinitesimal "displacement". The envelope can be found by inverting equation (3.27) with boundary conditions (3.26):

$$\frac{x}{l} = \frac{1}{\sqrt{D}}\int_0^{\phi/\phi_0}\frac{du}{u}\left[\frac{1+Du^2}{1-u^2}\right]^{1/2} \tag{3.29}$$

where $D \equiv 2C\phi_0^2$, $\phi_0 \equiv (2A/B)^{1/2}$, and $\phi(0) \equiv \phi_0$. The dependence of D on Hamiltonian parameters and the explicit evaluation of (3.29) for arbitrary D are not simple. One can easily deal with the interesting limiting cases: when $D=0$, as in the strict continuum limit, we recover the earlier solution

$$\rho(x) = |\alpha(x)|^2 = \left(\frac{\kappa}{2}\right)^{1/2} \operatorname{sech}(\kappa x) \tag{3.30}$$

as in (2.11). When D is large ($D \approx G\sqrt{\pi J l}$) the role of discreteness becomes important and the envelope acquires a completely different form:

$$\rho(x) = \begin{cases} \left(\frac{2}{\pi l}\right)^{1/2} \cos\left(\frac{x}{l}\right) & , \quad |x| < \frac{\pi l}{2} \\ 0 & , \quad |x| > \frac{\pi l}{2} \end{cases} . \tag{3.31}$$

One can see that the envelope now resists collapse to physical dimensions smaller than a lattice constant even when $\upsilon \to \upsilon_a$ and when $J \to 0$. This self-limiting property of the modified system represents a significant improvement over the singular behavior of the nonlinear Schrödinger equation in these limits.

We complete this discussion by noting several other measures of the improvement provided by retaining some discreteness corrections. Had we *not* truncated at second order in the lattice constant, then with the single approximation $\cos\omega_q t \approx \cos(\upsilon_a |q| t)$ we would have found, instead of (3.21), the envelope equation

$$\frac{\partial^2\phi}{\partial x^2} - A\phi + B\phi \cosh^2\left(\frac{l}{2}\frac{\partial}{\partial x}\right)\phi^2 = 0 \tag{3.32}$$

where $\cosh^2$ of an operator is to be understood in terms of its series expansion. The error involved in truncating (3.32) can be estimated for a typical envelope of width λ as follows:

$$\begin{aligned} \cosh^2\left(\frac{l}{2}\frac{\partial}{\partial x}\right)\phi^2 \approx \cosh^2\left(\frac{l}{\lambda}\right)\phi^2 &\equiv \Delta \qquad \textit{(untruncated)} \\ \approx \left[1 + \left(\frac{l}{\lambda}\right)^2\right]\phi^2 &\equiv \Delta_{MNLS} \qquad \textit{(modified nonlinear Schrödinger equation)} \\ \approx \phi^2 &\equiv \Delta_{NLS} \qquad \textit{(nonlinear Schrödinger equation)} . \end{aligned} \tag{3.33}$$

For sufficiently large λ the three Δ's all agree; on the other hand, for $\lambda = 2l$ we find

$$\frac{\Delta - \Delta_{NLS}}{\Delta} = 27\% , \qquad \frac{\Delta - \Delta_{MNLS}}{\Delta} = 1.6\% \tag{3.34}$$

and hence the retention of only one correction term in Δ_{MNLS} provides a great improvement. A second indicator is the degree to which discrete and continuum normalizations agree. Both the nonlinear Schrödinger soliton and the modified one are normalized to unity in the continuum; however, for the nonlinear Schrödinger soliton when $J \to 0$

$$l \sum_n |\alpha(R_n, t)|^2 \to \infty \tag{3.35a}$$

while the modified equation gives

$$l \sum_n |\alpha(R_n, t)|^2 = 1.0083 . \tag{3.35b}$$

A third indicator is the binding energy of the excitation, defined as follows:

$$E_{bind} = E_{total} - E_{total,x=0}$$
$$= \langle D(t)|H|D(t)\rangle - \langle D(t)|H|D(t)\rangle|_{x=0} \; . \tag{3.36}$$

The binding energy of the nonlinear Schrödinger soliton diverges as $J \to 0$ (as does that of the exact polaron solution in the continuum limit). The binding energy of the modified soliton remains finite in the $J \to 0$ limit and is in good agreement with that of the exact polaron solution in the discrete lattice.

We note that Eqs. (2.8) and (3.19) admit solutions besides the soliton solutions that we have discussed. These other solutions include linear and nonlinear waves such as plane waves (uniform probability density) and cnoidal waves. These solutions also satisfy the D'Alembert condition used to construct the equations and they survive indefinitely. Whether the system chooses a soliton solution or another [even in the "preformed" case that allowed us to arrive at (2.8) and (3.19)] depends on the initial condition.

Let us now turn to the other initial condition that is of experimental interest, namely, the initially bare exciton. This discussion will only be presented in the strict continuum limit and hence we return to Eq. (3.10). In this case the medium is initially quiescent, *i.e.* $f(x,t) = 0$. Equation (3.10) can schematically be rewritten in the form of a Schrödinger equation with a time-dependent potential $U(x,t)$:

$$i\hbar\dot{\alpha}(x,t) = -\frac{\hbar^2}{2m}\frac{\partial^2}{\partial x^2}\alpha(x,t) + E(0)\alpha(x,t) + U(x,t)\alpha(x,t) \tag{3.37}$$

with

$$U(x,t) = 2\frac{\varepsilon^2}{\zeta}\left[\frac{\upsilon_a}{\upsilon_a+\upsilon}\rho(x+\upsilon_a t) - \frac{2\upsilon_a^2}{\upsilon_a^2-\upsilon^2}\rho(x-\upsilon t) + \frac{\upsilon_a}{\upsilon_a-\upsilon}\rho(x-\upsilon_a t)\right] \; . \tag{3.38}$$

First, we note that a solution of (3.35) with this "potential" will not be of the D'Alembert form assumed to obtain it, so that strictly speaking we should return to (3.9) for this discussion. This caveat should be kept in mind in not pushing our conclusions too far. Note that the "potential" $U(x,t)$ contains three contributions: one traveling backward with velocity $-\upsilon_a$, one forward with velocity υ and the third also forward with velocity υ_a. The $-\upsilon_a$ and υ portions overlap for a time $\lambda/(\upsilon_a+\upsilon)$ while the υ and υ_a portions overlap for a longer time $\lambda/(\upsilon_a-\upsilon)$. This second time is the same as τ_D, the temporal scale (3.16) that emerged in the preformed soliton. During the overlap time, the "potential" is asymmetric with respect to the would-be soliton solution and the would-be soliton is not exactly at the bottom of the potential well but rather on its leading slope. The excitation therefore experiences a resistive force which must result in its deceleration (*i.e.* υ can not be strictly constant) as it attempts to reach the bottom of the potential well. The fate of the excitation in this scenario depends on the comparison of the time scale τ_D and the time τ_U for this deceleration to be finalized. The time τ_U can be estimated as the time it would take the centroid of the wave packet to reach the bottom of the potential well if the well were stationary. For this estimate we use the classical formula for the period of a bound oscillation, estimating τ_U to be one quarter of this period:

$$\tau_U \approx \frac{1}{4}\oint\frac{m\,dx}{\sqrt{2m[E-U(x)]}} \; . \tag{3.39}$$

Using the soliton (2.11) we estimate $[E-U(x)] \approx \frac{1}{2} G_\upsilon \rho(0)$ and $\rho(0) \approx \kappa/2 = (m\, G_\upsilon/4\hbar^2)$ whence

$$\tau_U \approx \frac{\lambda\hbar}{G_\upsilon} \,. \tag{3.40}$$

When υ is small and $G_0/\hbar\upsilon_a \ll 1$, then $\tau_U \gg \tau_D$ and the soliton deformation is completed (*i.e.* the $\pm\upsilon_a$ pulses are shed and gone) before the slow wave packet has had an opportunity to deviate significantly from the fixed-υ D'Alembert form. Under these conditions the initially bare excitation can thus decay into a soliton of nearly the same speed and width as the initial wave packet. One the other hand, when υ is small and $G_0/\hbar\upsilon_a \gg 1$, then $\tau_U \ll \tau_D$ and the wave packet responds to changes in the deformation potential as fast as these changes occur. Throughout the process the excitation slows down, the change $\Delta\upsilon$ in its velocity being estimated as

$$\Delta\upsilon \approx \int_0^t dt' \left[-\frac{1}{m}\frac{\partial U(x,t')}{\partial x} \right]_{x=\upsilon t'} \tag{3.41}$$

where the force is evaluated at the centroid of the would-be soliton. Again using the soliton (2.11) we estimate

$$\frac{\Delta\upsilon}{\upsilon} \approx \frac{G_\upsilon^2}{\hbar^2(\upsilon_a^2-\upsilon^2)} \tag{3.42}$$

which clearly leads to $\Delta\upsilon > \upsilon$ when $G_0/\hbar\upsilon > 1$. This suggests dramatic changes (perhaps breakup or coming to a complete rest) occur in a time of $O\{\tau_U\}$. The distance that an excitation would travel before coming to rest would be

$$d \approx \upsilon\tau_U \approx \lambda \left(\frac{\upsilon}{\upsilon_a}\right)\left(\frac{G_\upsilon}{\hbar\upsilon_a}\right)^{-1} , \tag{3.43}$$

which is smaller than λ, the width of the initial wave packet. For small excitation velocities we thus conclude that soliton formation with speed and width near that of the initial excitation is possible if $G_0/\hbar\upsilon_a \ll 1$, but if $G_0/\hbar\upsilon_a \gg 1$, *i.e* if the exciton-phonon coupling is strong, then the excitation is rather likely to come to a complete rest or perhaps to break up into different forms. Initial excitations launched with higher velocities (υ/υ_a nonnegligible) impose stronger conditions on the existence of solitons.

We note that the condition $G_0/\hbar\upsilon_a \ll 1$ is essentially the same as the weak-coupling condition of polaron theory. We also repeat that τ_D is the minimum time for a soliton to form from a bare initial state and it is also the time over which lattice coherence must be maintained for the preformed soliton. We therefore conclude that τ_D is the natural generalization of the polaron formation time discussed elsewhere [15,16].

It is interesting to note the parameter values for the hydrogen-bonded backbone of the α-helix protein, a system for which the Davydov model has been thought to be appropriate. The values that have been quoted for this system are [38] $\chi = 0.62 \times 10^{-10} N$, $M = 114\, m_p$ (m_p = proton mass) and $w = 13\, N/m$, from which one obtains $G_0/\hbar\upsilon_a \approx 1.36$. Simulations with these parameters have produced excitations that settle into fragmented self-trapped states pinned to the lattice. A reduction of only 20% in χ in these simulations leads to depinning, and to the value $G_0/\hbar\upsilon_a \approx 0.87$. These results are fully consistent with our estimates.

We end this section with some further comments on the validity of the $|D_2\rangle$ Davydov Ansatz. We concluded earlier that the $|D_1\rangle$ Ansatz gives the exact evolution of the system when $J=0$ but that the $|D_2\rangle$ Ansatz does not (the exact solution when $J\neq 0$ is not known). Nevertheless the relative simplicity of the $|D_2\rangle$ state makes it more ubiquitous than the $|D_1\rangle$ state. We are now in a position to pinpoint more precisely where the $|D_2\rangle$ state "goes wrong."

In the $J=0$ limit one can calculate the following energies. The superscript *bare* refers to a bare lattice initial condition and *pol* refers to a preformed polaron. All the energies are calculated at zero temperature. The subscript *tot* indicates the total energy of the system, *i.e.* the expectation value of the full Hamiltonian (1.1), $\langle\psi(t)|H|\psi(t)\rangle$, with the appropriate state $|\psi(t)\rangle$; the subscript *ph* indicates the expectation value of the phonon portion of the Hamiltonian, *i.e.* $\langle\psi(t)|\sum_q \hbar\omega_q\, b_q^+ b_q|\psi(t)\rangle$. The exact energies are

$$E_{tot}^{bare} = E \quad , \tag{3.44a}$$

$$E_{ph}^{bare}(t) = K(0) - K(t) \quad , \tag{3.44b}$$

$$E_{tot}^{pol} = E - \tfrac{1}{2} K(0) \quad , \tag{3.44c}$$

$$E_{ph}^{pol} = \tfrac{1}{2} K(0) \quad . \tag{3.44d}$$

Here $K(t) \equiv K_{nn}(t)$ of Eq. (3.5) is independent of the site index. Note that these exact energies are independent of the distribution of the excitation. The same quantities calculated with the Davydov $|D_2\rangle$ Ansatz when $J=0$ are

$$E_{tot}^{bare} = E \quad , \tag{3.45a}$$

$$E_{ph}^{bare}(t) = \sum_{m,n} [K_{mn}(0) - K_{mn}(t)]\,|\alpha_m|^2 |\alpha_n|^2 \ , \tag{3.45b}$$

$$E_{tot}^{sol} = E - \tfrac{1}{2}\sum_{m,n} K_{mn}(0)\,|\alpha_m|^2|\alpha_n|^2 \ , \tag{3.45c}$$

$$E_{ph}^{sol} = \tfrac{1}{2}\sum_{m,n} K_{mn}(0)\,|\alpha_m|^2|\alpha_n|^2 \ , \tag{3.45d}$$

where the superscript *sol* indicates a preformed soliton. Clearly the results (3.45b)-(3.45d) differ from the exact ones (unless the excitation is completely localized on a single site) and depend on the distribution of the excitation. Note that the energy (3.45c) of the soliton is higher than that of the polaron, (3.44c).

The dependence of the energy on the shape of the single-excitation distribution is a manifestation of an unphysical *self-interaction* implicit (also when $J\neq 0$) in the Davydov Ansatz. The nonlinear terms in (3.45) represent one portion of an excitation (on site m) interacting with another portion of the same excitation (on site n). These nonlinear terms are responsible for the formation of the soliton in this model.

To sum up what we have found so far: We have analyzed the consequences of the $|D_2\rangle$ Ansatz state, giving special attention to the initial preparation of the system and to the effects of discreteness and we have identified the physical source of the errors in this state. To round out this picture we present in the next section a theory that rectifies some of the faults of the $|D_2\rangle$ Ansatz for single excitations while preserving the simplicity of its form (as opposed to resorting to the complexity of $|D_1\rangle$). Furthermore, the theory we present provides a natural framework for generalization to finite temperatures by dealing with density matrices instead of wave functions.

4. Nonlinear Density Matrix Equations [39-41]

In the previous sections we have elaborated the virtues and faults of Davydov's $|D_2\rangle$ Ansatz. We have extolled its relative simplicity over the $|D_1\rangle$ Ansatz state, which does give the correct dynamics when $J=0$, but is difficult to handle. In this section we use the natural language of the statistical mechanics of quantum systems and use the density matrix, rather than the wave function, to formulate a new theory for the Fröhlich Hamiltonian (1.1). This theory attempts to preserve the simplicity of the D_2-state while at the same time incorporating all known exact results. The resulting formulation should thus strike or balance between the simple but inaccurate D_2 theories and the complex but more accurate D_1 theories. In particular, we attempt to recuperate the correct (linear) polaron dynamics when $J=0$ and we also incorporate the known quantum fluctuation-dissipation relations that must hold for an ensemble of systems. In constructing this theory we take care to respect the appropriate quantum commutation relations. Our derivation is initially exact, but some phenomenology must enter when formal operator expressions are simplified to a useful form. Our final results contain only the parameters that enter the Hamiltonian: the phenomenology does not introduce any new parameters in the model.

We begin with the Liouville-von Neumann equation for the full exciton-phonon system and construct a reduction procedure to obtain a reduced density matrix for the excitation. The reduction procedure must respect the nonlinear character of the excitations we attempt to describe. The usual projection operator P collapses a full phase space operator onto a subspace and re-completes the space by appending to the reduced operator one in the complementary space. For an arbitrary operator $O(t)$ *(e.g.* the full density matrix)

$$PO(t) = W_{ph} \otimes T_{r_{ph}} O(t) \ . \tag{4.1}$$

Here $T_{r_{ph}}$ is a trace over the phonon variables, $\otimes$ denotes a direct product, and W_{ph} is a phonon density matrix which is usually chosen to be the equilibrium density matrix of the phonon bath at temperature T. With this latter "standard" choice one would retreat from Davydov's principal achievement, namely, that of producing a dynamical description in which the phonon system is held *away* from its usual equilibrium through a constraint supplied by the moving excitation. In order to accommodate this constraint in our projection procedure we choose, as an alternative to the usual equilibrium form of the complementary operator, the *time-dependent* operator

$$W_{ph}(t) \equiv |\beta(t)\rangle\langle\beta(t)| \tag{4.2}$$

constructed from the phonon component of the Ansatz state vector (2.6), *i.e.* $|\beta(t)\rangle$ is the multi-mode coherent state [42]

$$|\beta(t)\rangle = \exp\Big[-\sum_q(\beta_q(t)b_q^+ - \beta_q^*(t)b_q)\Big]|0\rangle \ . \tag{4.3}$$

This choice embodies the desired constraint and differs from the usual one in an additional way: $|\beta(t)\rangle\langle\beta(t)|$ represents a *single system* with specified initial conditions $\{\beta_q(0)\}$. To introduce the temperature, one must follow the evolution of each of these distinct systems and construct an ensemble in which each set of *initial conditions* is weighted according to a thermal ensemble. In the coherent state representation the equilibrium phonon density matrix $W_{ph}^{eq}(T)$ takes the form [40]

$$W_{ph}^{eq}(T) = \prod_q \int d^2\gamma_q \, \frac{e^{-|\gamma_q|^2/\langle n_q\rangle}}{\pi\langle n_q\rangle} |\gamma_q\rangle\langle\gamma_q| \tag{4.4}$$

where $|\gamma_q\rangle$ represents a coherent state of wave vector q having the property $b_q|\gamma_q\rangle = \gamma_q|\gamma_q\rangle$. The number of phonons in each state $|\gamma_q\rangle$ is indefinite and has a mean value given by $|\gamma_q|^2$. In a thermal ensemble of these states, these mean values are in turn distributed according to the Gaussian weights indicated in (4.4) and their average is given by the Bose distribution $\langle n_q\rangle = [\exp(\hbar\omega_q/k_BT)-1]^{-1}$. Thus, in order for the ensemble of β-states to be representative of a thermal ensemble *at the initial time*, it is necessary that the initial phonon operator $W_{ph}(0) = |\beta(0)\rangle\langle\beta(0)| = |\{\beta_q(0)\}\rangle\langle\{\beta_q(0)\}|$ be represented in the ensemble with the weight

$$g\{\beta_q(0)\} = \prod_{\{\beta_q(0)\}} \frac{e^{-|\beta_q(0)|^2/\langle n_q\rangle}}{\pi\langle n_q\rangle} \ . \tag{4.5}$$

The temperature that characterizes the ensemble therefore enters only through these weights. In this way we achieve a description of an ensemble that is initially thermally distributed, each system of the ensemble evolving in a constrained manner.

In the present formalism we have not yet specified the way in which this constrained evolution of the phonon state is to take place, *i.e.* we have not yet specified the time dependence of the coefficients $\beta_q(t)$: only when we have specified this evolution will we have completed the definition of our projection operator. Within the Davydov formalism the parameters $\beta_q(t)$ are the expectation values of the phonon operators, *i.e.* $\beta_q(t) = \langle D_2(t)|b_q|D_2(t)\rangle$, which leads to the relation [*cf.* (2.7b)]

$$\beta_q(t) = e^{-i\omega_q t}\beta_q(0) - i\int_0^t d\tau\, e^{-i\omega_q(t-\tau)}\sum_n \chi_n^q \omega_q |\alpha_n(\tau)|^2 \ . \tag{4.6}$$

In the present development we no longer have the Davydov Ansatz at hand and so we must redefine the $\beta_q(t)$. Since our motivation is to retain the constraining aspects of the Davydov theory which "force" the lattice to recognize the presence of the excitation, we *define* $\beta_q(t)$ to be the *exact* quantum expectation value of the phonon operator b_q:

$$\beta_q(t) \equiv Tr\, W(t) b_q \ , \tag{4.7}$$

where $W(t)$ is the full density matrix of the exciton-phonon system and Tr is a trace over all variables.

We can make (4.7) more explicit by noting that for a single excitation the matrix elements of the reduced density operator

$$\rho(t) \equiv Tr_{ph}\; W(t) \tag{4.8}$$

are given by the single-particle relation

$$\rho_{mn}(t) = Tr\; W(0)a_n^+(t)a_m(t)\;. \tag{4.9}$$

We then can show that *without approximation*

$$\beta_q(t) = Tr\; W(0)b_q(t)$$

$$= e^{-i\omega_q t}\beta_q(0) - i\int_0^t d\tau\, e^{-i\omega_q(t-\tau)}\sum_n \chi_n^q\, \omega_q\, \rho_{nn}(\tau)\;. \tag{4.10}$$

The remaining unknowns are then only the reduced density matrix elements $\rho_{mn}(\tau)$, and these are fully and exactly determined (in principle) from the Liouville-von Neumann equation (*cf.* below).

Before proceeding to the determination of $\rho(t)$, we note the formal similarity of the Davydov result (4.6) and the exact relation (4.10). These two would in fact be entirely equivalent if one were to identify $\rho_{nn}(\tau)$ with $|\alpha_n(\tau)|^2$. We have already demonstrated that the Davydov procedure leads to an equation for $\alpha_n(\tau)$ that can not lead to the correct reduced density matrix. Our intention here is to provide a corrected procedure to evaluate this density matrix. Were we able to solve the full Liouville equation exactly, our task would be complete. Since this is not possible, we must introduce reasonable approximations. The Davydov Ansatz provides one such approximation; we introduce an alternative that has the virtue of exactly reproducing the known $J=0$ behavior (and that preserves the known quantum fluctuation-dissipation relation for the system).

The definition of our time-dependent projection operator (4.1) with (4.2),

$$P(t)O(t) = |\beta(t)\rangle\langle\beta(t)| \otimes Tr_{ph}\; O(t)\;, \tag{4.11}$$

together with (4.3) and (4.10) is now complete. A further average over the initial weights (4.5) need not yet be carried out. We stress that as a consequence, our reduced operator represents a *single system* in an ensemble, or alternatively that the reduced operator can be interpreted as a *stochastic* as well as a *quantum* operator.

Let L represent the Liouville operator that corresponds to the full Fröhlich Hamiltonian (1.1),

$$LO(t) = -\frac{i}{\hbar}[H, O(t)]\;, \tag{4.12}$$

and let us begin with the Liouville-von Neumann equation

$$\frac{d}{dt}W(t) = LW(t)\;. \tag{4.13}$$

From here we can construct an equation for the projected quantity

$$P(t)W(t) = |\beta(t)\rangle\langle\beta(t)| \otimes \rho(t) \tag{4.14}$$

using a somewhat more complicated version of standard projection operator techniques. The complications arise from the time dependence of $P(t)$, which forces us to retain contributions containing $\dot{P}(t)$ and observe standard time-ordering rules. To extract the reduced density operator $\rho(t)$, we must trace the projected equation over the phonon variables using the fact that by construction $Tr_{ph}\,|\beta(t)\rangle\langle\beta(t)|=1$. We obtain the formal expression for the evolution of the "stochastic" reduced density operator

$$\dot{\rho}(t) = Tr_{ph}\, L\,|\beta(t)\rangle\langle\beta(t)| \otimes \rho(t)$$

$$- Tr_{ph}\, L\int_0^t dt' \exp_T\Big[\int_{t'}^t d\tau\, \Omega(\tau)\Big]\, \Omega(t')\,|\beta(t')\rangle\langle\beta(t')| \otimes \rho(t') \,. \tag{4.15}$$

It is understood that the β-parameters implicit in (4.15) are to be eliminated using the exact relation (4.10) which introduces both stochasticity through the initial values $\beta_q(0)$ and nonlinearity through the dependence of $\beta_q(t)$ on the exciton probabilities. The subscript T denotes time ordering, and the super-operator $\Omega(t)$ is given by

$$\Omega(t) = -\dot{P}(t) + [1-P(t)]L \,. \tag{4.16}$$

In arriving at (4.15) we have set an initial condition term to zero, a term that vanishes identically if the initial density operator satisfies $P(0)W(0) = W(0)$; in our case this implies the form $W(0) = |\beta(0)\rangle\langle\beta(0)| \otimes W_{ex}(0)$. It is well known that an initial direct product form $W_{ph}(0) \otimes W_{ex}(0)$ insures that the ubiquitous and otherwise bothersome initial condition term in projected equations vanishes; As mentioned earlier, the usual projections involve a thermal ensemble whence $W_{ph}(0) = W_{ph}^{eq}$; in our case $W_{ph}(0) = |\beta(0)\rangle\langle\beta(0)|$ represents not a thermal ensemble but a single system chosen from a thermal ensemble. Since there is a one-to-one correspondence between the set of coherent state amplitudes $\{\beta\}$ and the set of classical coordinate pairs $\{q,p\}$, the product operator $|\beta(0)\rangle\langle\beta(0)|$ subsumes all classically describable initial conditions of the medium. However, since the operator $|\beta(0)\rangle\langle\beta(0)|$ represents only one of the many quantum states corresponding to each classical configuration (that of lowest energy), the initial operators we use constitute a sampling of all possible initial quantum states. While (4.4) shows that this sampling is wholly sufficient to compute any thermal expectation value, we have no guarantee that the quantum state implicit in our choice is actually representative of an ensemble of classically equivalent quantum states. We proceed with the expectation that, any biases introduced by the incomplete sampling of quantum ensembles are small and short-lived. We stress that the use of coherent state products in the initial operator *does not* imply a preexisting cooperation of normal modes nor special deformations of the medium at t=0.

Equation (4.15) is exact (subject to the initial condition just discussed) but useless in its present general form. A useful form is recovered only when approximations are made, and some of these are necessarily "blind" in that one really does not know the error made when implementing them. We discuss two such approximations: one corresponds to a generalization of the Davydov Ansatz factorization. The second uses external information to modify the equation and bring it into agreement with the known exact results.

Let us begin with the *factored approximation*

$$W(t) \approx |\beta(t)\rangle\langle\beta(t)| \otimes \rho(t) \tag{4.17}$$

which includes as a special case the operator $W(t) = |\beta(t)\rangle\langle\beta(t)| \otimes |\alpha(t)\rangle\langle\alpha(t)|$ which may be considered a density matrix version of the Davydov Ansatz. This latter form implies $W^2 = W(t)$ and hence represents only pure states, whereas (4.17) can represent certain statistical mixtures. On the other hand, any possible phase mixture between the exciton and phonon systems is lost in the approximation (4.17) and therefore so is the main purpose of the projection procedure. Nevertheless, let us pursue (4.17) for a moment and further on reinstate effects of the lost phase relations.

With the factored form (4.17) the intregral term in (4.15) vanishes identically and one obtains

$$\dot{\rho}(t) = -\frac{i}{\hbar}[H_{ex},\rho(t)] - \frac{i}{\hbar}[\bar{H}_{ex-ph}(t),\rho(t)] \tag{4.18}$$

wherein H_{ex} is the free-exciton part of the Hamiltonian, and, with H_{ex-ph} denoting the exciton-phonon coupling contribution,

$$\bar{H}_{ex-ph}(t) = Tr_{ph}\,|\beta(t)\rangle\langle\beta(t)|\,H_{ex-ph} \tag{4.19}$$

$$= \langle\beta(t)|\,H_{ex-ph}\,|\beta(t)\rangle\ . \tag{4.19}$$

For a single excitation, noting that $\langle\beta(t)|b_q|\beta(t)\rangle = \beta_q(t)$, and with $\beta_q(t)$ given by (4.10), we obtain for the matrix elements of $\rho(t)$

$$\dot{\rho}_{mn}(t) = -\frac{i}{\hbar}[H_{ex},\rho(t)]_{mn} - i[f_m(t) - f_n(t)]\rho_{mn}(t)$$

$$-i\int_0^t d\tau \sum_l [\dot{K}_{ml}(t-\tau) - \dot{K}_{nl}(t-\tau)]\rho_{ll}(\tau)\rho_{mn}(t)\ . \tag{4.20}$$

Here $K_{mn}(t)$ and $f_m(t)$ correspond exactly to the functions (3.5) and (3.4) encountered earlier.

Equation (4.20) is interesting in a number of respects. First, it is "stochastic" since it contains the initial conditions $\beta_q(0)$ in the $f_m(t)$. These latter functions can thus be regarded as (multiplicative) "fluctuations". The density matrix equation (4.20) can therefore only be compared with more standard ones after averaging over these fluctuations, a step made difficult by the nonlinear structure of (4.20). An average over the initial weights (4.5) indicated by a bracket yields

$$\langle f_m(t) f_n(\tau)\rangle = 2\sum_q \chi_m^q \chi_n^{-q}\,\omega_q^2[\cos\omega_q(t-\tau)]\langle n_q\rangle\ , \tag{4.21}$$

a result which would correspond to the quantum fluctuation-dissipation relation between $f_m(t)$ and the kernels $K_{mn}(t)$ if $\langle n_q\rangle$ were replaced by $\langle n_q\rangle + ½$ [40]. The second point to notice in (4.20) is the unusual nonlinear structure of the integral term. We know from the arguments of the previous section that the form (4.20) can not be entirely correct because it does not yield the correct J=0 limit. We also know that part of the problem lies with an inappropriate self-interaction contribution which, if implicit in the Davydov $|D_2\rangle$ Ansatz state, is surely implicit in the factorization (4.17) as well. One might be tempted to conclude that the term quadratic in ρ in (4.20) arises entirely from this self-interaction but this is not the case as will be seen subsequently. Part of the nonlinear contribution in (4.20) is correct and arises from the coherent response of the medium to the presence of the excitation.

If Eq. (4.20) had been our ultimate goal, it would not have been necessary to develop the full density matrix machinery; instead we would have started with (4.17) directly to arrive at this density matrix generalization of the Davydov $|D_2\rangle$ Ansatz in a more straightforward manner. Our aim is rather to preserve the simplicity of the $|D_2\rangle$ assumption without incurring the loss of phase coherence between the excitons and phonons which is implicit in (4.17). Let us therefore return to the exact equation (4.15) and construct an approximate procedure that retains contributions from the integral term.

The simplest approximation that does not altogether neglect the integral term in (4.15) is to replace the exponential operator kernel with a δ-function kernel [39-41],

$$\exp_T \left[\int_{t'}^{t} d\tau \Omega(\tau) \right] \approx \tau_M \, \delta(t-t') \, . \tag{4.22}$$

The resulting equation is ostensibly local in time, but memory terms such as appear in (4.20) are reintroduced upon elimination of the $\beta_q(t)$ using (4.10). In order to proceed further we simply (and "blindly") treat the super-operator τ_M as a scalar. The errors introduced by these assumptions are unknown *a priori*, but with these assumptions and with rather extensive (but otherwise exact) manipulations Eq. (4.15) reduces to

$$\begin{aligned} \dot{\rho}_{mn}(t) = & -\frac{i}{\hbar}[H_{ex},\rho(t)]_{mn} - i[f_m(t) - f_n(t)]\rho_{mn}(t) \\ & - i\int_0^t d\tau \sum_l [\dot{K}_{ml}(t-\tau) - \dot{K}_{nl}(t-\tau)]\rho_{ll}(\tau)\rho_{mn}(t) \\ & - (\lambda_{mm}^2 + \lambda_{nn}^2 - 2\lambda_{mn}^2)\tau_M \rho_{mn}(t) \, . \end{aligned} \tag{4.23}$$

The only new quantities appearing in (4.23) are the damping coefficients

$$\lambda_{mn}^2 = \sum_q \chi_m^q \chi_n^{-q} \omega_q^2 \, . \tag{4.24}$$

Equation (4.23) differs from the approximation (4.20) in the terms proportional to these quantities, which attempt to embody effects of the exciton-phonon phase relations. Note that the λ^2 terms enter (4.23) independently of the fluctuations $f_m(t)$ and that they are independent of J_{mn}. The form of these terms is reminiscent of the effect of the contributions that would arise from a full operator treatment of the problem as we now indicate.

In a fully quantum mechanical treatment one obtains the *exact* Heisenberg operator equation

$$\begin{aligned} \frac{d}{dt} a_n^+(t) a_m(t) = & \frac{i}{\hbar}[H_{ex}, a_n^+(t) a_m(t)] - i[F_m(t) - F_n(t)] a_n^+(t) a_m(t) \\ & - i\int_0^t d\tau \sum_l [\dot{K}_{ml}(t-\tau) - \dot{K}_{nl}(t-\tau)] \, a_l^+(\tau) a_l(\tau) a_n^+(t) a_m(t) \end{aligned} \tag{4.25}$$

(the transcription of this operator equation to a scalar equation is far from trivial and therefore it is difficult to compare it with the density matrix equations discussed in this section). Products of the quantum mechanical operator fluctuations

$$F_m(t) = \sum_q \chi_m^q \, \omega_q [e^{i\omega_q t} b_q^+(0) + e^{-\omega_q t} b_{-q}(0)] \tag{4.26}$$

when averaged over a thermal distribution of initial bath operators are related to the memory kernels $K_{mn}(t)$ via the quantum-mechanical fluctuation-dissipation relation [43]

$$\mathrm{Re}\,\langle F_m(t)F_n(\tau)\rangle = 2\sum_q \chi_m^q\,\chi_n^{-q}\,\omega_q^2\cos[\omega_q(t-\tau)](\langle n_q\rangle+\tfrac{1}{2}) \tag{4.27}$$

which incorporates the correct commutation properties of the operators $F_m(t)$. It would seem that the $\lambda^2\tau_M$ terms in (4.23) embody some of the effects of the nonthermal zero-point (T-independent and J-independent) contribution to (4.27) not present in (4.21). In fact, we have argued that the coefficients λ_{mn}^2 and the Markov time scale may originate from an approximation

$$\mathrm{Re}\,\langle F_m(t)F_n(\tau)\rangle\,\Big|_{T=0} \approx \lambda_{mn}^2\,2\tau_M\,\delta(t-\tau)\ , \tag{4.28}$$

implicit in our Markov approximation (4.22).

Although Eq. (4.23) goes beyond the factored approximation (4.20), it can easily be verified that it still does not give the correct density matrix when J_{mn}=0. It is not difficult to identify the simplest ways in which (4.23) would have to be modified in order to reproduce this behavior correctly. These arguments, which we have detailed elsewhere [40] and do not repeat here, are phenomenological, but leads to a substantially improved equation that contains no new parameters. We cite the final result, which we propose as an alternative to the foregoing theories [39-41]:

$$\dot{\rho}_{mn}(t) = -\frac{i}{\hbar}[H_{ex},\rho(t)]_{mn} - i\,[f_m(t)-f_n(t)]\rho_{mn}(t) \tag{4.28}$$

$$-\,i\int_0^t d\tau\sum_l [K_{ml}(t-\tau)-K_{nl}(t-\tau)\dot{\rho}_{ll}(\tau)\rho_{mn}(t)$$

$$-\frac{1}{2}\,[\dot{ff}_{mm}(t)+\dot{ff}_{nn}(t)-2\dot{ff}_{mn}(t)]\rho_{mn}(t)\ . \tag{4.28}$$

In going from (4.23) to (4.28) we have dropped surface terms that arise from an integration by parts and we have replaced the $\lambda^2\tau_M$-terms with ones containing the functions

$$ff_{mm}(t) = 2\sum_q \chi_m^q\,\chi_n^{-q}(1-\cos\omega_q t)\ , \tag{4.29}$$

due to the vibrational relaxation of the initially bare state. This new equation for the density matrix is in general nonlinear but reduces to the correct form when $J_{mn}=0$.

Equation (4.28) is a new starting point for the analysis of the Fröhlich Hamiltonian (1.1). Definite results at finite temperatures are not yet available; whether (4.28) supports solitons or other coherent solutions, and whether such structures may be stable at finite temperatures remains to be seen.

5. Future Directions

In addition to the analysis of the new density matrix equation (4.28), a number of new directions are currently under active pursuit in our group. Perhaps the most intriguing of these arises from the recognition that in many cases the excitation of interest is an intramolecular vibration (vibronic excitation) whose number of quanta need not be conserved. Indeed we believe that number non-conservation may be an important aspect of the behavior of systems with nonlinear vibron-phonon interactions. In particular, even at low levels of excitation there exists a finite probability that the vibronic excitation is multiply excited.

In order to address this and other questions, we have begun a thorough study of a Hamiltonian first proposed by Takeno as an alternative to the Fröhlich Hamiltonian [22,44]:

$$H_{Takeno} = H_{vib} + H_{ph} + H_{vib-ph} \ . \tag{5.1}$$

Here H_{vib} is the Hamiltonian for a set of bilinearly coupled local oscillators having local potentials $v(\hat{q}_n)$,

$$H_{vib} = \sum_n \left[\frac{\hat{p}_n^2}{2m} + v(\hat{q}_n) \right] - \sum_{m,n} L_{mn} \hat{q}_m \hat{q}_n \ . \tag{5.2}$$

$\hat{q}_n$ is the displacement operator of the oscillator on the nth molecule and $\hat{p}_n$ is the conjugate momentum operator. H_{ph} in (5.1) describes the acoustic phonons,

$$H_{ph} = \sum_n \left\{ \frac{\hat{P}_n^2}{2M} + \frac{w}{2} (\hat{Q}_n - \hat{Q}_{n-1})^2 \right\} = \sum_q \hbar\omega_q \, (b_q^+ b_q + ½) \ , \tag{5.3}$$

where $\hat{Q}_n$ and $\hat{P}_n$ are respectively the displacement and momentum operators of molecule n. These operators are related to the phonon creation and annihilation operators b_q^+ and b_q via standard relations. The interaction between the local oscillators and the acoustic phonons is contained in

$$H_{vib-ph} = \sum_n (\hat{Q}_{n+1} - \hat{Q}_{n-1}) g(\hat{q}_n) \ . \tag{5.4}$$

The potential $v(\hat{q}_n)$ in (5.2) and the force $g(\hat{q}_n)$ in (5.4) are in general analytic functions of at least second order in $\hat{q}_n$ but otherwise arbitrary. Most frequently they are taken to be of second order:

$$v(\hat{q}_n) = \frac{1}{2} m \omega_v^2 \hat{q}_n^2 \ , \qquad g(\hat{q}_n) = g \, \hat{q}_n^2 \tag{5.5}$$

where ω_v is the frequency of the local oscillator.

The Fröhlich Hamiltonian (1.1) can be viewed as a number-conserving approximation to the number-non-conserving special case of the Takeno Hamiltonian. Let us rewrite (1.1) as

$$H_{Fr\ddot{o}hlich} = H_{ex} + H_{ph} + H_{ex-ph} \tag{5.6}$$

with

$$H_{ex} = \sum_n E \, a_n^+ a_n - \sum_{m,n} J_{mn} \, a_m^+ a_n \ , \tag{5.7}$$

H_{ph} as given in (5.3), and

$$H_{ex-ph} = \sum_n \chi (\hat{Q}_{n+1} - \hat{Q}_{n-1}) a_n^+ a_n \ . \tag{5.8}$$

With the associations

$$\hat{q}_n = \left(\frac{\hbar}{2m\omega_v}\right)^{½} (a_n^+ + a_n) \ , \tag{5.9a}$$

$$\hat{p}_n = i\,(\frac{m\hbar\omega_v}{2})^{1/2}\,(a_n^+ - a_n)\ , \qquad (5.9b)$$

$$L_{mn} = \frac{m\,\omega_v}{\hbar}\,J_{mn}\ ,\quad E = \hbar\omega_v\ ,\quad \chi = \frac{\hbar g}{m\,\omega_v} \qquad (5.10a)$$

and the quadratic forms (5.5) one finds that the Takeno Hamiltonian contains the Fröhlich Hamiltonian and, in addition, number non-conserving contributions of the form a^+a^+ and aa. Neglect of these (*e.g.* a rotating wave approximation in the Hamiltonian) might seem appropriate if $\omega_v \gg \omega_q$ for all q, but turns out to affect the system dynamics in important ways. In particular, the Takeno Hamiltonian leads to a greater likelihood of soliton formation at low levels of excitation and to greater soliton stability against temperature fluctuations. The detailed analysis of the Takeno Hamiltonian will be published elsewhere [44].

We end this review with a few words about the experimental situation *vis a vis* the problems we have discussed. It is difficult to design an experiment that will unequivocally distinguish the precise nature of an excitation from among various possibilities. Spectral measurements may identify red-shifts (*i.e.* energy lowering mechanisms), but it is difficult to say whether the collective excitation being observed is a polaron or soliton or yet another structure. Such red-shifts have been observed, for instance, in ACN crystals [9,10,45], and have been used to infer the presence of a soliton. Subsequently the observations have been re-interpreted as arising from a polaron (as must be the case since the dressed excitation is known to be immobile) [46]. Less equivocal information could be obtained from transport measurements of mobile excitations, a goal that has proven elusive. Current measurements on *l*-alanine crystals [47] may offer the most optimistic outlook on the experimental verification of the phenomena we have discussed in this review.

Acknowledgement

This research is supported in part by DARPA Grant No. DAAG 29-85-K-0246 and NSF grant No. DMR 86-19650-A1.

References

1. Davydov, A.S., Kisluka, N. I.: Zh. Eksp. Teor. Fiz. **71**, 1090 (1976) [Sov. Phys. - JETP **44**, 571 (1976)].
2. Davydov, A.S.: Phys. Scr. **20**, 387 (1979).
3. Davydov, A.S.: Zh. Eksp. Teor. Fiz. **78**, 789 (1980) [Sov. Phys. - JETP **51**, 397 (1980)].
4. Davydov, A.S.: Usp. Fiz. Nauk. **138**, 603 (1982) [Sov. Phys. Usp. **25**, 898 (1982)].
5. Davydov, A.S.: Biology and Quantum Mechanics. New York: Pergamon, 1982.
6. Davydov. A.S.: Solitons in Molecular Systems. Boston: Reidel, 1985.
7. Davydov. A.S.: in Solitons. Edited by S.E. Trullinger, V.E. Zhakarov, and V. I. Pokrovsky. New York: North-Holland, 1986.
8. Scott, A.C.: Phys. Rev. A**26**, 578 (1982).
9. Careri, G., Buontempo, U., Galluzzi, F., Scott, A.C., Gratton, E., Shyamsunder, E.: Phys. Rev. B**30**, 4689 (1984).

10. Eilbeck, J.C., Lomdahl, P.S., Scott, A.C.: Phys. Rev. B**30**, 4703 (1984).
11. Fröhlich, H,: Proc. R. Soc. London Ser. A**215**, 291 (1952); Adv. Phys. **3**, 325 (1954).
12. Holstein, T.: Ann. Phys. (NY) **8**, 325 (1959); 343, (1959).
13. Lee, T.D., Low, F.E., Pines, D.: Phys. Rev. **90**, 297 (1953).
14. Grover, M., Silbey, R.: J. Chem. Phys. **54**, 4843 (1971); Rackovsky, S., Silbey, R.: Mol. Phys. **25**, 61 (1973).
15. Brown, D.W., Lindenberg, K., West, B.J.: J. Chem. Phys. **84**, 1574 (1986).
16. Brown, D.W., Lindenberg, K., West, B.J., Cina, J., Silbey, R.: J. Chem. Phys. **87**, 6700 (1987).
17. Lomdahl, P.S., Kerr, W.C.: in Physics of Many Particle Systems. Edited by A. S. Davydov. Kiev: Naukova Dumka, 1988.
18. Kerr, W.C. Lomdahl, P.S.: Phys. Rev. B**35**, 3629 (1987).
19. Venzl, G. Fischer, S.F.: J. Chem. Phys. **81**, 6090 (1984); Phys. Rev. B**32**, 6437 (1985).
20. Lomdahl, P.S., Kerr, W.C.: Phys. Rev. Lett **55**, 1235 (1985).
21. Lawrence, A.F., McDaniel, J.C., Chang, D.B., Pierce, B.M., Birge, P. R.: Phys. Rev. A**33**, 1188 (1986).
22. Takeno, S,: Prog. Theor, Phys. **69**, 1798 (1983); **71**, 395 (1984); **73**, 853 (1985).
23. This picture is due to A. Migliori.
24. Swofford, R.L., Long, M.E., Albrecht, A.C.: J. Chem. Phys. **65**, 179 (1976).
25. Swofford, R.L., Burberry, M.S., Morrell, J.A., Albrecht, A.C.: J. Chem. Phys. **66**, 5245 (1977).
26. Bruinsma, R., Maki, K., Wheatley, J.: Phys. Rev. Lett. **57**, 1773 (1986).
27. Scott, A.C., Gratton, E., Shyamsunder, E., Careri, G.: Phys. Rev. B**32**, 5551 (1985).
28. Kenkre, V.M., Reineker, P.: Exciton Dynamics in Molecular Crystals and Aggregates. Vol. 94, Springer Tracts in Modern Physics. Edited by G. Höhler. Berlin: Springer-Verlag, 1982.
29. Brown, D.W., West, B.J., Lindenberg, K,: Phys. Rev. A**33**, 4110 (1986).
30. Cruzeiro, L., Halding, J., Christiansen, P.L., Skφvgaard, O., Scott, A.C.: Phys. Rev. A**37**, 880 (1988).
31. Škrinjar, M.J., Kapor, D.V., Stojanović, S.D.: Phys. Rev. A**37**, 639 (1988).
32. Brown, D.W., West, B.J., Lindenberg, K.: Phys. Rev. A**37**, 642 (1988).
33. Brown, D.W. Lindenberg, K., West, B.J.: Phys. Rev. A**33**, 4104 (1986).
34. Brown, D.W.: to appear in Phys. Rev. A (Rapid Communications).
35. Zhang, Q., Romero-Rochin, V., Silbey, R., to appear in Phys. Rev. A.
36. Škrinjar, M.J., Kapor, D.V., Stojanović, S.D.: to appear in Phys. Rev. B.
37. Wang, X., Brown, D.W., Lindenberg, K., West, B.J.: Phys. Rev. A**37**, 3557 (1988).
38. Scott, A.C.: Philos. Trans. R. Soc. London Ser. A **315**, 423 (1986).
39. Brown, D.W., Lindenberg, K., West, B.J.: Phys. Rev. Lett. **57**, 2341 (1986).
40. Brown, D.W., Lindenberg, K., West, B.J.: Phys. Rev. B**35**, 6169 (1987).
41. Brown, D.W., Lindenberg, K., West, B.J.: Phys. Rev. B**37**, 2946 (1988).
42. Glauber, R.J.: Phys. Rev. **131**, 2766 (1963).
43. West, B.J., Lindenberg, K.: J. Chem. Phys. **83**, 4118 (1986).
44. Wang, X., Brown, D.W., Lindenberg, K.: in preparation.
45. Careri, G., Gratton, E., Shyamsunder, E.: Phys. Rev. A**37**, 4048 (1988).
46. Alexander, D.M.: Phys. Rev. Lett **54**, 138 (1985).
47. Migliori, A., Maxton, P.: in progress.

FLUCTUATIONS IN THE TRANSIENT DYNAMICS OF NONLINEAR OPTICAL SYSTEMS

M. San Miguel

Departament de Física, Universitat de les Illes Balears

07071 Palma de Mallorca, Spain

1.- INTRODUCTION

The laser threshold instability can be considered a prototype among the many physical situations exhibiting transitions in states far from equilibrium. The analogies between equilibrium phase transitions and nonequilibrium instabilities were understood rather early in this context.[1] Differences between equilibrium and nonequilibrium transitions and also the general validity of mean-field approaches for the latter, have also been discussed in the context of optical systems.[2] Besides the laser threshold instability, other nonequilibrium transitions are common in nonlinear optical systems. For example transitions to chaos in the laser itself,[3] or instabilities in passive systems, as optical bistability[4] which has close analogies to a first order phase transition.

A question of general interest in nonequilibrium statistical mechanics is the role played by fluctuations. This role is crucial in dynamical processes involving a phase transition.[5] Generally speaking these processes involve the relaxation of metastable (nucleation) or unstable states (spinodal decomposition). They are initiated by fluctuations which dominate the early stages of evolution. Likewise, transient dynamical processes involving relaxation from states which have lost global stability in optical systems are also dominated by fluctuations. These processes generally involve the passage through a nonequilibrium instability point. The general problem addressed in these lectures is precisely the description of the transient relaxation in nonlinear optical systems following a change of a control parameter through an instability point. A difference with situations more akin to traditional statistical mechanics is that the final steady state is a nonequilibrium state, while, for example, in nucleation or spinodal decomposition the final state has well defined equilibrium thermodynamics. A simplification occuring in optical systems is that they often admit zero-dimensional modeling.

The approach followed here to the general problem stated above consists in reviewing a few representative examples in which the main ideas, physical mechanisms and mathematical techniques can be illustrated. First, questions related to the laser switch-on will be discussed. This refers to the decay of the unstable state leading to laser radiation. The simplest basic theory, appropriate for He,Ne lasers, will be first reviewed.[6-13] Modifications of this theory to account for situations of more recent interest will be discussed in the case of the existence of pump-noise[14-17] (dye lasers) and in connection with the concept of delayed bifurcation[18] when sweeping through the laser instability at a finite rate. The second problem to be discussed is relaxation in optical bistability when the system is placed close the end-points of the hysteresis cycle.[19-20] It corresponds to relaxation from spinodal points. These two first problems involve different relaxation mechanisms but do not include any space-dependence complication. The final problem addressed considers fluctuations in the transient dynamics of an optical instability leading to a spatial pattern.[21] The unifying theme in the discussion of these representative examples is the calculation, through similar stochastic methods, of the characteristic time scale in which the system relaxes from its original state and the fluctuations occurring during the relaxation process. The time scale is determined by passage-time techniques.

2.- LASER SWITCH-ON PROBLEMS

2.1 Basic theory

The build-up of the laser radiation when the pump parameter becomes larger than its threshold value is an example of decay of an unstable state. Such a decay can be generally described by two alternative approaches. In the laser problem the first approach focuses on the time dependence of the moments of the laser intensity. The classical experiments of Arecchi[6] and Mandel[7] show the existence of anomalously large intensity fluctuations during the transient. A second more recent approach[8-10] addresses directly the question of when the laser is switched-on: The time at which the laser intensity reaches an observable value is a stochastic quantity determined by a passage-time distribution. The mean passage time is identified with the lifetime of the unstable zero-intensity state.

The basic physical mechanisms involved in the decay of an unstable state are rather well understood. They have been formulated in essentially equivalent ways by several authors.[11-13] A key idea in these theories is the existance of two different stages of evolution. A first one is dominated by noise and linear dynamics. In the second regime saturation nonlinear terms become important and fluctuations are not essential. A final stage in which the system fluctuates around the equilibrium state is not considered here. During the first stage the system leaves the vicinity of the unstable state, so that the lifetime of the unstable state gives the time scale wich determines the separation of the two stages of evolution. The transient anomalous fluctuations occur during the nonlinear regime. They appear as an amplification by nonlinear deterministic dynamics of the initial fluctuations triggering the decay process. A summary of the application of these ideas to the laser problem is given below in the framework of the Q.D.T. (Quasideterministic theory) formulation[13].

As a starting pont I consider the appropriate laser model equation for a single mode on resonance, near threshold and in the good-cavity limit:[22]

$$\partial_t E = a\,E - |E|^2 E + \sqrt{\varepsilon}\,\xi(t) \tag{2.1}$$

E is the electric field complex amplitude, $E = E_1 + i\,E_2$, and $a = \Gamma - \kappa$ where Γ and κ are, respectively, the gain and loss parameters. The complex random term $\xi(t) = \xi_1(t) + i\,\xi_2(t)$ models spontaneous emission fluctuations of strength ε. It is taken as Ganssian white noise of zero mean and correlation

$$< \xi_i(t)\,\xi_j(t') > = \delta_{ij}\,\delta(t-t'),\ i = 1, 2 \tag{2.2}$$

The Q.D.T. approximation consists in replacing the actual process (2.1) by a process obtained from the nonlinear deterministic solution of (2.1) replacing the initial condition in this solution by the stochastic process which appears in the solution of the linear approximation to (2.1). In this way one profits the existance of the two stages of evolution mentioned before: The initial fluctuating regime is propagated in time by the nonlinear deterministic mapping. For the laser intensity $I = |E|^2$, the Q.D.T. approximation is given by

$$I(t) = |h|^2(t)\, e^{2at} \Big/ \left[1 + \frac{|h|^2(t)}{a} (e^{2at} - 1) \right] \tag{2.3}$$

$I(t)$ is a functional of the process $h(t)$. The functional form is given by the deteministic nonlinear solution of (2.1) and the process $h(t)$ enters as an effective initial condition in the solution of the linear problem:

$$I(t) = |h|^2(t)\ \exp(2at) \tag{2.4}$$

The complex process $h = h_1 + i h_2$ is defined by

$$h_i(t) = \sqrt{\varepsilon} \int_0^t dt'\, e^{-at'} \xi_i(t') + E_i(0)\ , i = 1, 2 \tag{2.5}$$

It is Gaussian bivariate and the variance of the modulus is

$$< |h|^2(t) > = (2\varepsilon / a)(1 - e^{-2at}) + < I^2(0) > \tag{2.6}$$

The initial intensity fluctuations are determined by the linear dynamics around the stable state $I = 0$ associated with an initial value of the pump parameter $a_0 < 0 : < I^2(0) > = \varepsilon / |a_0|$.
The intensity moments $< I^n(t) >$ can be calculated from (2.3) by Gaussian averaging over the realizations of the process $h(t)$. Two important consequences of this treatment are the following. First, (2.3) and (2.6) imply that for times $at \gg 1$, the time dependance of $< I^n(t) >$ is given through a scaling variable $\tau = \varepsilon\ e^{2at}$. This dynamical scaling has been verified in available data.[13] Secondly, $\delta(t) = < I^2(t) > - < I(t) >^2$ calculated from (2.3) exhibits the characteristic peak associated with transient anomalous fluctuations. At its peak value $\delta(t) \gg \delta(0), \delta(\infty)$.

The statistics of the passage time to an observable value of the intensity can be calculated neglecting the saturation term in (2.3). For times $at \gg 1$, the process $h(t)$ becomes a time-independent random variable, so that (2.4) can be inverted giving t as a random function of a fixed prescribed value I. The statistics of t are determined by the random variable $|h|^2(\infty)$. The generating function of the passage time distribution is easily calculated as

$$W(\lambda) = < e^{-\lambda t} > = \Gamma((\lambda / a) + 1)\, (I / < |h|^2(\infty) >)^{-\lambda/2a} \tag{2.7}$$

so that

$$< t > = (1/2a) \ln (aI / < |h|^2(\infty) >) - (\psi(1) / 2a) \tag{2.8}$$

$$(\Delta t)^2 = < t^2 > - < t >^2 = (1/4a^2)\, \psi'(1) \tag{2.9}$$

where ψ is the digamma function. An important feature of (2.8) – (2.9) is that $< t >$ diverges as $\ln \varepsilon^{-1}$ when $\varepsilon \to 0$, while $(\Delta t)^2$ is independent of the noise strength. The consistency of (2.8) with the assumption on the time scale of interest ($at \gg 1$) implies that the calculation is valid in the asymptotic limit $(\varepsilon / a) \ll 1$. Nonlinearities do not modify the essential ε–dependance in (2.8) – (2.9). An interesting consequence of (2.8) is that the time scale needed for the system to leave the unstable state is basically given by the time at which the scaling variable τ is of order unity. The scaling regime connects then the linear and nonlinear stages of evolution.

These general ideas for the laser switch-on problem are applied below to two particular situations which require the consideration of some new ingredients.

2.2 Transient statistics of dye lasers

Dye laser light exhibits anomalous statistical properties reviewed elsewhere.[14, 15] It has been established that these anomalies can be traced back to the existence of pump fluctuations with a finite correlation time.[16, 23] The appropriate model to study transient dynamics [10, 17] of a single-mode dye laser is obtained from (2.1) replacing the gain parameter Γ by $\Gamma + \eta(t)$, where $\eta(t)$ models pump fluctuations. Fluctuations of the saturation term are also needed when studying steady-state properties.[23] The noise source $\eta(t)$ is taken to be Gaussian, of strength Q and correlation time τ:

$$<\eta(t)\,\eta(t')> = (Q/\tau)\exp - |t-t'|/\tau \tag{2.10}$$

The new feature of this problem is the competing effects of quantum fluctuation modeled by $\xi(t)$ and pump noise modeled by $\eta(t)$. The main idea here [10, 17] is that the initial decay is triggered by $\xi(t)$, while in the nonlinear regime $\eta(t)$ becomes preponderant. This suggests the way to modify the previous Q.D.T. approximation: Instead of using a deterministic mapping to propagate the initial fluctuating regime one uses here the stochastic nonlinear mapping which incorporates pump noise when $\varepsilon = 0$

$$I(t) = |h|^2(t)\, e^{2(at+\omega(t))} / [1 + 2|h|^2(t) \int_0^t dt'\, e^{2(at'+\omega(t'))}] \tag{2.11}$$

where $\omega(t) = \int_0^t dt'\, \eta(t')$. Equation (2.11) is formally obtained replacing the initial condition $I(0)$ by $|h|^2(t)$ in the solution of the stochastic nonlinear equation for $\varepsilon = 0$. The process $\omega(t)$ is Gaussian with zero mean and variance

$$Q(t) = <\omega^2(t)> = Dt - D\tau(1 - e^{-t/\tau}) \tag{2.12}$$

The denominator in (2.11) can be neglected when describing the initial regime in which the system leaves the unstable state. The calculation of the passage time distribution requires the inversion of the formula $I(t) = |h|^2(t)\, e^{2(at+\omega(t))}$ to give t as a stochastic function of I. In order to do this $\omega(t)$ can be replaced by $\mu Q(t)$ where μ is a Gaussian random variable of unit variance. The inversion is now trivially done using two approximations. First one takes, as before, $at \gg 1$ so that $|h|^2(t) \simeq |h|^2(\infty)$. Second, for $t \gg \tau$, $Q(t) \simeq D(t-\tau)$. Proceeding now as in the case $Q = 0$ above, but taking a double average over the distributions of the random variables h and μ, one obtains for asymptotically small ε and $I(0) = 0$:

$$W(\lambda) = \frac{W_{Q=0}(\lambda)}{(1 + \lambda Q/a^2)^{\frac{1}{2}}} \exp\{\lambda^2 Q(T_0 - \tau)/[2a^2(1 + \lambda Q/a^2)]\} \tag{2.13}$$

where $W_{Q=0}(\lambda)$ is given by (2.7) and T_0 is the dominant term in (2.8).

$$T_0 = (1/2a)\ln(aI/2\varepsilon) \tag{2.14}$$

From (2.13) one obtains

$$<t> = T_0 - (\psi(1)/2a) + (Q/2a) \tag{2.15}$$

$$(\Delta t)^2 = (1/4a^2)\psi'(1) + Q(T_0 - \tau)/a^2 + Q^2/2a^4 \tag{2.16}$$

The result for $<t>$ indicates that the lifetime of the unstable state is independent of τ and essentially unaffected by pump noise since $T_0 \gg (Q/2a)$. However the variance of the passage-time distribution has an important enhancement due to pump noise through the term $Q T_0/a^2$. This enhancement is partially reduced by the finite correlation time τ. The independence of $<t>$ on pump noise and the enhancement of $(\Delta t)^2$ are in agreement with experimental results.[10] Equation (2.13) displays cleary the role of the pump-noise parameters in the passage-time distribution, while other available theories based on the same starting model do not account for the effect of the parameter τ.[14] It must be also noted that (2.13) is only valid for reference values I up to 0.5 - 0.7 the stationary value of the intensity. Beyond these values the passage time has not a clear physical meaning as a lifetime of an unstable state, and the system enters the nonlinear regime.

The nonlinear regime is better described by the time dependence of $\delta(t)$. A calculation[17] based on (2.11) shows that pump-noise produces a large increase of $\delta(t)$ at its peak value associated with the large transient fluctuations. It is also seen that $\delta(\infty)$ has a much larger value than in the case $Q = 0$ due to the preponderance of pump-noise in the final approach to equilibrium. The anomalous fluctuation defined as the difference between peak and final values of $\delta(t)$ tends to dissappear for large Q. This effect is partially compensated when increasing the value of τ.

2.3 Delayed laser instability

Another situation related to the laser switch-on problem concerns the general problem of describing an instability in which the appropriate control parameter is swept in time through the instability point. Theoretical studies of such dynamical bifurcation are reviewed by P. Mandel in these Proceedings.[18] For the laser threshold instability experimental results are available for Argon,[24] semiconductor[25] and CO_2 lasers.[26] I will consider here[27] a model directly connected with the specific case of Ref. 24. This is given by (2.1) in which the loss parameter κ is continuously swept in time at a rate α driving the laser form below to above threshold. As a consequence the parameter a in (2.1) becomes time dependent as

$$a(t) = \begin{cases} a_0 + \alpha t, & 0 \leqq t < t_1 \\ a, & t > t_1 \end{cases} \tag{2.17}$$

where $a_0 < 0$ is the initial value of a below threshold. The point of interest in this general physical situation[18] is illustrated by a deterministic linear stability analysis of (2.1) with (2.17). Defining a dynamical instability point $a(t^*)$ by the time t^* at which $I(t)$ starts growing exponentially, one finds that t^* is twice the time t at which the instability $a(t) = 0$ is crossed. This indicates a dynamical stabilization which occurs because the solution does not follow adiabatically the variation of the control parameter. The point of view adopted here is that stability in a dynamical sense is associated with the concept of lifetime and thus determined by fluctuations. The lifetime is defined as a mean passage-time. The identification of the mean passage-time with the time at which the instability is observed gives an operational definition of the dynamical bifurcation point and of the observed delay which has unambiguous experimental meaning.

The calculation of the passage time statistics follows the same steps that in Sect 2.1 with eqs. (2.4) and (2.5) replaced respectively by

$$I(t) = |h|^2(t) \exp 2\int_0^t a(t')\,dt' \tag{2.18}$$

$$h_i(t) = \sqrt{\varepsilon}\int_0^t dt' \exp\left(\int_0^{t'} dt''\, a(t'')\right) + E_i(0) \tag{2.19}$$

The calculation is carried out in the limit $\varepsilon \ll \sqrt{\alpha} \ll a$. The first inequality represents the small noise limit and the second one corresponds to slow sweeping. The times of interest are in this case always smaller than t_1. The results obtained for the first two moments of the passage time distribution are

$$\langle t\rangle - t_0 = (\alpha^{-1} T)^{\frac{1}{2}} + O(T^{-\frac{1}{2}}) \tag{2.20}$$

$$(\Delta t)^2 = (1/4\alpha)\, T^{-1}\, \psi'(1) \tag{2.21}$$

where $T \equiv \ln (I / < | h^2 | (\infty) >) \sim \ln \varepsilon^{-1}$. The right hand side of (2.20) gives the delay in the observation of the instability. It is clear in the stochastic framework considered here that such delay is determined by the noise intensity ε . The dependence on ε of $< t >$ and $(\Delta t)^2$ are clearly different than the corresponding ones given by (2.8) and (2.9) for the case of an instantaneous change of the parameter a .Here $< t > \sim (\ln \varepsilon^{-1})^{\frac{1}{2}}$ and $(\Delta t)^2$ does depend on ε . The dynamical bifurcation point can now be defined by the mean value of $a(t)$ to reach an observable value I . From (2.20) this gives

$$< a(t) > = (\alpha T)^{\frac{1}{2}} + 0 (T^{-\frac{1}{2}}) \tag{2.22}$$

The $\alpha^{\frac{1}{2}}$ dependence of (2.22) coincides with the one found in Ref. 24. However, the identification of the mean passage-time with the time of maximum rate of growth of the output intensity monitored by Scharpf et al[24] is not obvious, since the latter seems to be beyond the linear regime considered here. A stringent test of (2.20) and (2.21) is given by a direct measurement of passage time statistics as done by Spano et al.[25] In particular the α^{-1} dependence of $(\Delta t)^2$ is well confirmed. In addition, (2.20) - (2.21) give an overall good description of earlier numerical studies[28] and analogic simulations.[29]

The calculation of passage time statistics can also be carried out in the limit $\varepsilon \ll a \lesssim \sqrt{\alpha}$ of small noise and fast sweeping.[27] In this limit one finds the modifications, for a finite α , to the results (2.8) - (2.9) found in the instantaneous limit $\alpha \to \infty$. One obtains that $(\Delta t)^2$ remains independent of ε and unchanged while

$$< t > = T - (\psi(1) / 2a) + (a + | a_0 |)^2 / 2a\alpha \tag{2.23}$$

T has the same formal expression than before, but $< | h |^2 (\infty) >$ is now obviously different.

3. RELAXATION FROM MARGINALITY IN OPTICAL BISTABILITY

A relaxation problem, different from the laser switch-on problem, but which also occurs in nonlinear optical systems is associated with the decay from a state of marginal stability. For a simple potential dynamical system a possible state of marginal stability corresponds to an inflection point of the potential. A characteristic example of such marginal points is given by the end points of the hysteresis cycle occuring in optically bistable devices. In a typical experiment[19,20] involving relaxation from marginality the incident intensity is suddenly changed from a value for which the system is in equilibrium outside the hysteresis cycle to a final value close to the one associated with the end point of the hysteresis cycle. A quantity of interest is then the time that the system takes to reach the equilibrium state in the other branch of the cycle.

The normal form of a potential close to a marginality of this type is given by

$$V(x) = a x^3 + \beta x \qquad a > 0 \tag{3.1}$$

The marginal point occurs at $x = 0$ for $\beta = 0$. The control parameter β measures the distance to the point in parameter space for which the potential has a marginal point. In optical bistability β is the difference between the incident intensity and its value at the end point of the hysteresis cycle. For $x < 0$ the marginal point behaves as locally stable and for $x > 0$ as locally unstable. In the absence of fluctuations and for $\beta \leq 0$ a trajectory starting at $x_0 < 0$ never reaches the region of $x > 0$ and the marginal state $x = 0$ has infinite lifetime. A main difficulty associated with marginal relaxation is the different mechanisms acting for $\beta >< 0$. For $\beta < 0$ relaxation occurs via an activation mechanism of escape through a barrier. For $\beta > 0$ deterministic relaxation is possible and for $\beta = 0$ relaxation is triggered by diffusion in a locally flat potential.

In spite of different models and experimental situations, relaxation from marginality has some universal model-independent features. They can be manifested through a calculation of the passage-time statistics associated with the time scale in which the system leaves the vicinity of the marginal state.[30] The calculation is based on the normal form equation

$$d_t x = -\partial V/\partial x + \sqrt{2\varepsilon}\,\xi(t) \tag{3.2}$$

where $\xi(t)$ is the real part of $\xi(t)$ in (2.1) and $V(x)$ is given by (3.1) for $x < R_0$. For $x > R_0$, $V(x) = \infty$ to have global stability. Passage-times to $x = R_0$ will be calculated. An exact analytic expression exists for the mean first passage time of a process involving a single degree of freedom like (3.2).[31] This expression is given in terms of a double integral of the stationary probability distribution associated with (3.2). (Note that this result cannot be used in more complicated situations as the ones discussed in Sects 2.2 and 2.3). The asymptotic evaluation for small ε and β of that double integral for the passage-time to R_0 with initial condition x_0 gives.[30]

$$\overline{T} \equiv \langle t \rangle = (a^2\varepsilon)^{-1/3}\,\varnothing(k) + A \tag{3.3}$$

$\varnothing(k)$ is a universal function independent of x_0 and R_0;

$$\varnothing(k) = \sum_{n=0}^{\infty} B_n(-1)^n k^n \;;\; B_n = (\pi/27)^{1/2}\, 2^{2n+1}\,\Gamma(\,(2n+1)/6\,) \tag{3.4}$$

where k is the parameter that measures deviation from marginality

$$k = (\beta/a)\,(a/\varepsilon)^{2/3} \tag{3.5}$$

The constant A includes the dependence on x_0, R_0 but it is independent of a, ε and β.

There are several important consequences of (3.3). First $\overline{T}$ scales with $\varepsilon^{-1/3}$ which is essentially different from the result for the decay of an unstable state. In fact the plot of $\overline{T}$ given by (3.3) vs. $\ln \varepsilon^{-1}$ gives characteristic s - shaped curves observed in experiments.[19] More important is the structure of the series (3.4). The result for $\beta = 0$ is well known.[32] The validity of keeping one or two terms in the series is given by the smallnes of k. The series explains the different behavior of $\overline{T}$ for $\beta >< 0$. For $\beta < 0$ all the terms in the series are positive and one gets a large value of $\overline{T}$. In fact for $\beta < 0$, $|k| \gg 1$, $\varnothing(k) \sim \exp(\,(4/3\sqrt{3})\,|k|^{3/2}\,)$ which has the characteristic Kramers form for escape through a barrier. For $\beta > 0$ (3.4) is an alternating series giving a small value of $\overline{T}$ which, as $\varepsilon \to 0$, tends to the deterministic value $T_d = \int_{x_0}^{R_0} dx\,(V'(x))^{-1}$. The accuracy and universality of (3.4) is seen in Fig. 1. Fig. 1a shows a comparison of the results based on the normal form eq. (3.2) with a direct numerical simulation of the prototype model of Bonifacio and Lugiato of Optical Bistability:[4]

$$\partial_t I = I - I_I - (\,2CI/(1+I^2)\,) + \sqrt{2\varepsilon}\,\xi(t) \tag{3.6}$$

C is the cooperativity parameter and I_I the intensity of the incident field. The qualitative behavior in Fig. 1a reproduces the one of experiments in nonlinear passive systems[19] and analogic simulations.[33] The change of behavior of $\overline{T}$ from $\beta < 0$ to $\beta > 0$ indicates the change from activated relaxation to deterministic relaxation. The comparison between (3.3) and the numerical simulation is made fitting the constat A. This constant carries the model dependence of the results and it is the same for a given model independently of values of a, β and ε. The good agreement found indicates the accuracy of (3.3) and also that the same result for $\overline{T}$ holds for different models up to an adjustable additive constant A. The universal scaling form of (3.3) a)

b)

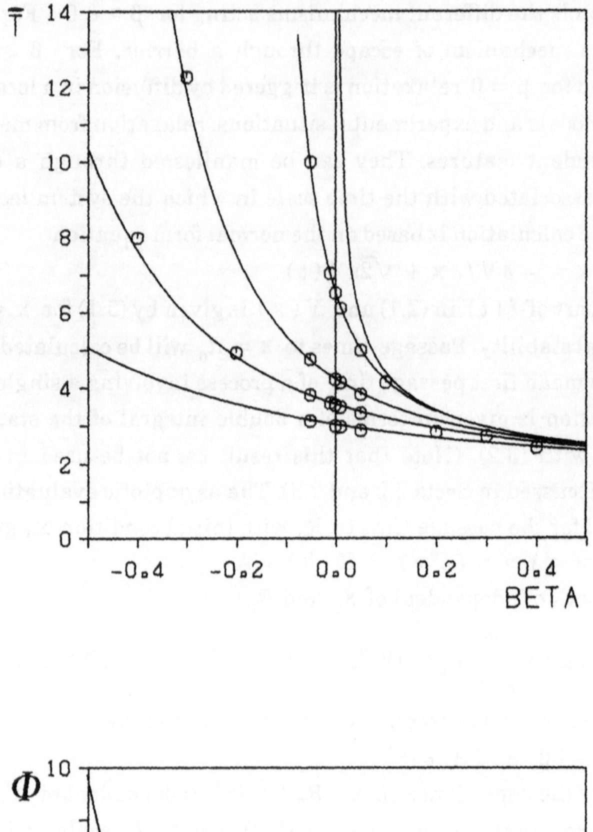

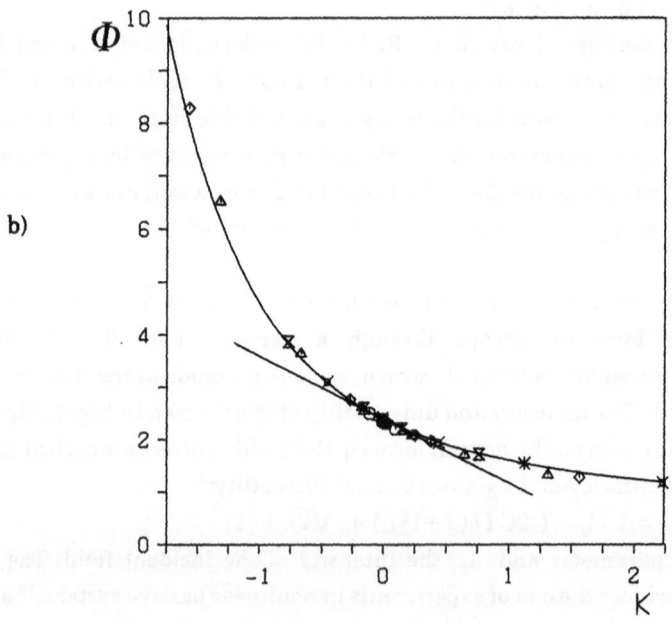

Fig. 1: a) T as a function of β for different noise intensities. From top to bottom $\varepsilon = 0$, 0.01, 0.05, 0.1, 0.2. Continuous lines correspond to (3.3) and small squares to simulation of (3.6) with $C = 20$, $I(0) = 0$, $R_0 = 10$. The parameter β for the simulation results is $\beta = I_I - I_{I,M}$ with $I_{I,M} = 21.026$. b) Scaling function $\emptyset(k)$ as a function of k.

implies that there is no essential difference in the different curves of Fig. 1a. An appropriate rescaling of axis makes them collapse into a single curve. This scaling form is shown in Fig. 1b where $\varnothing(k)$ is plotted vs. k together with simulation points in the rescaled variables. The straight line around $\beta = 0$ is obtained keeping only the terms $n = 0, 1$ in (3.4). It shows the domain of validity of the first order approximation when β is close to $\beta = 0$. Characteristic features associated with the marginality have to be observed for any small k. The scaling form of $\overline{T}$ described here has not yet been studied experimentally.

A similar universal form is found for the variance of the passage-time distribution. The asymptotic calculation of a four-fold integral[31] gives[30]

$$\sigma_T^2 \equiv (\Delta t)^2 = \sum_{n=0}^{\infty} \overline{B}_n (-1)^n k^n \tag{3.7}$$

with known coefficients $\overline{B}_n$. Using the first order approximation to (3.3) and (3.7) one finds a linear relation between σ_T and $\overline{T}$ close to $\beta = 0$:

$$\sigma_T = m + n\overline{T} \tag{3.8}$$

where n is independent of ε. As a consequence, a plot of σ_T vs. $\overline{T}$ for different ε gives a collection of parallel lines. This result accounts for the experimental observation in optical bistability of a CO_2 laser with saturable absorber.[20]

4. DYNAMICS OF PATTERN FORMATION IN THE TURING OPTICAL INSTABILITY

Rather recently,[21] instabilities leading to spontaneous pattern formation in nonlinear optical systems have been predicted and investigated. The mathematical formulation is analogous to the one describing the well known Turing instability in nonlinear chemical systems. The difference is that the role played by diffusion in these systems is played in optical systems by diffraction associated with transverse effects. This formally amounts to the consideration of a pure imaginary diffusion coefficient. Pattern formation has been studied in different geometries for nonlinear passive systems as well as laser systems.[34] Here only a passive externally driven system in a rectangular geometry is considered. The system is a Fabry - Perot cavity containing a Kerr medium and illuminated by an incident laser. The model equation in appropriate units and in the selffocusing case is[21]

$$\partial_t E(x,t) = -E + E_I + iE(|E|^2 - \theta) + ia\,\partial_x^2 E, \quad 0 \le x \le 1 \tag{4.1}$$

where x is the transverse coordinate, E the envelope amplitude of the field in the cavity, E_I the incident field and θ a detuning parameter which involves cavity and atomic detunings. Equation (4.1) is known to have homogeneous stationary solutions exhibiting optical bistability for $\theta > \sqrt{3}$. An instability[21] of the homogeneous solution E_s leading to pattern formation appears for $\theta < \sqrt{3}$. Linear stability analysis shows that the homogeneous mode $n = 0$ remains stable, but unstable modes $\cos n\pi x$ appear for $|E_s|^2 \ge 1$ and $2|E_s|^2 > \theta$. The instability first appears for $|E_s|^2 = 1$ and $a(\pi n_c)^2 = 2 - \theta$. The steady state solution describing the pattern has been calculated by methods of bifurcation theory close to the instability point.[21] A numerical study of the deterministic dynamics of pattern formation is also available.[34] The aim of the present study is to give an analytical description of the transient dynamics leading to pattern formation in a stochastic framework. The idea is that when the incident field is suddenly changed from a value smaller than the critical one E_{I_C} (associated with the instability point) to a value $E_I > E_{I_C}$, the system finds itself in an unstalbe homogeneous state E_s. This state decays through a transient dynamical process during which the pattern is formed. The decay of the unstable state is triggered by fluctuations and falls into the same general category of the problems studied in previous sections. It is here analyzed within the same stochastic scheme using similar techniques.

The inclusion of noise terms in (4.1) is essential in describing the transient process. They are responsible for the initial decay, determine the time scale in which the relaxation occurs and describe fluctuations during the transient. In addition they have in this case an important effect in the final steady state: The deterministic solutions of (4.1) imply symmetry breaking in two senses. First, the emergence of the pattern breaks the continous space translational symmetry. Secondly, there are two solutions for the intensity $I_\pm(x)$ which have no definite parity when $\Delta \to -\Delta$, being $x = \frac{1}{2} + \Delta$. This broken reflection symmetry is restored when a small noise term is included, because fluctuations mix the two solutions. This symmetry restoring effect is preserved if the noise intensity is taken to zero once the steady state is reached. This fact reflects the noncommutativity of the deterministic limit and the limit $t \to \infty$. Possible sources of noise to be included in (4.1) are quantum noise, parametric or external noise and thermal noise. Quantum noise will not be considered here. In order to introduce parametric noise associated with phase and intensity fluctuations of the incident field a frame of reference selecting the phase difference between E and E_I is introduced:

$$E_I(t) = |E_I(t)| \exp i\psi(t) \quad ; \quad \bar{E}(x,t) = E(x,t) \exp -i\psi(t) \tag{4.2}$$

The chosen stochastic model for $\bar{E}(x,t)$ is

$$\partial_t \bar{E}(x,t) = -\bar{E} + |E_I| + i(|\bar{E}|^2 - \theta)\bar{E} + i a\, \partial_x^2 E - i\dot{\psi}(t)\bar{E} + \omega(t) + \eta(x,t) \tag{4.3}$$

$\dot{\psi}(t)$ and $\omega(t)$ are associated with frequency and amplitude fluctuations of E_I respectively. They are real and space-independent stochastic processes which are taken to be Gaussian. $\dot{\psi}(t)$ models equally well fluctuations of the detuning parameter.[35] The complex and space dependent process $\eta = \eta_1 + i\eta_2$ models thermal fluctuations. It is taken to be Gaussian white noise with correlation

$$\langle \eta_\alpha(x,t)\eta_\beta(x,t')\rangle = \varepsilon_T\, \delta(x-x')\, \delta(t-t')\, \delta_{\alpha\beta} \quad , \quad \alpha,\beta = 1,2 \tag{4.4}$$

The strategy now is to reduce (4.3) to a stochastic amplitude equation for a space independent real variable describing the dynamics close to the instability.[36] One writes $\bar{E}_\alpha(x,t) = E_{s,\alpha} + q_\alpha(x,t)$ and a solution for the deviation from the homogeneous solution is seeked in the form

$$q_\alpha(x,t) = \sum_{kj} \zeta_{k,j}(t)\, O_\alpha^j(k) \cos kx \tag{4.5}$$

where $O_\alpha^j(k)$ is the α - component of the j eigenvector $(j = 1, 2)$ associated with the eigenvalue $\lambda_j(k)$ obtained linearizing the deterministic part of (4.3) around $E_{s,\alpha}$. The eigenvalues and eigenvectors are obtained for each wavenumber $k = \pi n$. Dynamical equations for the amplitudes $\zeta_{k,j}(t)$ are found[37] using the completeness of the eigenvectors $O_\alpha^j(k) \cos kx$ and introducing left-hand eigenvector $\bar{O}_\alpha^j$ satisfying

$$\sum_\alpha \bar{O}_\alpha^j(k)\, O_\alpha^l(k) = \delta_{jl} \tag{4.6}$$

The idea of the reduced dynamics is to single out the evolution of the amplitude of the unstable mode k_c in the direction of instability, $j = 1$, associated with the single positive eigenvalue $\lambda_1(k_c) > 0$, and to express the remaining amplitudes in terms of $\zeta_1(k_c)$. A useful notation that makes clear the difference between stable and unstable modes is $u = \zeta_1(k_c)$ and $s_{k,j} = \zeta_j(k)$ for $k = k_c$ and $k = k_c, j = 2$. To lowest nontrivial order in u one finds[38]

$$\partial_t u = \lambda_1(k_c)u + A_u^{(3)}u + \sum_{kj} A_u^{(2)} s_{k,j} u + \psi(t) \sum_j B_u \zeta_j(k_c) + \eta_{kc,1}(t) \tag{4.7}$$

$$\partial_t s_{k,j} = \lambda_j(k) s_{k,j} + A_{s_k}^{(2)} u^2 + \psi(t) \sum_k B_s s_k + \eta_{k,j}(t) + \delta_{k,0}\, O_1^j(0)\,\omega(t) \tag{4.8}$$

where $\eta_{k,j}(t)$ are the amplitudes of $\eta(x,t)$ in an expansion as (4.5).

The coefficients A and B are known in terms of the eigenvectors $O_\alpha^{\,l}$ and $O_\beta^{\,m}$. Mode coupling selection rules are such that $A_u^{(2)}$ vanishes except for $k = 0, 2k_c$. In addition $A_{s_k}^{(2)} = 0$ except for $k = 0$, $2k_c$ so that the unstable mode is coupled to the homogeneous and first harmonic, while the remaining modes relax to zero. Amplitude noise only couples to the homogeneous mode. Frequency noise couples to all the modes and introduces a nondiagonal linear term in (4.7). This is so because (4.5) is based on the linearization of the deterministic part of (4.3). A first approximation to the noise terms in (4.7) - (4.8) is introduced as follows: The important fluctuations are those connected with the grow of the unstable mode and therefore noise terms in (4.8) are neglected. Also, since the main interest is here in the early stages of decay, multiplicative frequency noise in (4.7) can be neglected in front of $\eta_{kc,1}$. In fact, B_u is such that strictly at the instability point, $\psi(t)$ does not couple to u. Note that within this scheme the most relevant source of noise in (4.3) influencing the decay process is found to be $\eta(x,t)$. Standard adiabatic elimination of the stable modes leads from (4.7) - (4.8) to

$$u(t) = \lambda_1(k_c)u + Cu^3 + \eta_{k_c,1}(t) \tag{4.9}$$

where C is known in terms of the previous coefficients and $\eta_{k_c,1}(t)$ is Gaussian white noise with correlation

$$< \eta_{k_c,1}(t)\, \eta_{k_c,1}(t') > = \varepsilon\, \delta(t-t') \quad ; \quad \varepsilon = 2\varepsilon_T \sum_\alpha O_\alpha^{\,1}(k_c)\, O_\alpha^{\,1}(k_c) \tag{4.10}$$

The stochastic amplitude equation (4.9) for a single space-independent variable is the reduced form of the stochastic dynamics problem posed by (4.3) for two real space-dependent variables. Eq. (4.9) can be handled with the general techniques discussed in Sect. 2.1 to study transient dynamics. If necessary, multiplicative noise terms neglected here can be taken into account as done in Sect. 2.2. From the solution of (4.9) a solution for the original problem is recovered from (4.5) recalling that only $k = 0, k_c, 2k_c$ survive and that $\zeta_{k=0,2k_c}$ are given in terms of u by adiabatic elimination.

Important qualitative facts concerning noise effects can be already discussed without further explicit calculation: For $\varepsilon = 0$ the laser intensity $I(\Delta, t)$ $(x = \frac{1}{2} + \Delta)$ calculated from (4.5) and (4.9) has the general form

$$I(\Delta,t) = I_{st} + a_1(\Delta)\, u(t) + a_2(\Delta)\, u^2(t) + a_3(\Delta)\, u^3(t) + a_4(\Delta)\, u^4(t) \tag{4.11}$$

where $I_s = E_{1,s}^{\,2} + E_{2,s}^{\,2}$ and $a_i(\Delta)$ are known coefficient with the symmetry properties $a_{1,3}(\Delta) = -a_{1,3}(-\Delta)$ and $a_{2,4}(\Delta) = a_{2,4}(-\Delta)$. In the steady state $t = \infty$, $u(t)$ reaches either of two values giving two solutions $I_\pm(\Delta)$. These two solutions only coincide at the center of the cavity $\Delta = 0$. For small ε the qualitative stochastic picture is then that for $\Delta \neq 0$ an intensity probability distribution initially peaked around I_s will evolve splitting in two peaks that will approach positions centered around $I_\pm(\Delta)$. The peak splitting will be more noticeable close to the

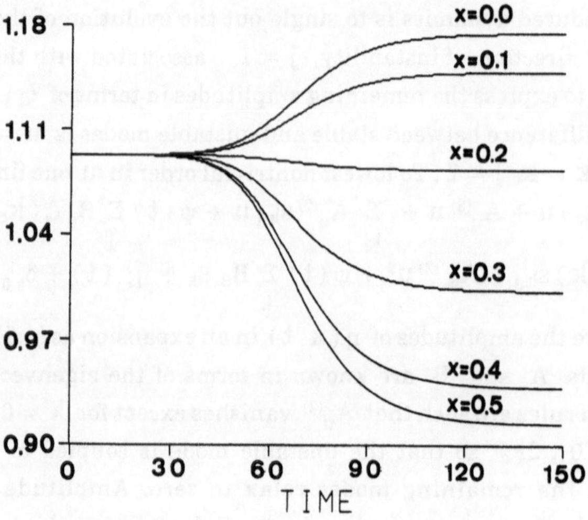

Fig. 3: Mean intensity for different space points **x**. Parameters used: $\theta = 1$, $E_I = 1.05$, $\varepsilon = 10^{-6}$

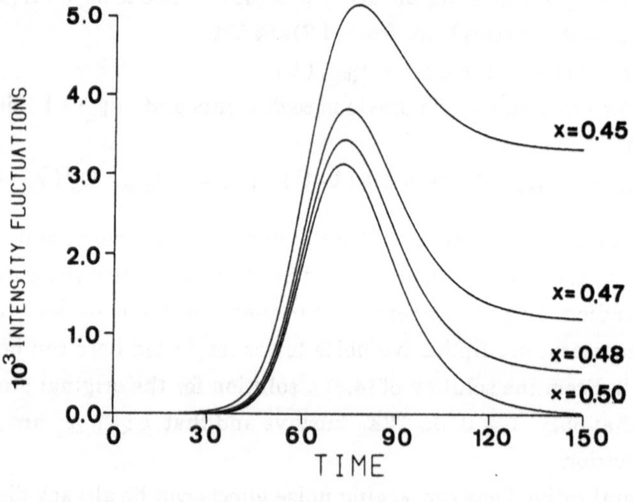

Fig. 4: Intensity fluctuations for different space points **x**. Same parameters than in Fig. 3.

walls of the cavity, while in the middle of the cavity the probability distribution remains single peaked for all times, the position of the peak moving from I_s at $t = 0$ to $I(\Delta = 0)$ for $t = \infty$. The symmetry restoring mentioned before is now obvious and it is due to mixing of the two deterministic solutions by fluctuations. Indeed $u(t)$ in (4.11) is now a stochastic process such that $< u^{2n+1}(t) > = 0$ for all times. Then $< I(\Delta, t) > = < I(-\Delta, t) >$ because only the a_2 and a_4 terms in (4.11) remain after averaging. Of course, a given realization approaches at intermediate times one of the two deterministic solutions, but at late times the two solutions are mixed by an escape mechanism activated by fluctuations.

The second important noise effect concerns intensity fluctuations in the transient evolution. One finds

$$\delta(x,t) = < (I - <I>)^2 > =$$
$$= a_1^2(x) < u^2(t) > + 2a_1(x)a_3(x) < u^4(t) > + a_3^2(x) < u^6(t) > + \Sigma(x,t) \quad (4.12)$$

where

$$\Sigma(x,t) = a_2^2(x) < (u^2(t) - < u^2(t) >)^2 > + a_4^2(x) < (u^4(t) - < u^4(t) >)^2 >$$
$$+ 2a_2(x)\, a_4(x) < (u^2(t) - < u^2(t) >)\,(u^4(t) - < u^4(t) >) > \quad (4.13)$$

$\delta(x,t)$ has two contributions: $\Sigma(x,t)$ involves variances and at steady state $t = \infty$ vanishes as $\varepsilon \rightarrow 0$. The remaining part $\delta - \Sigma$ grows with time and reaches a final value which remains finite for $\varepsilon = 0$. At $\Delta = 0$, $\delta = \Sigma$. Essentially, $\delta - \Sigma$ is associated with fluctuations occuring in two-peaked probability distributions. These remain in the limit of vanishing fluctuations in which the probability distribution becomes a superposition of two delta functions. On the other hand Σ is associated with the transient anomalous fluctuations which occur in the evolution of a single-peaked distribution. Those are the anomalous fluctuations discussed in Sect. 2 which now depend on the space point x. They are masked by the part of δ which grows monotonously in time except close to the center of the cavity ($\Delta = 0$) in which $\delta - \Sigma$ becomes small, because no peak-splitting occurs. Only in these space points the characteristic peak of the transient anomalous fluctuations will become apparent in $\delta(t)$.

An explicit calculation[38] of the time evolution of the mean intensity and intensity fluctuations is shown in Figs. 2 and 3. The calculation is based on the solution of (4.9) using the Q.D.T. approximation discussed in Sect. 2.1 with initial condition $u(0) = 0$. Note that in Sect. 2.1 the Q.D.T. is directly applied to the laser intensity associated with (2.1) while here the same development is applied to the amplitude variable u. The mean intensity evolves in time taking values smaller or larger than the original I_s value depending on the space position. The time scale in which the pattern forms is the time at which $< I >$ becomes essentially different of I_s at all points simultaneously. This time scale is given by the first passage time for the amplitude u. A calculation like (2.8) for (4.9) gives a value $< t > \simeq 60$ for the parameter values in Fig. 2. This time scale is clearly the one seen in this figure. The intensity fluctuations shown in Fig. 3 make quatitative our previous discussion. At $x = 0.5$ only Σ survives in δ and a clear peak associated with transient fluctuations occurs. The peak becomes less defined when moving away from the center of the cavity and for $0.4 > x$ or $x > 0.6$, $\delta(t)$ grows monotonously in time to a final value which is larger as one moves closer to the cavity walls.

ACKNOWLEDGMENT:

Different aspects of original research work summarized in these lectures is the result of collaborations with M. Aguado, J. Casademunt, P. Colet, F. De Pasquale, R. F. Rodríguez, J. M. Sancho, P. Tartaglia and M. C. Torrent. Financial support from Dirección General de Investigación Científica y Técnica Project No. PB – 86 –0534 (Spain) is acknowledged.

REFERENCES

1.- R. Graham and H. Haken, Z. Phys. 213, 420 (1968), 237, 31 (1970); V. De Giorgio and M. O. Scully, Phys. Rev. A 2, 1170 (1970).
2.- R. Graham in *Order and Fluctuations in Equilibrium and Nonequilibrium Statistical Mechanims*. Eds. G. Nicolis, G. Dewel and I. Turner (Wiley, 1981).
3.- N. B. Abraham in this volume.
4.- L. A. Lugiato in *Progress in Optics XXI*. Ed. E. Wolf (North Holland, 1984).
5.- J. D. Gunton, M. San Miguel and P. S. Sahni in *Phase Transitions and Critical Phenomena*, vol. 8. Eds. C. Domb and J. Lebowitz (Academic Press, 1983).
6.- F. T. Arecchi, V. De Giorgio and B. Querzola, Phys. Rev. Lett. 19, 1168 (1967).
7.- D. Meltzer and L. Mandel, Phys. Rev. Lett. 25, 1151 (1970).
8.- F. T. Arecchi and A. Politi, Phys. Rev. Lett. 45, 1219 (1980).
9.- M. R. Young and S. Singh, Phys. Rev. A 35, 1453 (1987); J. Opt. Soc. Am. B 5, 1011 (1988).
10.- R. Roy, A. W. Yu and S. Zhu, Phys. Rev. Lett. 55, 2794 (1985); Phys. Rev. A 34, 4333 (1986).
11.- M. Suzuki, Adv. Chem. Phys. 46, 195 (1981).
12.- F. Haake, J. W. Haus and R. Glauber, Phys. Rev. A 23, 3235 (1981).
13.- F. De Pasquale and P. Tombesi, Phys. Lett. 72 A, 7 (1979); F. De Pasquale, P. Tartaglia and P. Tombesi, Physica 99 A, 581 (1979); Z. Phys. B 43, 353 (1981); Phys. Rev. A 25, 466 (1982).
14.- R. Roy, A. W. Yu and S. Zhu in *Noise in Nonlinear Dynamical Systems*. Eds. F. Moss and P. McClintock, (Cambridge University Press, 1988).
15.- M. San Miguel in *Instabilities and Chaos in Quantum Optics II*, Eds. N. Abraham, F. T. Arecchi and L. Lugiato, (Plenum Press, 1988).
16.- A. W. Yu, G. P. Agrawal and R. Roy, Opt. Lett. 12, 806 (1987).
17.- F. De Pasquale, J. M. Sancho, M. San Miguel and P. Tartaglia, Phys. Rev. Lett. 56, 2473 (1986).
18.- P. Mandel in this volume.
19.- R. Deserno, R. Kume, F. Mitschke and J. Mlyneck, SPIE Proc. 700, 83 (1986); W. Lange, F. Mitschke, R. Deserno and J. Mlyneck, Phys. Rev. A 32, 1271 (1985); W. Lange in *Noise in Nonlinear Dynamical Systems*. Eds. F. Moss and P. McClintock, (Cambridge University Press, 1988); in *Instabilities and Chaos in Quantum Optics II*, Eds. N. Abraham, F. T. Arecchi and L. Lugiato, (Plenum Press, 1988).
20.- E. Arimondo, D. Dangoisse, L. Fronzoni, O. Incani and N. K. Rahman in *Optical Bistability with External Noise* (AIP Conf. Proc., to be published); E. Arimondo, C. Gabanini, E. Menchi and B. Zambon, SPIE Proc. 667, 234 (1986).
21.- L. Lugiato and R. Lefever, Phys. Rev. Lett. 58, 2209 (1987).
22.- H. Haken, in *Encyclopedia of Physics*, vol. XXV / 2c, (Springer, 1970).
23.- M. Aguado, E. Hernández - García and M. San Miguel, Phys. Rev. A (1988).
24.- W. Scharpf, M. Squicciarini, D. Bromley, C. Green, J. R. Tredicce, L. M. Narducci, Opt. Comm. 63, 344 (1987).
25.- A. Mecozzi, S. Piazzola, A. D'Ottavi and P. Spano (unpublished); P. Spano, A. D'Ottavi, A. Mecozzi and B. Daino (unpublished).
26.- F. T. Arecchi, Private communication.
27.- M. C. Torrent and M. San Miguel, Phys. Rev. A (1988).
28.- G. Broggi, A. Colombo, L. Lugiato and P. Mandel, Phys. Rev. A 33, 3635 (1986).
29.- R. Mannella, F. Moss and P. V. McClintock, Phys. Rev. A 35, 2560 (1987).
30.- P. Colet, M. San Miguel, J. Casademunt and J. M. Sancho, (unpublished).
31.- C. W. Gardiner, *Handbook of Stochastic Methods*, (Springer, 1985).
32.- F. T. Arecchi, A. Politi, L. Ulivi, Nuovo Cimento 71 B, 119 (1982).
33.- F. T. Mitschke, R. Deserno, J. Mlyneck and W. Lange, IEEE – QE 21, 1435 (1985).
34.- L. Lugiato in this volume.
35.- R. Lefever, J. W. Turner and L. A. Lugiato, J. Stat. Phys. 48, 1045 (1987).
36.- P. Coullet in *Nonlinear Phenomena in Physics*, Ed. F. Claro, (Springer, 1985).
37.- H. Haken, *Light*, vol. II (North Holland, 1985).
38.- M. Aguado, R. F. Rodríguez and M. San Miguel (unpublished).

THEORETICAL METHODS IN PATTERN FORMATION IN PHYSICS, CHEMISTRY AND BIOLOGY

M.C. Cross
Division of Physics, Mathematics and Astronomy
California Institute of Technology
Pasadena, CA 91125

I. INTRODUCTION

In these lectures some simple ideas common to theories of pattern formation in diverse areas of Physics, Chemistry and Biology are reviewed. The unifying theme of these theoretical models is the existence of an instability (bifurcation) in a system driven far from equilibrium to a state with spatial variation at a non-zero wave vector. The basic pattern forming tendency is understood by a *linear stability analysis* of the spatially uniform system. The subsequent saturation of the exponential growth by *non-linearities* together with *boundary effects* (the influence of the finite size of the domain) leads to the complexity and subtlety in the understanding of non-equilibrium structures. Some elements of the theory of linear instability, non-linearity and boundary effects will be briefly reviewed in these lectures. The emphasis is on common features of the theoretical modeling of the phenomena. No attempt will be made to assess the detailed applicability of these models to the different fields, or the main aims of the study of structure formation in the different fields. There are clearly great differences between the carefully controlled laboratory models of the physicist studying hydrodynamic instabilities the chemist studying transient evolution of complicated chemical reactions, and the in vivo observations of the biologist studying developing organisms, differences in the ability to control the system, the understanding of the underlying microscopic processes, and particularly the goals of the endeavor. At the current elementary understanding of the general phenomena, these disparities are not very evident in the theoretical modeling.

In the second part of the lectures I will report on two examples of the application of some of the above methods.

These lectures are derived from a longer forthcoming review with P.C. Hohenberg. A general overview has been presented elsewhere (Hohenberg and Cross, 1986): The present lectures provide a more detailed presentation of a narrower range of topics.

II. BASIC FEATURES OF PATTERN FORMATION

A. PATTERN FORMING SYSTEMS

In this section we describe the elementary physical properties which lead to pattern formation in a few canonical examples

1. Convection

We will often use the language of convection in describing more general phenomena, referring for example to "rolls" rather than the more cumbersome "spatial periods".

In its idealized form Rayleigh-Benard convection (Busse, 1978) involves a fluid placed between flat horizontal plates which are infinite in extent and are perfect heat conductors. The fluid is driven by maintaining the lower plate at a temperature ΔT above the upper plate temperature. For small driving the fluid remains at rest and a linear temperature profile is set up interpolating between the upper and lower plate temperatures. This is the "conducting" or "uniform" solution. Due to the thermal expansion, however, the fluid near the lower plate is less dense, an intrinsically unstable situation in the gravitational field. Of course the fluid cannot rise as a whole since there would be no place for the fluid above it to go. Thus, due to a conservation law (mass in this case) we encounter an instability at a finite wavelength — a fundamental precursor of pattern formation. This instability occurs when the driving ΔT is strong enough to overcome the dissipative effects of thermal conduction and viscosity. The "control parameter" describing the instability, the Rayleigh number R, is the dimensionless ratio of the destabilizing buoyancy force $\rho_0 \alpha g \Delta T$ to the stabilizing dissipative force $\nu\kappa/d^3$

$$R = \frac{\rho_0 \alpha g \Delta T d^3}{\kappa\nu}, \tag{2.1}$$

where ρ_0 is the average mass density, α the thermal expansion coefficient, g the acceleration of gravity, ν the kinematice viscosity, κ the thermal diffusivity and d the plate separation. The instability occurs at the value $R = 1708$, independent of the fluid under consideration. The wave number q_0 of the instability can easily be seen to be of order d^{-1}, with d the plate separation, since this is the only length scale available in this ideal, static problem. We thus arrive at the picture of an instability towards a pattern in which the fluid rises in some regions and falls in others with a characteristic horizontal length scale d. The simplest manifestation of such a solution is the familiar convective roll pattern (Fig. 1a). However due to the rotational degeneracy in the plane more complex patterns are often seen.

If the bounding plates are made of poor thermal conductors (Busse and Riahi, 1980) (compared to the conductivity of the fluid) the critical wavenumber q_0 tends to zero, corresponding to very wide convecting rolls. Another case of interest is convection in a fluid mixture (Platten and Legros, 1984) rather than a pure system. There are now two diffusing fields, the concentration and the temperature. If, as usually happens, the concentration field acts as a stabilizing effect on the usual (static) convective instability, the latter is pushed up to larger values of R. In addition, there is now a new *oscillatory* instability where the concentration moves in opposition to the temperature in a sustained temporal oscillation, which occurs in conjunction with the spatial periodicity of wavevector q_0.

2. Taylor-Couette flow

The Taylor-Couette system (DiPrima and Swinney, 1981) is another hydrodynamical example analogous to Rayleigh-Benard convection, except that the buoyancy force is replaced by the centrifugal force due to rotation. The apparatus consists of two concentic circular cylinders with fluid confined to the gap between the cylinders. If the inner cylinder is rotated an azimuthal shear flow (Couette flow) is set up. However, the larger centrifugal force near the rotating cylinder leads to an instability above a critical rotation rate, towards circulating rolls (called Taylor vortices) perpendicular to the axis of the cylinder (Fig. 1b). The radial coordinate is analogous to the vertical coordinate in convection, and the azimuthal and axial directions correspond to the horizontal directions in the Rayleigh-Benard system. Note however that there is no symmetry between these two directions in the Taylor-Couette case: the first instability is to a state of azimuthal rolls, with no spatial variation around the cylinders. Until this azimuthal invariance is destroyed, Taylor-Couette flow provides a good laboratory example for studying "one-dimensional" pattern formation. Eventually, as the rotation rate is increased, a second instability occurs to a time-dependent flow in which first one and then a second wavy modulation of the Taylor vortices travel around the cylinder at independent velocities. The behavior is even richer if the outer cylinder is rotated in the reverse direction to the inner cylinder: now the first transition may be to a spiral (barber's pole) pattern with the rolls simultaneously traveling up (or down) and around the cylinder.

3. Reaction-diffusion systems

Forces and flows are central to fluid systems; chemical systems are dominated by reaction and diffusion. In a remarkable paper Turing (1952) showed that these two simple ingredients could lead to a wide range of pattern forming instabilities. This paper opened up an enormous range of study spanning the fields of developmental biology (Turing's main interest), laboratory chemistry, applied mathematics and engineering.

The general feature of these systems is the competition between different temporal growth rates and spatial ranges of diffusion for the different chemicals in the system. For example the very simple linear equations for the concentrations $u_1(x,t)$ and $u_2(x,t)$ of two reacting and diffusing chemicals in one dimension

$$\partial_t u_1 = D_1 \partial_x^2 u_1 + a_1 u_1 + b_1 u_2, \tag{2.2}$$

$$\partial_t u_2 = D_2 \partial_x^2 u_2 + a_2 u_2 + b_2 u_1, \tag{2.3}$$

lead to an instability to a state with a wavenumber

$$q_0 = \left[\frac{1}{2}\left(\frac{a_1}{D_1} - \frac{a_2}{D_2}\right)\right]^{1/2}. \tag{2.4}$$

These equations represent some sort of superficial description of a complicated set of reactions. For

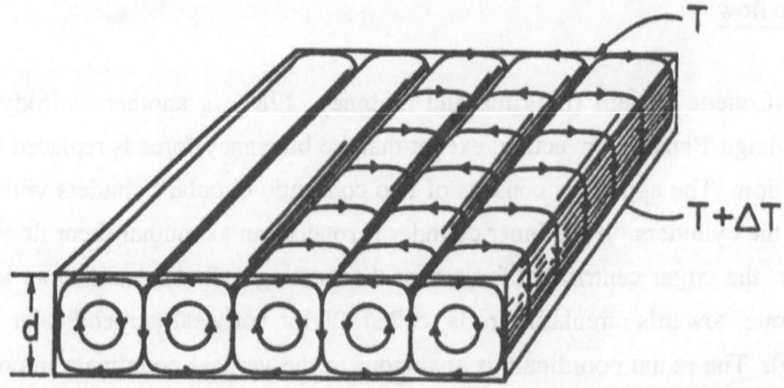

Fig. 1a) Schematic diagram illustrating Rayliegh-Benard convection. A fluid is placed between horizontal plates and heated from below. When the temperature difference ΔT exceeds a critical value ΔT_c the heat can no longer be carried up by conduction alone and the fluid is set into motion with flow in the form of convective rolls whose characteristic spacing is of order d, the plate separation.

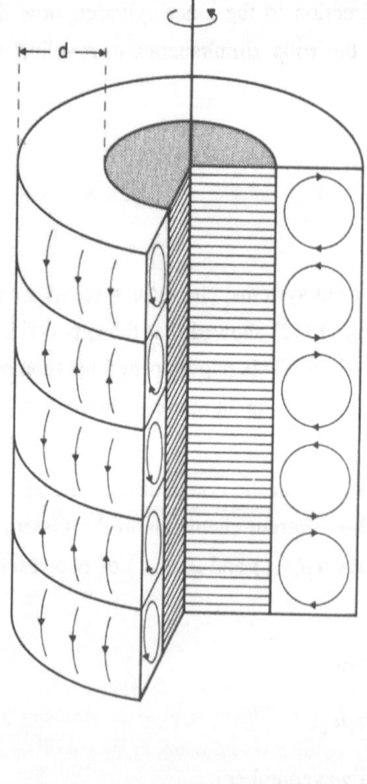

Fig. 1b) Schematic diagram illustrating Taylor-Couette flow. A fluid is placed between concentric cylinders and the inner cylinder is rotated. When the angular frequency Ω exceeds a critical value Ω_c, the flow is no longer purely azimuthal. An instability occurs to a pattern of Taylor vortices with an axial component of flow and a characteristic separation of order d, the distance between cylinders.

example, we have not discussed the mechanism for the production of u_1 and u_2. Often, more detailed models attempt to describe the competition between a slowly diffusing self-catalytic substance and a rapidly diffusing inhibiting chemical. Equations (2.2-3) then represent the deviations of the concentrations from a spatially homogeneous (unstable) steady state. More complicated sets of equations can lead to spatially periodic instabilities at finite frequencies.

A closed chemical system, just as a closed fluid system, ultimately must come to equilibrium. Nonequilibrium phenomena of interest to us often occur as a transient — maybe over long times. Recently, an experimental configuration with spatially uniform chemical pumping has been developed, so that the patterns can be investigated in steady state (Tam *et al.*, 1988).

4. Discrete Systems

In biological systems discrete models are often more natural, describing the states of discrete cells or neurons interconnected in prescribed ways. Oster (1987) recently has emphasized this point of view. If the connections are arranged to lead to *short range activation* and *long range inhibition*, instabilities at finite wavelength (of order the scale of the change over) are typical. Gierer and Meinhardt (1972) had previously emphasized the importance of such a competition, implementing the ideas in terms of coupled reaction diffusion equations for activators and inhibitors with different diffusion rates. There is a natural mapping between the cell models and discrete approximations to the differential operator of diffusion and the phenomena are closely related ; which description is the more fundamental hinges on the length scale of the instability compared to the size of the discrete elements, and will depend on the particular system.

B. LINEAR INSTABILITIES

A unifying theme for pattern forming systems such as the ones introduced above is given by linear stability analysis. We consider systems with equations that can be written in the rather general form

$$\partial_t \underset{\sim}{U}(\mathbf{x},t) = \underset{\sim}{G}[\underset{\sim}{U}, \nabla \underset{\sim}{U}, \ldots] , \tag{2.5}$$

where $\underset{\sim}{U}$ denotes the n functions $u_1(\mathbf{x},t),\ldots,u_n(\mathbf{x},t)$, and the functional $\underset{\sim}{G}$ in general involves the field $\underset{\sim}{U}$ as well as its derivatives $\nabla \underset{\sim}{U}$, $\nabla^2 \underset{\sim}{U}$, etc. We suppose that $\underset{\sim}{G}[0] = 0$ so that the uniform state $\underset{\sim}{U} = 0$ is a stationary solution of (2.5). In order to define the problem mathematically we must also specify boundary conditions and initial values. The basic instability of (2.5) is found by *linearizing* $\underset{\sim}{G}[\underset{\sim}{U}]$ about $\underset{\sim}{U} = 0$ and studying the evolution of modes of given wavevector $\mathbf{q}$

$$u_j(\mathbf{x},t) = u_{j0} e^{i\mathbf{q}\cdot\mathbf{x}+\lambda t} . \tag{2.6}$$

The ensuing *linear* equations have a set of eigenvalues $\lambda_\alpha(q)$ and we choose to focus on the one with the largest real part, which we denote as $\lambda(q)$. It is interesting to remark that in most pattern forming systems the wavevector **q** lies in a restricted dimension (1 or 2). For example in convection the periodicity is in the horizontal plane: the vertical structure is completely determined by the boundary conditions at the plates. Similarly in biology pattern formation seems to largely occur on surfaces or membranes.

Now suppose that $\underset{\sim}{G}$ depends on *a control parameter R*, such that for $R < R_c$, $\mathrm{Re}\lambda(q) < 0$, and for $R = R_c$, $\mathrm{Re}\lambda(q = q_0) = 0$, for some q_0. We introduce the *reduced* control parameter

$$\varepsilon = (R - R_c)/R_c \tag{2.7}$$

(assuming $R_c \neq 0$), and show in Fig. (2I) the dependence of $\mathrm{Re}\lambda(q)$ on ε. For $\varepsilon < 0$ the uniform state is stable and $\mathrm{Re}\lambda < 0$, whereas for $\varepsilon = 0$ the instability sets in ($\mathrm{Re}\lambda = 0$ at a wavevector $q = q_0$. For $\varepsilon > 0$ there is a *band* of wavevectors $q_- < q < q_+$ (in the infinite systems we are considering), for which the uniform state is unstable. The instability of Fig. (2I) can be of two types; either *stationary* if $\mathrm{Im}\lambda(q_0) = 0$, or *oscillatory* if $\mathrm{Im}\lambda(q_0) \equiv \omega_0 \neq 0$ for $\varepsilon = 0$. We denote these as Types Is and Io.

Another class of instability occurs if for some reason (usually a conservation law) $\mathrm{Re}\lambda(q = 0) = 0$ for all ε. We then have the situation depicted in Fig. (2II). The critical wavevector is $q_0 = 0$ and the unstable band for $\varepsilon > 0$ is $0 \leq q \leq q_+$, with $q_+ \sim \varepsilon^{1/2}$, so that the pattern occurs on a long length scale near threshold. Once again there are two possible cases (IIs and IIo), steady $[\mathrm{Im}\lambda(q = 0) = 0]$ or oscillatory $[\mathrm{Im}\lambda(q = 0) \neq 0]$.

In the case depicted in Fig. (2 III) both the instability and the maximum growth rate occur at $q_0 = 0$. Here there is no intrinsic length scale. The structure will presumably occur on a scale defined by the system size, or by the dynamics. As in the other two cases, this situation can correspond to either a steady $[\mathrm{Im}\lambda = 0]$ or an oscillatory $[\mathrm{Im}\lambda \neq 0]$ instability (Types IIIs and IIIo).

We are thus led to a classification of pattern forming systems into classes depending purely on the *linear* instabilities of the system, and in particular whether the characteristic wavenumber and frequency at the linear instability are zero or not. The ultimate long-time state depends on the nonlinearities of the equations. The simplest situation is when the nonlinearities act to saturate the exponential growth of the unstable solution. In this case, sufficiently close to the instability threshold the effect of the nonlinearities on the local structure of the pattern are weak and the final state will reflect the linear classification scheme. In this way diverse pattern forming systems tend to show similar behavior within the weakly non-linear regime.

C. NON-LINEAR STATES.

Here we will qualitatively discuss ideal non-linear states, i.e., those characterized by certain simple *symmetries* reflecting the nature of the transition in the laterally infinite system. In subsequent sections we discuss perturbation methods to calculate these simple states and the more complicated patterns encountered in realistic situations.

We are focusing on transitions at finite wavevector **q**. In systems with rotational invariance the wavevector may point in any direction in the plane; and as far as the linear analysis is concerned any superposition of these degenerate states also grows at the same rate. The non-linearity selects between different superpositions, typically leading to states of different symmetries.

For the stationary instability, patterns in the form of rolls, squares or hexagons can be constructed by superposing states of a given wavenumber. Which one is stable is determined by the non-linearity. The hexagonal state is particularly interesting since it breaks the $\underset{\sim}{U} \rightarrow -\underset{\sim}{U}$ symmetry: small first order transitions where this symmetry is broken will always favor a particular state. All three states (rolls, hexagons and squares) are observed near threshold in different convection systems.

In the case of oscillatory instabilities the basic solution is a wave. Obvious possibilities are traveling waves, or standing waves if the $\mathbf{q}$ and $-\mathbf{q}$ instabilities occur together. In addition mixed solutions may be possible.

In addition we can ask about the magnitude of the wavevector. In fact above but near the threshold non-linear spatially periodic states can be found over the whole band of unstable wavenumbers given by the linear stability analysis of the uniform state (Fig. 2). However the band of useful states is further restricted by the *stability* of these states. For the example of stationary convection the linear stability analysis of the non-linear steady states has been analyzed in some detail (Fig. 3). Away from threshold numerical methods are needed; Busse (1974) and coworkers have found a large number of characteristic instabilities, leaving however a finite *band* of stable wavenumbers for Rayleigh numbers not too far away from threshold. Two instabilities are quite generally seen in such systems, and survive to limit the band approaching threshold. These are the Eckhaus instability, a long wavelength longitudinal (compressional) instability and the Zig-Zag instability, a long wavelength transverse instability. Long wavelength here means that the instability first occurs as a distortion over arbitrarily long lengthscales. Further from threshold the instabilities (knot, oscillatory etc.) are short wavelength, and are specific to the physics of the fluid system. The Skew-Varicose instability is a mixed (neither purely longitudinal or transverse) long wavelength instability that has a certain degree of generality, as will be seen below.

III. SIMPLE THEORIES

A. AMPLITUDE EQUATIONS

In most cases of interest near the threshold for pattern forming instabilities, i.e., in the weakly non-linear regime, it is possible to reduce the complicated equations leading to the instability to a *universal form*, which goes under the name of "*amplitude equation*" (Newell and Whitehead, 1969; Segel, 1969). This approach has been rediscovered in many different specific contexts, and it bears a strong relationship to the mean-field Landau description of equilibrium phase transitions. In fact, for stationary instabilities, the resulting description is formally identical to statistical mechanics near equilibrium, and many questions of interest can in this *restricted range*, often be answered by minimizing a "potential" or "Lyapunov functional". This feature dramatically aids the analysis, but equally eliminates from consideration some subtle and interesting "non-potential" phenomena. Although the mathematical approximation leading to amplitude equations is easy to state, the precise physical conditions of applicability of the resulting theories remain unclear in many cases. (For example when do non-perturbative "non-adiabatic" effect become important?)

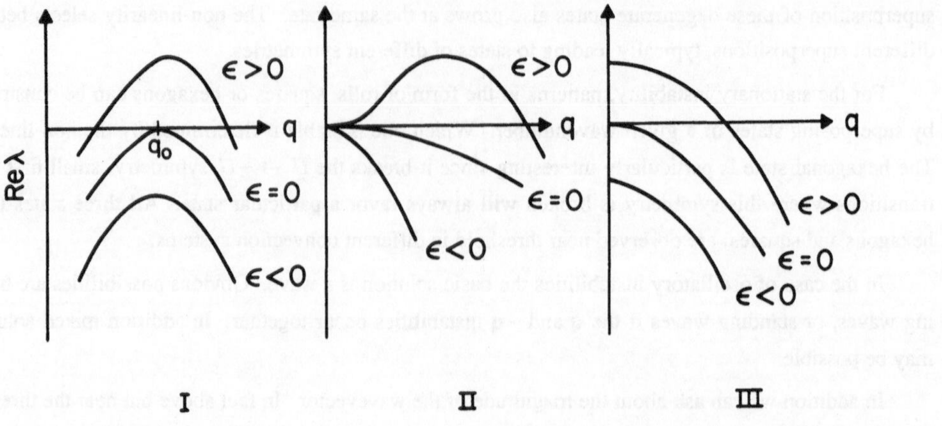

Fig. 2. The linear dispersion relation for pattern forming instabilities. The growth rate Reλ for small disturbances of a reference state with wavevector q, is plotted vs. q for three different types of linear instabilities. In case (I) the reference state is stable for all q when $\varepsilon = (R - R_c)/R_c < 0$, where R is the control parameter. When the latter reaches its critical value R_c, i.e., when $\varepsilon = 0$, the growth rate is zero at $q = q_0$ and a finite wavelength disturbance becomes marginal. For $\varepsilon > 0$ there is a continuous band of wavevectors for which the reference state is unstable (Re$\lambda > 0$). In case (II) the growth rate vanishes at $q = 0$ for all ε, and the initial instability band has a width of order $\delta q \sim \varepsilon^{1/2}$ for $\varepsilon > 0$. In case (III) the maximum growth rate is always at $q = 0$ and no intrinsic length scale is singled out by the linear dynamics.

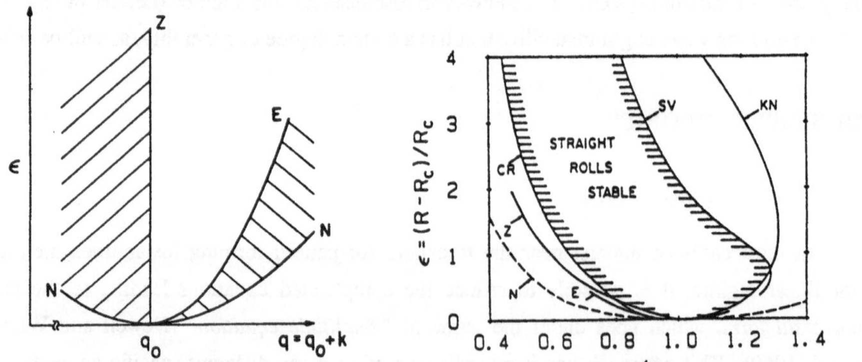

Fig. 3. The stability balloon for roll solutions of (a) the amplitude equation and (b) the hydrodynamic equations for convection in water. In part (a) we show the neutral curve N and Eckhaus (E) and zig-zag (Z) stability boundaries. Roll convection is only linearly stable between curves Z and E near threshold. In part (b) we show the results of Busse and coworkers for roll convection in water over a larger range of Rayleigh numbers. In addition to the boundaries calculated near threshold in part (a), the system has a long-wavelength instability [skewed-varicose SV] and short-wavelength instabilities (knot KN, and cross-roll CR) away from threshold.

1. Stationary periodic instability

Let us consider the plane-wave growing solution above the threshold for a steady instability ($q_0 \neq 0$, $\omega_0 = 0$ type Is)

$$\underset{\sim}{U}(\mathbf{x},t) = \underset{\sim}{U}_0 e^{iq_0 \cdot x + \lambda_0 t} \ . \tag{3.1}$$

For values of the control parameter close to threshold ($|\varepsilon|) << 1$), the structure on short length scales will be insensitive to ε, but a slow modulation in space and time is possible and the linear growth (3.1) is likely to saturate due to non-linear effects. This behavior can be analyzed by writing

$$\underset{\sim}{U}(\mathbf{x},t) = [\underset{\sim}{U}_0(z)\,A(x,y,t)e^{iq_0 x} + cc] + O(\varepsilon) \ , \tag{3.2}$$

where we have assumed a two-dimensional pattern ($\mathbf{x} = x, y$) consisting of one-dimensional rolls perpendicular to the x-direction, and cc denotes the complex conjugate. The complex amplitude $A(x,y,t)$ satisfies the equation

$$\tau_0 \partial_t A = \varepsilon A + \xi_o^2(\partial_x - \frac{i}{2q_o}\partial_y^2)^2 A - g_0|A|^2 A \ , \tag{3.3}$$

which is obtained by inserting the ansatz (3.2) into the "microscopic" equations of motion (2.5) and requiring a balance at each order in $\varepsilon^{1/4}$. Alternatively the equation in Fourier space can be derived as the projection of the full equations onto the "slow mode" corresponding to the marginally stable eigenvector (Cross, 1980). The form of the amplitude equation is quite general, and depends on the *symmetry* of the fundamental roll pattern (3.1). The linear terms in particular are a direct reflection of the instability spectrum. Newell (1974) has presented a formal derivation of amplitude equations in various situations incorporating this idea. The difference in scaling in the two directions in (3.3) reflects the inherent symmetry breaking of the roll pattern, which was here chosen with wavevector in the x-direction. This form is easily understood from the Fourier expansion for $\mathbf{q} = q_o\hat{x} + \mathbf{k}$

$$\tau_0\lambda(\mathbf{q}) = \varepsilon - \tfrac{1}{2}\lambda''(q - q_o)^2 \simeq \varepsilon - \tfrac{1}{2}\lambda''(2q_o\, k_x + k_y{}^2 + ..)^2 \tag{3.4}$$

where lowest order terms in k_x and k_y are retained. The form of the non-linear term may often be written down by inspection. Sometimes symmetry properties (e.g., $A \to -A$, which is in fact a symmetry of the Boussinesq convection equations) can be used to eliminate certain hypothetical non-linearities: this has been assumed in (3.3). In very complicated situations (e.g., degenerate bifurcations) the more formal method of "normal forms" can be used to construct the complete set of non-linear terms.

The detailed properties of individual systems undergoing the steady instability are entirely contained in the real constants in (3.3) τ_0, ξ_o, q_o and g_0, which set the scales of variation in time, space and amplitude. Equation (3.3) correctly describes the variations of the pattern on the slow time scale εt and the slow spatial scales $\varepsilon^{1/2}x$ perpendicular to the rolls and $\varepsilon^{1/4}y$ parallel to the rolls.

We may use the amplitude equation to describe many properties of patterns near threshold, such as pattern stability, competition and dynamics (Ahlers *et al.*, 1981). An important property of (3.3) is that the time evolution is "potential" in that it can be written in the form $\tau_0 \partial_t A = -\delta F/\delta A^*$ with the potential functional $F\{A, A^*\}$ given by

$$F = \iint dxdy\, [-\varepsilon |A|^2 + (g_0/2)|A|^4 + |\xi_o(\partial_x - \frac{i}{2q_o}\partial_y^2)A|^2]\,. \tag{3.5}$$

The equation of motion (3.3) then implies

$$d_t F = -2\tau_0 \iint dxdy\, |\partial_t A|^2 \le 0\,, \tag{3.6}$$

so that F is a "Lyapunov function", decreasing in any dynamics. Since F is bounded below, (3.6) implies that at long times the dynamics must eventually cease. This considerably aids in the analysis of (3.3). On the other hand, the validity of the amplitude equation is clearly shown to be restricted to the range of driving for which persistent motion is absent.

Another important limitation of the amplitude equation is that it only describes situations in which the rolls are everywhere almost normal to a particular direction, labeled the x direction [more precisely, the roll orientation may only vary by an angle of $O(\varepsilon^{1/4})$]. The slow reorientation of the rolls over large angles commonly observed, cannot be accounted for by the present theory.

Finally it should be remarked that implicitly a translational invariance of the underlying microscopic system is assumed in supposing a *complex* amplitude (a phase change then corresponds simply to a translation and the amplitude equation contains no terms coupling directly to the phase). For discrete models where the instability wavelength is comparable to the size of the discrete elements there will be strong pinning effects breaking this translational freedom.

We can easily use the amplitude equation to consider the linear stability of the non-linear steady state solutions of (3.3)

$$A_k(x) = g_0^{-1/2}(\varepsilon - \xi_o^2 k^2)^{1/2} e^{ikx}\,, \tag{3.7}$$

which corresponds to a roll solution (3.1) with wavenumber $q = q_o + k$. It is a straightforward matter to calculate the *linear stability* of (3.7) by inserting the solution $A(x) = A_k(x) + \delta A_k(x,t)$ with

$$\delta A_k(x,t) = e^{ikx}\,[\delta a_+(t)e^{i\mathbf{Q}\cdot\mathbf{x}} + \delta a_-(t)e^{-i\mathbf{Q}\cdot\mathbf{x}}] \tag{3.8}$$

into (3.3) and linearizing in $\delta a(t)$. The ensuing "stability balloon" is shown in Fig. 3, with its three stability boundaries. The neutral stability curve N given by $k_N^2 = \xi_o^{-2}\varepsilon$ marks the limit of existence of the finite solution (3.7) and the limit of instability of the "conducting" solution $A \equiv 0$. The Eckhaus boundary **E** corresponds to a longitudinal instability with $\mathbf{Q} = Q\hat{x}$ and is given by $k_E^2 = (1/3)k_N^2 = (1/3)\xi_o^{-2}\varepsilon$, and the Zig-Zag boundary Z given by $k_Z = 0$ corresponds to a growing modulation with wavevector along the rolls $\mathbf{Q} = Q\hat{y}$. Within the balloon the spatially periodic solutions are stable to all infinitesimal perturbations. Both the Eckhaus and Zig-Zag instabilities occur at long wavelength (i.e., a perturbation $\cos \mathbf{Q}.\mathbf{r}$ with $|Q| \to 0$). They are both given by the relevant "phase diffusion constant" passing through zero, and this can be used to continue their characterization into the strongly non-linear regime (see below).

Additional predictions can be made for the fastest growing mode (**Q**) for a given unstable **k**. Also the time evolution following the growing instability can be investigated numerically. For example, the complete process by which an Eckhaus instability grows to large amplitude eventually eliminating a number of rolls to yield a final wavenumber within the stable band has been followed in fascinating detail (Kramer and Riecke, 1985).

In (3.7) the assumption of a single roll solution was made. We can also look for the growth of solutions that are the superposition of many sets $i = 1$ to N of rolls at various orientations.

$$\underset{\sim}{U} = [\underset{\sim}{U}_o \sum_{i=1}^{N} A_i(x,y,t)\, e^{i\mathbf{q}_i \cdot \mathbf{x}} + c.c.] + O(\varepsilon) \tag{3.9}$$

with $|\mathbf{q}_i| = q_0$. If we leave out spatial variation of the A_i, in general (3.3) is replaced by a set of N equations

$$\tau_0 \partial_t A_i = \varepsilon A_i - \sum_{j=1}^{N} g_{ij} \; |A_j|^2 A_i \tag{3.10}$$

where the constants $g_{ij} = g(\theta_{ij})$ depends on the angles $\hat{q}_i \cdot \hat{q}_j = \cos(\theta_{ij})$. These equations can be used to investigate the competition between rolls and squares for example, with the latter described by two amplitudes A_1, A_2 and $\hat{q}_1 \cdot \hat{q}_2 = 0$. The relative stability of single rolls or squares depends on the function $g(\theta)$, which must be calculated for each system. An example where squares are in fact the preferred solution is convection between poor conductors. A particularly interesting case is that of three wavevectors mutually at an angle of 120°. In this case, because $\mathbf{q}_1 + \mathbf{q}_2 + \mathbf{q}_3 = 0$ an additional quadratic nonlinearity occurs, e.g.,

$$\tau_0 \partial_t A_1 = \varepsilon A_1 - \gamma A_2 A_3 - [g_0 |A_1|^2 + g_1(|A_2|^2 + |A_3|^2)]\, A_1 . \tag{3.11}$$

This extra nonlinearity is dominant near onset. It is easily seen that the solution $A_1 = A_2 = A_3 = A_h$ corresponding to a pattern of hexagonal cells develops by a discontinuous (backward) bifurcation, and that near threshold it is stable (and the roll solution developing continuously is unstable). This is quite generally true if $\gamma \neq 0$. This condition requires, however, that the operation $A_i \rightarrow -A_i$ (i.e., $\mathbf{U} \rightarrow \mathbf{U}$) is not a symmetry of the original system. In the case of convection this corresponds to the up-down symmetry being broken, which is only the case if typically small non-Boussinesq effects are included. The original experiment of Bénard in a dish of fluid with a free surface is an example where the symmetry is clearly broken and a hexagonal pattern was seen. Sometimes dynamic effects may lead to the asymmetry, and (3.10) can be used to follow the dynamics of the roll-hexagon transition (Ahlers *et al.*, 1981).

Having considered the existence of nonlinear solutions consisting of superpositions of different periodic solutions, we may also consider the instability of the plane wave solutions to such perturbations. In general the stability diagram becomes changed by extra instability lines, again varying as $k \propto \varepsilon^{1/2}$ corresponding to "cross-roll" instabilities. The proportionality constant depends on the details of the nonlinear couplings. For particular system parameters these new instabilities may bound the stable region.

2. Other instabilities

The same expansion about threshold can be done for the other classes of linear instability. Again, in these simple cases the results can be written down by inspection. We will describe just some of the more useful ones here.

i) Type Io

Near an oscillatory spatially periodic instability [Type Io, $\omega_0 \neq 0$, $q_0 \neq 0$] the analogue of (3.1) is

$$\underset{\sim}{U}(\mathbf{x},t) = \underset{\sim}{U}_0 [A_R(x,y,t) e^{i(q_0 x - \omega_0 t)} + A_L(x,y,t) e^{i(-q_0 x - \omega_0 t)}] + c.c. + O(\varepsilon) \tag{3.12}$$

where $A_R(x,y,t)$ and $A_L(x,y,t)$ are right- and left- traveling wave amplitudes, respectively. For the one-dimensional case the amplitude equations then read (Brand *et al.*, 1987)

$$\begin{aligned}\tau_0\{\partial_t A_R + s[\partial_x - \frac{i}{2q_0}\partial_y{}^2]A_R\} &= (1+ic_0)\,\varepsilon A_R \\ &+ \xi_0^2(1+ic_1 - \frac{is}{2q_0\xi_0^2})(\partial_x - \frac{i}{2q_0}\partial_y{}^2)^2 A_R + \frac{is}{2q_0}\partial_x{}^2 A_R \\ &- g_0(1+ic_2)|A_R|^2 A_R - g_1(1+ic_3)|A_L|^2 A_R\end{aligned} \tag{3.13a}$$

$$\begin{aligned}\tau_0\{\partial_t A_L - s[\partial_x - \frac{i}{2q_0}\partial_y{}^2]A_L\} &= (1+ic_0)\,\varepsilon A_L \\ &+ \xi_0^2(1+ic_1 + \frac{is}{2q_0\xi_0^2})(\partial_x - \frac{i}{2q_0}\partial_y{}^2)^2 A_L - \frac{is}{2q_0}\partial_x{}^2 A_L \\ &- g_0(1+ic_2)|A_L|^2 A_L - g_1(1+ic_3)|A_R|^2 A_L\end{aligned} \tag{3.13b}$$

Note that for a one dimensional situation and with only a single wave present (e.g., $A_L = 0$), (3.13) for the oscillatory periodic case can be reduced to the form of the time dependent complex Ginzburg-Landau equation by the introduction of a moving coordinate $\bar{x} = x - st$. In the limit $c_1^{-1}, c_2^{-1}, c_3^{-1} \to 0$, the imaginary terms in (3.13) dominate, and the amplitude equation reduces to the non-linear Schrodinger equation, an integrable differential equation which has been studied in great detail.

A major question for wave instabilities is the nature of the spatially homogeneous solutions: traveling or standing. Using (3.13) for solutions at wavevector $\mathbf{q}_o$ and frequency $\omega = \omega_0 + \Omega$ (Coullet *et al.*, 1985) we have

a) traveling waves:

$$A_R = a\exp(-i\Omega t + \phi); \; A_L = 0 \tag{3.14}$$

with $a^2 = \varepsilon/g_0$, and $\Omega = (c_2 - c_0)\,\varepsilon/\tau_0$ corresponding to waves traveling to the right, or the alternate solution with A_R and A_L interchanged corresponding to left moving waves

b) standing waves:

$$A_R = A_L = b\exp(-i\Omega t + \phi) \tag{3.15}$$

with $b^2 = \varepsilon/(g_0 + g_1)$ and $\Omega = [(c_2 g_0 + c_3 g_1)/(g_0 + g_1) - c_0]\varepsilon/\tau_0$. Traveling waves are stable for $g_1 > g_0 > 0$; standing waves are stable for $-g_0 < g_1 < g_0$, $g_0 > 0$; and there is no saturation for $g_0 < 0$ or $g_1 < -g_0$.

The important instability in non-linear wave systems is the Benjamin-Feir instability (Benjamin and Feir, 1967; Newell, 1974; Stuart and DiPrima, 1978). This corresponds to the instability of a wave at $(\mathbf{q}, \omega)$ by resonant excitation of sidebands with wavevectors $\mathbf{q}_1, \mathbf{q}_2$, and frequencies ω_1, ω_2 with

$$\tfrac{1}{2}(\mathbf{q}_1 + \mathbf{q}_2) = \mathbf{q} \quad ; \quad \tfrac{1}{2}(\omega_1 + \omega_2) = \omega . \tag{3.16}$$

The analysis of the one dimensional situation from the amplitude equation was discussed by Newell (1974) and in more detail by Stuart and DiPrima (1978). We assume a base state of a traveling wave

$$A_k(x) = a_k \exp[i(kx - \Omega t)] \tag{3.17}$$

and seek an instability in the form

$$\delta A_k(x) = \exp[i(kx - \Omega t)]\,[\delta a_+(t)e^{iQx} + \delta a_-(t)e^{-iQx}] \tag{3.18}$$

with Q the wavevector of the perturbation. In many ways this is the analogue of the Eckhaus instability of the stationary case, although because of the larger parameter space a full analysis becomes quite complex. In addition the instability is more potent, rendering all plane wave solutions near onset unstable for $1 + c_1 c_2 < 0$. Before this full instability limit is reached a band of stable solutions is found with

$$-\Lambda k_N < k < \Lambda k_N \tag{3.19}$$

with $k_N{}^2 = \xi_o^{-2}\,\varepsilon$ the neutral stability limit and with Λ a complicated function of the parameters. For many parameter values, but not all, the instability first occurs for long wavelength disturbances ($Q \to 0$), and may be calculated from the phase diffusion equation (see below).

For transverse perturbations varying as $\exp(iQy)$ the diffusive restoring forces are absent, and the condition for the critical wavenumber to be unstable is simply $c_2 s < 0$ (Brand *et al.*, 1987; Ohta and Kawasaki, 1987).

ii) Type IIIo

For the oscillatory uniform instability [Type IIIo $\omega_0 \neq 0$; instability and maximum growth rate at $q_0 = 0$] we again need a complex amplitude, with the phase describing the phase of the basic oscillator:

$$\underline{U}(x,t) = \underline{U}_0[A(x,y,t)e^{-i\omega_0 t} + c.c.] + O(\varepsilon) \tag{3.20}$$

The amplitude equation is similar to (3.13) with $A_L = 0$, except that now the x and y coordinates play a symmetric role:

$$\tau_0 \partial_t A = (1 + ic_0)\varepsilon A + \xi_0^2 (1 + ic_1)(\partial_x^2 + \partial_y^2) A - g_0 (1 + ic_2) |A|^2 A . \tag{3.21}$$

For one dimensional situations the analysis of (3.21) corresponds to a special case of (3.13). The longitudinal stability analysis can be derived in this way. For transverse instabilities we must remember the different form of the y-derivatives: it turns out that the transverse instability is never more important than the longitudinal one for (3.21). Finally, if the non-linear coefficient g_0 is positive, leading to saturation near onset, standing wave solutions are always unstable to traveling waves.

iii) Type IIs

For the stationary instability with zero growth rate at zero wavenumber a real amplitude is sufficient

$$\underset{\sim}{U} = \underset{\sim}{U}_0 \, \psi(x, y, t) . \tag{3.22}$$

If the non-linear terms of the basic equations are also zero at long wavelengths, a consistent long-wavelength expansion of the equations can be made. The most general amplitude equation then takes the form

$$\partial_t \psi = -\varepsilon \nabla^2 \psi - \nabla^4 \psi + g_1 \nabla . [(\nabla\psi)^2 \nabla\psi] + g_2 (\nabla\psi)^2 \nabla^2 \psi \tag{3.23}$$

where we have assumed a situation with $\psi \to -\psi$ symmetry so that quadratic non-linearities are absent. Equation (3.23) with g_2 identically zero, describes convection between infinitely poorly conducting plates (Gertsberg and Sivashinsky, 1981). In this case (3.23) is potential, whereas the general equation is not.

B. PHASE EQUATIONS

1. Steady instabilities

The amplitude equation (3.3) for the stationary instability describes the dynamics of both the magnitude $|A|$ and phase ϕ of the complex amplitude. Consider a small perturbation from the solution $A_k(x)$ describing a periodic state of wavenumber $q = q_0 + k$:

$$A_k(x) = (|A_k| + \delta |A|) e^{i(kx + \delta\phi)} \tag{3.24}$$

with $g_0 |A_k|^2 = (\varepsilon - \xi_0^2 k^2)$. From (3.3) we see that the perturbation $\delta|A|$ in the amplitude relaxes in a time of order $\varepsilon^{-1} \tau_0$, which is a slow rate near threshold, but one that remains finite for a fixed control parameter. On the other hand a homogeneous phase perturbation $\delta\phi$ does not relax at all — it is simply a uniform shift of all the rolls in the x direction. A very slow perturbation, e.g., $\delta\phi = \delta\phi_0 \cos(Qx)$ with $Q \to 0$, will relax arbitrarily slowly (on a time scale typically of order Q^{-2}). For long wavelength perturbations, with $\xi_0 Q < \varepsilon^{1/2}$, we can assume that the magnitude adiabatically follows any phase (actually any wavenumber) variation. We implement this by substituting (3.24) into (3.3) and multiplying through by $e^{-i(kx + \delta\phi)}$. The real part gives us the adiabatic amplitude change

$$g_0|A_k|\ \delta|A| = -\xi_0^2 k\partial_x\phi \tag{3.25}$$

where we neglect time and space derivatives of $\delta|A|$, and also higher order derivatives. The imaginary part gives the phase variation

$$\tau_0\partial_t\phi = \xi_0^2(\partial_x{}^2\phi + (k/q_0)\partial_y{}^2\phi) + 2\xi_0^2 k\partial_x\delta|A|/|A_k|\,. \tag{3.26}$$

Eliminating $\delta|A|$ we can derive a single equation for the phase dynamics for slow, long wavelength perturbations:

$$\partial_t\phi = D_{\|}(k)\partial_x{}^2\phi + D_{\perp}(k)\partial_y{}^2\phi \tag{3.27}$$

with

$$D_{\|} = \frac{\varepsilon - 3\xi_0^2k^2}{\varepsilon - \xi_0^2k^2}\ \frac{\xi_0^2}{\tau_0} \tag{3.28a}$$

$$D_{\perp} = \frac{k}{q_0}\ \frac{\xi_0^2}{\tau_0} \tag{3.28b}$$

This "phase diffusion" equation was derived in the context of convection by Pomeau and Manneville (1981). Notice that the zeroes of the diffusion constants $D_{\|}(k)$, $D_{\perp}(k)$ give the long wavelength phase instabilities (Eckhaus and Zig-Zag respectively) bounding the stable balloon defined by $D_{\|}$, $D_{\perp} > 0$ as in Fig. 3.

It may be noted that the description of the validity of the phase equation is not restricted to the vicinity of the threshold, but is in fact a fortiriori true away from threshold (Cross and Newell, 1984). Here "magnitude" perturbations (i.e., perturbations of the local structure) relax on a rapid timescale. Long-wavelength phase perturbations again relax arbitrarily slowly. The phase variable can now be defined more generally: if $\underset{\sim}{U}_\infty(\mathbf{q}\,.\,\mathbf{x})$ is the fully non-linear singly periodic solution with constant wavevector $\mathbf{q}$ then the solution with slow changes (on a scale η^{-1}) in the magnitude or direction of the *local* wavevector $\mathbf{q}(x,t)$ is

$$\underset{\sim}{U}(x,t) = \underset{\sim}{U}_\infty(\phi) + O(\eta)\,, \tag{3.29}$$

where $\nabla\phi(x,t) = \mathbf{q}$ and gradients of $\mathbf{q}$ are $O(\eta)$. With this general definition we are no longer restricted to small perturbations $\delta\phi$: the solution can describe the variation of the direction of the rolls through large angles, providing this takes place slowly, i.e., over many of the basic periods. The formulation of the problem in terms of a phase variable defined by (3.29) is reminiscent of the WKB approach in linear waves, and indeed this kind of approach in non-linear systems was used first in non-linear wave systems by Whitham (1970) and later by Howard and Kopell (1977). The derivation for slow distortions of stationary state was discussed by Cross and Newell (1984) for various model equations. Again, expanding to lowest order in η, the equation often takes a universal form reflecting the symmetries of the problem and certain smoothness assumptions: (i.e., distortions of a stationary, locally periodic rotationally degenerate pattern)

$$\tau(q)\,\partial_t\phi = -\nabla\,.\,[\mathbf{q}\,B(q)] \tag{3.30a}$$

$$\mathbf{q} = \nabla \phi \tag{3.30b}$$

where $\tau(q)$ and $B(q)$ are functions of the wavenumber that depend on the specific system under study.

The parameters $\tau(q)$ and $B(q)$ can be related to the diffusion constants for small perturbations from uniform rolls.

$$D_{\parallel} = -\tau^{-1}\,\partial_q (qB)\ ; \qquad D_{\perp} = -\tau^{-1} B \tag{3.31}$$

Thus the long wavelength instabilities of Fig. 3 are contained in the simple phase equation (3.30) even into the highly non-linear regime. These instabilities (Eckhaus and Zig-Zag) continue to have a generality of behavior away from threshold. Other, short wavelength instabilities depend sensitively on the specific details of the system.

It turns out for Rayleigh-Benard convection (Cross, 1983) and other similar fluid systems (Hall, 1984) that the smoothness assumption used in deriving the general form (3.30) breaks down, and in fact the expansion in the small wavenumber Q of the distortion depends on quantities such as $Q_x/|Q|$ that are not analytic as $|Q| \to 0$. A simple way of incorporating this effect is to include a coupling to a slow divergence-free "mean drift" flow $\mathbf{V}$, averaged over the depth of the cell. This gives an additional advection term,

$$[\partial_t \phi + \mathbf{V}\,.\,\nabla\phi] = -\tau^{-1}(q)\nabla\,.\,[\mathbf{q}B(q)] \tag{3.32}$$

and $\mathbf{V}$ in turn is driven by spatial inhomogeneities of the wavenumber $\mathbf{q}$ (e.g., by curvature). With this addition the "skew varicose" instability, an important long wavelength instability in convection, may be calculated within the phase formulation. The existence of "mean drift" effects is often an important additional effect in non-linear systems, and must be carefully sought. Sometimes they can be traced back to an additional conserved quantity, but this is not always the case: it is all to easy to miss such terms in developing the perturbation expansion.

The form of the phase equation is also changed if other slow modes exist. The slow phase equation must then be coupled to the dynamical equation for the slow mode, leading to higher order dynamical equations, and often propagating, rather than diffusing, solutions (Brand and Cross, 1983). This may occur, for example, if we have an additional conserved physical quantity, such as the horizontal momentum for convection between free slip boundaries (Siggia and Zippelius, 1981). Alternatively it may occur because of an additional symmetry. An example is where we look at the long wavelength dynamics of a spatially periodic solution to a phase equation $\phi_0(qx + \xi_0)$, with the new "phase" ξ_0 giving translations of the pattern (Shraiman, 1986). There are now two spatially uniform perturbations that do not relax (i.e. are at zero frequency):

$$\phi = \phi_0(x) + \delta\phi\ ; \qquad \xi = \xi_0 + \delta\xi \tag{3.33}$$

and coupled dynamics for slowly varying $\delta\phi$ and $\delta\xi$ must be considered. An alternative scheme for this problem is to introduce a derived "velocity" field $u = \partial_x \phi$ rather like we do in superfluid Helium. Then we have

$$u_t = \partial_x(\phi_t) = f(u, u_x ...) \tag{3.34}$$

where f does not depend on ϕ itself, only its derivatives u, u_x.... The quantity u is conserved (i.e., a constant u does not relax), and coupled equations for u and ξ are considered as in our first example. Note that the need for coupled equations is implied by the conservation of u. The field u need not be an actual velocity implying Galilean invariance $u \rightarrow u + c$, $x \rightarrow x + ct$. The attribution of propagating dynamics (Coullet *et al.*, 1985; Shraiman, 1986) to Galilean invariance is too restrictive: only the conservation (3.34) is needed. Galilean invariance does however give additional restrictions on the *parameters* of the coupled equations, e.g., in the equation

$$\xi_t = \alpha u + ... \tag{3.35}$$

the coefficient α is unity if Galilean invariance holds.

2. Control parameter ramps.

A variation on the idea of the phase equation is to investigate a system with a slow spatial variation ("ramp") of a quantity p that contributes to the control parameter. A particularly interesting case is where p is a function of one space dimension x, and $p(x)$ interpolates slowly between a value yielding a control parameter $\varepsilon(x)$ below threshold for $x < 0$, to a constant value p_+ giving $\varepsilon_+ > 0$ above threshold for large positive x. Using the same methods as in deriving the phase equation Kramer *et al.* (1982) showed how to derive an equation for the stationary solution for the wavenumber q for slow ramps, which is a natural generalization of the phase equation:

$$f_1(q,p)\,\partial_x q + f_2(q,p)\,\partial_x p = 0 \tag{3.36}$$

where f_1 and f_2 are functions that can be calculated for each particular system and quantity p. Integrating from the linear state for $x < 0$ where a unique solution exists yields a precise wavenumber $q_p(\varepsilon_+)$ in the constant region. For a given quantity p the wavenumber $q(\varepsilon_+)$ is unique, independent of the functional form of $p(x)$ for slow enough spatial variation. However different quantities varied yield *different* selected wavenumbers for the same final ε_+. Thus slow control parameter ramps to subthreshold values provide a precise and selectable wavenumber as opposed to the wide band consistent with stability in a periodic system. This has been beautifully exploited experimentally in the Taylor-Couette (Ahlers *et al.*, 1986), with quantitative theoretical predictions by Riecke and Paap (1987). In this system choosing p to be inner or outer cylinder radii selects different wavenumbers.

3. Oscillatory instabilities.

We can introduce phase equations for the oscillatory systems in the same way as above starting from the Amplitude equations.

For a single wave solution to the oscillatory periodic case we use the analogue to (3.24):

$$A(x,t) = (|A_k| + \delta|A|)\, \exp\,[i(kx - \Omega_k t + \delta\phi)] \tag{3.37}$$

with Ω_k the frequency of the plane wave solution. Eliminating magnitude perturbations can be done, with more algebra, as in the stationary case, and we arrive at the equation for small phase variations on long length scales $>> \varepsilon^{-1/2}\xi_0$

$$\partial_t \delta\phi + \bar{s}\, \partial_x \delta\phi = D_\parallel \partial_x^{\,2} \delta\phi + D_\perp \partial_y^{\,2} \delta\phi \tag{3.38}$$

with

$$\bar{s} = s + 2k(c_1 - c_2)\, \tau_0^{-1} \xi_0^2 \tag{3.39a}$$

$$D_\parallel = (1 + c_1 c_2) \left[\frac{\varepsilon - \xi_0^2 k^2 \left(\dfrac{3 + c_1 c_2 + 2c_2^{\,2}}{1 + c_1 c_2} \right)}{\varepsilon - \xi_0^2 k^2} \right] \tau_0^{-1} \xi_0^2 \tag{3.39b}$$

$$D_\perp = \frac{c_2 s}{2q_0} + \frac{k}{q_0} (1 + c_1 c_2)\, \tau_0^{-1} \xi_0^2 \tag{3.39c}$$

The zeroes of $D_\parallel$ and $D_\perp$ again delineate the stability boundaries, here the Benjamin-Feir instability (see above). We see that the whole band becomes longitudinally unstable for $c_1 c_2 < -1$, the classic balance of dispersive and diffusional effects. The transverse instability at $k = 0$ simply requires $c_2 < 0$, and may preempt the more familiar longitudinal instability (Kawasaki, 1986; Brand *et al.*, 1986). More generally, away from the weakly non-linear regime the phase equation takes the form

$$\partial_t \phi + \Omega(\mathbf{q}) = -\tau^{-1}(q)\, \nabla \,.\, [\mathbf{q}\, B(q)] \tag{3.40}$$

where $\Omega(q)$ is the dispersion relation of the fully non-linear but undistorted plane wave state, and $\tau(q)$ and $B(q)$ are functions of this state.

For the oscillatory uniform case we can follow the same approach to give

$$\partial_t \phi + \omega_0 = -\,\alpha \nabla^2 \phi - \beta (\nabla \phi)^2 \tag{3.41}$$

with

$$\omega_0 = \varepsilon (c_2 - c_0) \tau_0^{-1} \tag{3.42a}$$

$$\alpha = -\xi_0^2 (1 + c_1 c_2) \tau_0^{-1} \tag{3.42b}$$

$$\beta = \xi_0^2 (c_1 - c_2) \tau_0^{-1} \,. \tag{3.42c}$$

Away from threshold the phase equation takes the same form, with α, β given by the fully non-linear solution. Note β is just given by the dispersion. In one spatial dimension, this equation becomes more familiar in the form

$$u_t + \alpha\, u_{xx} + \beta u u_x = 0 \tag{3.43}$$

(with $u = \sqrt{2}\, \phi_x$) which is Burgers' equation.

4. Higher order phase equations

If the lowest order, diffusive terms in the phase equations pass through zero into the unstable regime it becomes necessary to add higher order terms to control the dynamics. (The dynamics may not however always remain within the range of validity of the phase equations: then a more complete description would be needed).

The higher order equations depend more specifically of the problem addressed than the lower order equations. Often it is not convenient to maintain the rotationally invariant description, since a different scaling of the spatial derivatives is needed to give a suitable balance. Kuramoto (1984) has written down a classification of higher order equations for small deviations from plane wave states, steady or oscillatory. The symmetries implied by the different cases may be used to restrict the possible terms. In addition relationships exist between some non-linear terms in the phase gradients and the linear dispersion relation. The higher order equations involve a balance between non-linear terms and higher order gradient terms when the coefficient of the diffusion term becomes small. The choice of balancing terms, given by suitably scaling space and non-linearities, amongst possible "higher order" ones is not always unique — often numerical work is needed to test whether the evolution remains withing the validity of the chosen scaling, or is robust to the addition of ignored terms (e.g., adding dissipative terms to otherwise conservative equations will change the long time behavior). Analysis along these lines remains in its early stages. In certain simple cases the equations reduce to well studied equations.

A simple example is given by considering the phase equation derived from the amplitude equation for the stationary-periodic case. If our reference state is at the critical wavenumber ($k = 0$ in 3.24), the state is Zig-Zag unstable and the coefficient of the ∂y^2 disappears. It is straightforward to repeat the derivation of (3.27) keeping higher order terms. If we continue to restrict ourselves to a linear equation we find

$$\partial_t \phi = \partial_x{}^2\phi + (k/q_0)\,\partial_y{}^2\phi - \frac{1}{4}q_0^{-2}\,\partial_y{}^4\phi \tag{3.44}$$

where we are imagining $\partial_x \sim \eta$, $\partial_y \sim \eta^{1/2}$, $k \sim \eta$. More generally see Cross and Newell (1984) for non-linear terms.

Other examples come from studying the variation in one spatial dimension x for our three major classes of instability. Kuramoto has also considered transverse variations. To specify the symmetries we will assume the fundamental equations are symmetric under $x \to -x$. Since we are usually dealing with dissipative systems, no time reversal symmetry is assumed.

The stationary-periodic system (Type Is) is invariant under $(\phi \to -\phi,\, x \to -x)$. This allows the equation

$$\phi_t = \alpha\phi_{xx} - \beta\phi_{xxxx} + \gamma\,\phi_x\,\phi_{xx} + \ldots \tag{3.45}$$

with $\alpha \to 0$ signaling the diffusive (Eckhaus) instability. Scaling $\partial_x \to (\alpha/\beta)^{1/2}\partial_X$, $\phi \to (\alpha/\gamma)^{1/2}\bar{\phi}$, $\partial_t \to \alpha^2\partial_T$ yields

$$\bar{\phi}_T = \bar{\phi}_{XX} - \bar{\phi}_{XXXX} + \bar{\phi}_X\,\bar{\phi}_{XX}\,. \tag{3.46}$$

Interestingly, this equation derives from a potential

$$U = \tfrac{1}{2}\int dX\ \{\alpha\bar{\phi}_X^2 + 1/3(\bar{\phi}_X)^3 + (\bar{\phi}_{XX})^2\}$$

which however is not bounded above or below, so that higher order terms may be needed to understand the dynamics. Empirically the dynamics subsequent to the Eckhaus instability is "catastrophic" evolving outside the slow variation assumed, eventually to "unwind" the phase to give a new wavenumber.

The oscillatory-uniform (Type IIIo) system is invariant under simply $x \to -x$. Thus we expect

$$\phi_t = \alpha\,\phi_{xx} - \beta\,\phi_{xxxx} + \gamma(\phi_x)^2 + \ldots \tag{3.47}$$

Note that the non-linear coefficient is given by the dispersion relation of the oscillations $\omega = \omega_0 + \gamma q^2$. Rescaling $\partial_x \to (\alpha/\beta)^{1/2}\,\partial_X$, $\phi \to (\alpha/\gamma)\bar{\phi}$, $\partial_t \to \alpha^2\partial_T$ and defining $u = \sqrt{2}\,\bar{\phi}_X$ yields at (α^3) the Kuramoto-Sivashinsky equation (Kuramoto, 1978; Sivashinsky, 1979)

$$u_T = u_{XX} - u_{XXXX} + uu_X\ . \tag{3.48}$$

The same result is expected for *transverse* variations in the stationary periodic case.

The oscillatory-periodic phase equation (Type Io) has no symmetry restrictions and is

$$\phi_t = \alpha\,\phi_{xx} + \beta\,\phi_{xxx} + \gamma(\phi_x)^2 + \ldots \tag{3.49}$$

after using a Galilean transformation to eliminate the term in ϕ_x. The scaling $\partial_x \to (\alpha/\beta))^{1/2}\,\partial_X$, $\phi \to (\alpha/\gamma)^{1/2}\bar{\phi}$, $\partial_t \to \alpha^{3/2}\partial_T$ and again defining $u = \bar{\phi}_X$ yields

$$u_T - u_{XXX} + uu_X = 0(\alpha^{1/2})\,\{u_{XX}, u_{XXX}, (uu_X)_X\} \tag{3.50}$$

Equating the left hand side to zero gives the Kortveg de Vries equation, balancing the dispersive and non-linear terms. The $(\alpha^{1/2})$ diffusive correction terms will destroy the integrability of the lowest order equation.

All these results rely on the apparently innocuous assumption of the symmetries and a smooth expansion in slow spatial gradients (compared with the characteristic universe length scales of the unperturbed pattern). However, as we have already seen in the lowest order phase equation, these assumptions may break down. Finally, to make progress it is assumed that the higher order terms have the "right sign" to saturate the dynamics on the weakly non-linear, slowly varying realm for which the equation is valid. In specific situations one must be aware of these potential difficulties.

C. TOPOLOGICAL METHODS

The introduction of a slow phase variable has implications beyond dynamical equations. The global constraints implied by the requirement that the phase field lead to single-valued microscopic fields (e.g., fluid velocity) can be used to give a topological classification of possible patterns. The importance of such mathematical methods increases with the complexity of the patterns, and they have played an important part in studies of wave instabilities in three-dimensional excitable media (Winfree, 1984).

D. ROTATIONALLY INVARIANT AMPLITUDE EQUATIONS.

The phase equation is valid over a wide range of control parameters and for arbitrarily large reorientations of the rolls, provided the rate of spatial variations is small compared with the local wavenumber. The whole approach is based on the *slow* variations of a structure that locally has the simple, singly periodic spatial structure. This method does not give a complete description of the whole pattern in typical situations, since defects — where no locally periodic structure can be identified — are common. The amplitude equation allows for more general modulations of the pattern, including amplitude modulations. For example, the properties of *dislocation* defects may be completely studied near threshold using the amplitude equation (Siggia and Zippelius, 1981), and boundary effects are easily included. However large changes in the direction of the wavevector over the box are not treated. The question arises: Can we find a treatment that includes both options?

Near threshold the "order parameter" equation introduced by Swift and Hohenberg (1977) seems a possible candidate. The form of this equation can be motivated by returning to the characteristic spectrum of the linear instability Is of Fig. (2). Let us define ψ_q to be simply the full amplitude of the plane wave eigenvector at **q**. To get the correct spectrum (3.4) must be satisfied in the linear regime. The non-linear terms may be developed perturbatively by assuming that the modes away from the critical wavenumber q_o adiabatically follow the slow time dependence. The equation becomes

$$\partial_t \psi_{\mathbf{q}} = [\varepsilon - (q - q_o)^2]\, \psi_{\mathbf{q}} + \sum_{\mathbf{q}'\mathbf{q}''\mathbf{q}'''} g(\mathbf{q}', \mathbf{q}'', \mathbf{q}''')\, \psi_{\mathbf{q}'} \psi_{\mathbf{q}''} \psi_{\mathbf{q}'''} \tag{3.51}$$

where g is a complicated function that can be calculated (Cross, 1980). We now define the real space, real order parameter as

$$\psi(r) = \sum_{\mathbf{q}} \psi_{\mathbf{q}}\, e^{i\mathbf{q}\cdot\mathbf{r}} \tag{3.52}$$

and Fourier transform (3.51) to give a real space equation. There are three difficulties. The first is that the $(q - q_o)^2$ terms does not give a nice expression. Instead we use the idea

$$(\nabla^2 + q_o{}^2)^2 \psi(r) \rightarrow (q^2 - q_o{}^2)^2 \psi_{\mathbf{q}} \simeq 4q_o{}^2(q - q_o)^2 \tag{3.53}$$

Secondly, the non-linear terms gives a short range, but non-local interaction term. The structure is quantitatively important when considering states which locally are the superposition of a number of singly periodic states (hexagons, grain boundaries, etc.) but otherwise may be replaced by a local interaction $g\psi^3(r)$. Thirdly, the full effect of boundaries is not included in the projection onto the unstable eigenvector, since the "fast" modes are forced by the boundary conditions.

However the equation for $\psi(r)$:

$$\partial_t \psi = \varepsilon\psi - (\nabla^2 + q_o{}^2)^2\, \psi - g\psi^3 \tag{3.54}$$

together with conditions

$$\psi = \hat{n} \cdot \nabla\psi = 0 \tag{3.55}$$

at boundaries with normal $\hat{s}$ reduces to the amplitude equation (3.3) to lowest order for nearly parallel rolls and reproduces the correct boundary conditions on the amplitude A. Thus the Swift-Hohenberg equation seems to have accomplished our goal near threshold.

We have discussed the derivation of (3.54) in some detail because we now have the following remarkable fact: equation (3.54) is potential, i.e.,

$$\partial_t \psi(r) = -\frac{\delta F}{\delta \psi} \tag{3.56}$$

with

$$F = \iint dxdy \, \{ -\frac{1}{2}\varepsilon\psi^2 + \frac{1}{4}\psi^4 + \frac{1}{2}[(\nabla^2 + q_o{}^2)\psi]^2 \} \tag{3.57}$$

and

$$\partial_t F = -\iint dxdy \, (\partial_t \psi)^2 \leq 0 \, . \tag{3.58}$$

The analysis of the properties are then vastly simplified, since we now have a global principle — the reduction of F — with which any dynamics must be consistent. We also arrive at a statement that is apparently inconsistent with experiment in large Rayleigh-Benard boxes: namely that all motion is a transient and persistent dynamics is impossible. Thus, as with the usual amplitude equation, the domain of validity of these equations is not entirely clear.

IV. APPLICATIONS

In the second part of the lectures I will report on two examples of the application of some of the above methods. In particular I will discuss one dimensional pattern formation in the stationary periodic case (Type Is), emphasizing the questions of wavenumber selection and experiments on Taylor-Couette flow, and the spatial structure and dynamics of the amplitude of the disturbance in the oscillatory-periodic case (Type Io).

A. TAYLOR VORTICES

The azimuthally symmetric Taylor vortices provide an experimental configuration for investigating one-dimensional pattern formation. In a series of experiments reviewed in Ahlers *et al.* (1986) these authors have demonstrated quite subtle features. In many cases quantitative understanding of the results has been found, either using the amplitude equation description near threshold, or calculations based on the full fluid equations. This work provides perhaps the most dramatic comparison of experiment and theory in this area of pattern formation, where often the connection remains vague. Long Taylor-Couette cylinders, with amplitude or phase equation descriptions backed up by quantitative calculations on the full equations, provide a fertile area for developing and testing ideas on one dimensional pattern formation. Here I point out the main features of the work: the interested reader is referred to the excellent review by Ahlers *et al.* (1986) for a more complete discussion.

1. Wavenumber selection by Natural Boundaries

The first question that naturally arises is the question of the size of the Taylor vortices. This is an important parameter on which further calculations of the flow properties (e.g., torque on the cylinders) and subsequent instabilities strongly depend.

The ends of a Couette system strongly perturb the ideal Couette flow. Typically, rigid plates co-rotating with either inner or outer cylinder are used, at least at one end. These do not match to the Couette flow characteristic of the ideal infinite cylinder. Consequently the approach to this ideal situation as the cylinder length is increased is a question that has provoked considerable attention. Firstly, the end perturbation leads to an imperfect bifurcation: the end drives a strong Taylor vortex in its vicinity, even well below the co-system threshold (Eckman vortex) and as the rotation rate is increased the Taylor vortex state spreads from this localized disturbance. Secondly, the boundary effects strongly perturb the nature of the vortices (e.g., the wavelength) near the ends. Finally, even in long cylinders the strong Taylor vortex at the ends affects the possible wavenumbers of the vortices in the center of the cylinder where Taylor vortices may (once the wavelength is fixed) be well characterized by an infinite length calculation. Because of the strong boundary pinning, the wavenumber of the central rolls can be varied, over a band by changing the length of the cylinder with the number of rolls being fixed. Eventually, when the wavenumber reaches extreme limits time dependence develops first in the central rolls. This dynamics presumably well approximates the Eckhaus instability of the infinite system. The stable band agrees beautifully with this theoretically calculated Eckhaus boundary (Riecke and Paap, 1986) shown in Fig. 5. Notice that the amplitude equation prediction is good only for very small values of ε. The sharp up turn on the low q side is due to a non-linear resonance of the solution at wavenumber q with the $2q$ solution, which also becomes unstable here.

There are, of course, finite size corrections. These were investigated quantitatively using the complex amplitude equation. The boundary condition to be applied is not obvious since the Eckman vortex leads to a large ($O(1)$ not $O(\varepsilon^{1/2})$) flow amplitude near the ends. Hall (1980) and Graham and Domaradzki (1982) investigated this question, and suggest

$$A(0) = A(L) = A_b >> \varepsilon^{1/2} \tag{4.1}$$

with A_b a fixed number (chosen to be 1 by Ahlers *et al.*). Solving for the local wavenumber at the center of the cylinder for fixed mean wavenumber $\bar{q} = 2\pi N/L$ for a number of rolls $N = 10$ and various lengths L leads to the comparison in Fig. 16 of Ahlers et al. It is easy to see from (3.26) that in steady state

$$|A|^2 q = const \tag{4.2}$$

Thus variations in $q(x)$ are most evident at small ε where the amplitude is strongly varying in space.

Ahlers et al find a quantitative agreement between the predictions of the amplitude equation together with (4.1) and experiment, including the effect of the finite size on the Eckhaus instability, and its complete suppression for a small range of wavenumbers about q_0 so that at least one stable state always grows from threshold in the finite system.

2. Ramped Boundaries

The Eckhaus instability still leaves a wide band of possible solutions in long cylinders. The actual one realized must depend on initial conditions and the experimental protocol. A unique wavenumber determined purely by the system parameters has been long sought. Ahlers *et al.* implemented the idea of a slow control parameter ramp described in section II by axially varying the outer cylinder radius and hence the gap between the cylinders with various small ramp angles α (e.g., $\alpha = .015$ radians). They then measured the local wavenumber of the vortices well away from either end (ramp or fixed), as a function of the cylinder length L. A much narrow band of wavenumbers is seen, and becoming smaller for small ramp angles confirming the qualitative theory (Fig. 5). The increasing bandwidth at small ε can be understood as arising from pinning of the vortices at the corner of the ramp: this was modeled (Cross, 1984) by adding a localized forcing term to the amplitude equation:

$$\xi_o^2 \frac{d^2A}{dz^2} + \varepsilon A - g|A|^2 A + f(z) = 0 \quad (4.3)$$

$$f(z) = h\xi_o \delta(z) \quad (4.4)$$

The solutions of this equation with the single fit parameter h dependent on the ramp angle predicts the bandwidth and dependence of the wavenumber on the length, including histeretic transitions for small ε, very well. The increasing bandwidth at larger ε is attributed to higher order terms than contained in the amplitude equation description. The unique wavenumber found by extrapolation to very small ramp angles agrees with the predictions based on calculations of the Navier Stokes equations (Riecke and Paap, 1987) as shown in Fig. 5. This is one example where a unique wavenumber is predicted and quantitatively calculated, and moreover measured experimentally. One fascinating prediction is that the wavenumber selected depends on the nature of the ramp (inner cylinder or outer cylinder ramped etc.) This method can now be used to experimentally prepare and study a particular wavenumber in a controlled predictable way. If both the cylinders are ramped, Riecke and Paap predicted the unique wavenumber selected by the ramp may become Eckhaus unstable for some Reynolds numbers. Again this was quantitatively confirmed experimentally. Similarly, different ramps at either end may be used to prepare a dynamic state where the vortices are created at one ramp, drift along the cylinder, and are destroyed at the second ramp.

B. NON-LINEAR TRAVELING WAVE STATES IN FINITE GEOMETRIES.

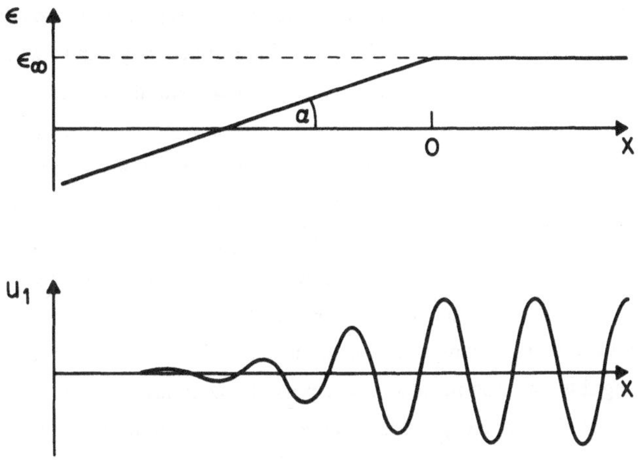

Fig. 4. Effect of a "soft boundary" on wavevector selection. The control parameter is here assumed to vary in space from a positive constant for $x > 0$ to negative values for $x < 0$, with a ramp rate α. The solution $u_1(x)$ has a well-defined wavevector at positive x. In the limit of slow ramp rates ($\alpha \rightarrow 0$) this wavevector is unique function of ε_∞.

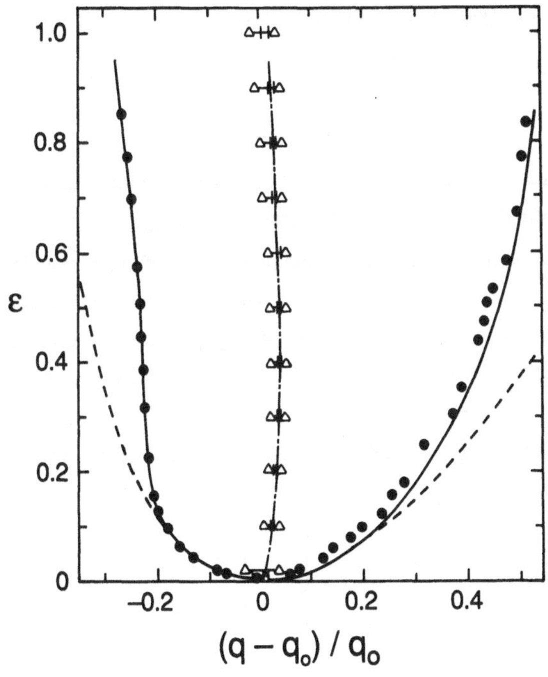

Fig. 5. Solid circles: experimental results for the Eckhaus boundary; solid line: numerical results for the Eckhaus boundary; dashed line: predicted Eckhaus boundary based on the amplitude equation; triangles($\alpha = .030$) and vertical bars ($\alpha = .015$) give bands of states selected by a ramp in the outer radius with angles α; dash-dotted line gives the theoretical prediction for slow variation of the outer cylinder radius. (From Ahlers et al. 1986 and Riecke and Paap 1986 and 1987)

In this section we discuss the question of the nature of the wave state in a one dimensional or quasi one dimensional system near an instability at finite frequency and wavevector (Type Io $\omega_0 \neq 0,\ q_0 \neq 0$). Again here the main results are summarized and the reader is referred to the literature for a more complete report. We will assume an inversion symmetry so that waves $q_0\hat{x}$ and $-q_0\hat{x}$ simultaneously go unstable. (Note this is not the case for the azimuthal waves on Taylor vortices, where $\theta \to -\theta$ is not a symmetry, but *is* the case for spiral waves, where $z \to -z$ is a symmetry.) In the nonlinear saturated state the superposition principle no longer applies and the question of standing or traveling waves immediately arises. This is easily answered in the laterally infinite geometry as discussed in section III and we will assume traveling waves to be stable there. We may then ask for the state in a finite one-dimensional geometry, where reflection at the ends might be expected to enhance the standing wave nature. Studying this question in its simplest possible form has turned out to lead to surprisingly rich answers (Cross, 1986) with the prediction of a number of static and dynamic states that have apparently been observed in binary fluid convection, waves on pure fluid convection rolls above the oscillatory instability and the spiral vortex states in Couette flow.

The full amplitude equations for this case (3.10) are still complicated, even just considering one dimensional states (i.e., $\partial_y \to 0$). The study so far has focussed on simplified equations corresponding to the case where the oscillatory effects are rather small (quantitatively $\omega_0\tau_0$ small). In this case all the imaginary terms c_0, c_1, c_2 may be ignored, and the effect of the propagation appears simply through the group advection term. Simplifying and rescaling we have

$$\partial_T A_R + \bar{s}\,\partial_X A_R = A_R - (|A_R|^2 + g\,|A_L|^2)A_R + \partial_X{}^2 A_R \tag{4.5a}$$

$$\partial_T A_L - \bar{s}\,\partial_X A_L = A_L - (|A_L|^2 + g\,|A_R|^2)A_L + \partial_X{}^2 A_L \tag{4.5b}$$

with $\bar{s} = s\varepsilon^{-1/2}\tau_0/\xi_0$ assumed to be $O(1)$, $g = g_2/g_1$ and we have introduced slow space $X = \varepsilon^{1/2}x/\xi_o$ and time $T = \varepsilon t/\tau_0$ coordinates and scaled system size $\bar{l} = \varepsilon^{1/2}l/\xi_o$.

This simplification was motivated by experiments on binary fluid convection, is necessary to get a fully consistent treatment of the end effects within the amplitude equation formulation, and has the advantage of allowing the study of the effect of propagation on one dimensional spatial structure separately from the complex Benjamin-Feir instabilities of the full equations. The qualitative results might hold in the general case however.

Equations (4.5) must be supplemented by boundary conditions. In the small frequency limit these are

$$\left.\begin{aligned} A_R - \bar{\alpha}_{\pm}\,\partial_X A_R - \bar{\beta}_{\pm}\,\partial_X A_L &= 0 \\ A_L - \bar{\alpha}_{\pm}^{*}\,\partial_X A_L - \bar{\beta}_{\pm}^{*}\,\partial_X A_R &= 0 \end{aligned}\right\} \quad \begin{aligned} x &= \pm\tfrac{1}{2}l \\ X &= \pm\tfrac{1}{2}\bar{l} \end{aligned} \tag{4.6}$$

Here $\bar{\alpha}_+ = -\bar{\alpha}_-^{*} = \bar{\alpha}$ and $\bar{\beta}_+ = -\bar{\beta}_-^{*} = \bar{\beta}$ are parameters that depend on the thermal properties of the endwall and should be taken as small compared to unity. The reflection coefficient for low amplitude (linear) waves is given by $r = -\bar{\beta}^{*}\bar{s}$ for small r.

A wide range of possible states are found solving these equations numerically. An example is shown in Fig. 6. First we consider the existence of steady state solutions to (4.5) corresponding to simply periodic solutions of the original equations, ignoring questions of stability, in a large cell and for small reflection coefficient (Figs. g,h). Here we find a dramatic transition around $\bar{s} = 2$. For $\bar{s} < 2$ the steady state consists of large amplitude traveling waves almost throughout the cell, except for short healing lengths, roughly independent of l, near the ends. Fig. h shows large amplitude right moving waves: the symmetry related state may also be found. On the other hand for $\bar{s} > 2$ the steady state in Fig. g surprisingly consists of the right hand end of the cell with saturated right moving waves, with the left hand end remaining in the quiescent, unstable state. The symmetry related state may also be obtained. The fraction of the cell containing large amplitude cells is asymptotically independent of the length of the system. The steady state is maintained by the small amplitude right moving waves produced by reflection at the right end. In a semi-infinite system (e.g., remove the right hand end wall of the system in Fig. 6), a non-linear steady state of left moving traveling waves growing out of the left end may be found for $\bar{s} \leq 2$; for $\bar{s} > 2$ no steady state solution is found, with the excited region propagating away to $+\infty$ leaving the unstable quiescent state. The evidence for the transition at $\bar{s} = 2$ rests largely on numerical work. However, for small $\bar{\alpha}, \bar{\beta}$ it can be understood analytically in terms of the change in the number of degrees of freedom at $\bar{s} = 2$ in the approach to zero of the amplitude of left moving waves for $x \to 0$. The value $\bar{s} = 2$ also corresponds to the transition point between convective and absolute instability in the infinite geometry (Deissler, 1987): the difference g-h can be thought of as the manifestation of this change in weakly non-linear states in finite geometries.

The range of solutions found numerically, including dynamics, is shown in Fig. 6. For ease of comparison with experimental trends the sequence is shown for fixed physical parameters s, l and increasing $\varepsilon/\varepsilon_c$ with ε_c the control parameter at onset shifted by finite size effects. Two combinations of system parameters play an important role: the shift in onset ε_c given by

$$\varepsilon_c = \frac{s}{l} \ln\left|\frac{1}{r}\right| \tau_0 \tag{4.7}$$

and the value of the scaled group speed at threshold $\bar{s}_c$ given by

$$\bar{s}_c = \left[\frac{sl}{\ln\left|\frac{1}{r}\right|} \frac{\tau_0}{\xi_o^{\,2}} \right]^{1/2} \tag{4.8}$$

Then above threshold the scaled group speed is

$$\bar{s} = \bar{s}_c (\varepsilon_c/\varepsilon)^{1/2} \tag{4.9}$$

always decreasing below its threshold value, and the degree of non-linearity is given by $\varepsilon/\varepsilon_c$. For the system in Fig. 6 $\bar{s}_c$ takes on the value 4.7, and ε_c is about .05.

We now describe the states (a-h). State (a) is the state found just above threshold: it consists of a stationary amplitude of counterpropagating waves, with large amplitude right moving waves in the right end of the cell, and large amplitude left moving waves in the left end of the cell. This state can be understood as the weak non-linear saturation of the linear transient (Cross, 1988). For slightly larger ε

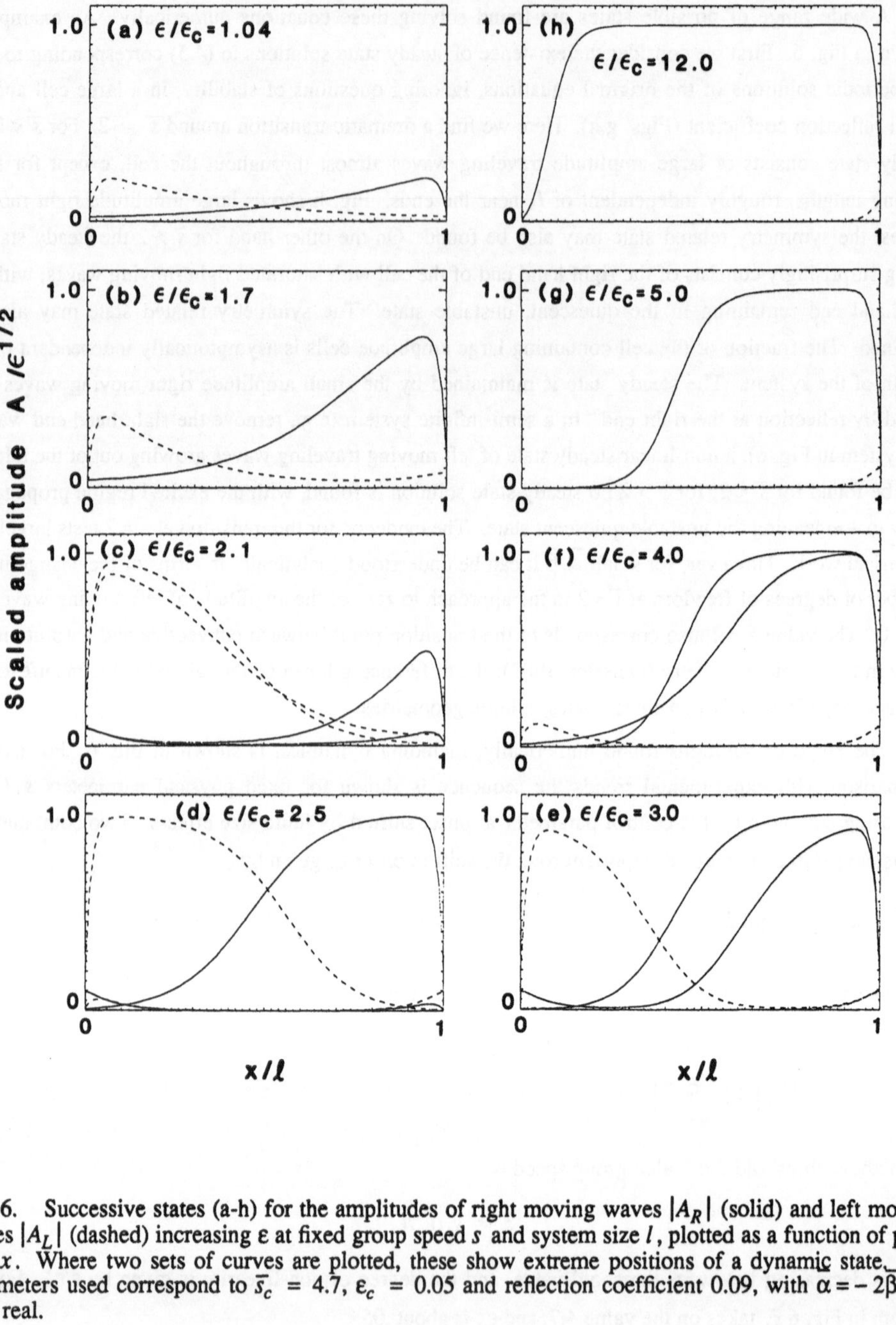

Fig. 6. Successive states (a-h) for the amplitudes of right moving waves $|A_R|$ (solid) and left moving waves $|A_L|$ (dashed) increasing ε at fixed group speed s and system size l, plotted as a function of position x. Where two sets of curves are plotted, these show extreme positions of a dynamic state. The parameters used correspond to $\bar{s}_c = 4.7$, $\varepsilon_c = 0.05$ and reflection coefficient 0.09, with $\alpha = -2\beta$ and both real.

(b) the symmetry is broken, and one wave dominates. Then (c) a periodic modulation develops, which grows in amplitude to give a symmetric oscillating state ("blinking" or "sloshing") in which large amplitude right moving waves at the right end of the cell are replaced half a cycle later by large amplitude left moving waves at the left end of the cell (d). For higher ε (e-f) the symmetry is again broken, and a modulated confined state is seen, in which the amplitude in the active region becomes saturated at $\varepsilon^{1/2}$ during parts of the cycle. The period in (c-f) is comparable to l/s, but longer by factors of order 2-4. (Specifically, we have times of 4.8, 3.8, 5.0, 4.0 l/s for the to and fro motion in (c), (d), (e) and (f) respectively, although we have not looked at very long time sequences to check the periodicity.) Note that periodic states here correspond to quasiperiodic states in the original variables. Finally, the dynamics ceases, to give the confined state (g) and subsequently the filling state (h) as $\bar{s}$ decreases below 2. The sequence of states depends on the system parameters. Most importantly, for "shorter" systems (ones with larger ε_c and smaller $\bar{s}_c$, e.g., $\varepsilon_c \simeq .12$, $\bar{s}_c \simeq 3.0$) the dynamic states (c-f) may be absent, and if $\bar{s}_c < 2$ the non-linear confined state (g) is also absent. There may perhaps be other states, for example, non-linear defect states, in longer systems: Fig. 6 does not represent a complete search of parameter space.

Remarkably many states strongly resembling those in Fig. 6 have been observed experimentally in binary fluid convection, Taylor-Couette spirals and the oscillatory waves on pure fluid convection rolls. A quantitative comparison has not yet been made. For summaries of the experimental results, see the reports in the Proceedings of the 1988 Los Alamos Conference on Advance in Fluid Turbulence (Physica D, to be published).

REFERENCES

Ahlers G., M.C. Cross, P.C. Hohenberg and S. Safran, (1981), J. Fluid Mech. *110*, 297.

Ahlers G., D.S. Cannell and M.A. Dominguez-Lerma, (1986), Physica *23D* 202.

Benjamin, T.B. and J.E. Feir, (1967), J. Fluid Mech. *27*, 417.

Bernoff, A.J., (1987), Physica *29D*.

Brand, H.R., P.S. Lomdahl and A.C. Newell, (1986), Physica *23D*, 345.

Brand. H.R. and M.C. Cross, (1982), Phys. Rev. A*27*, 1237.

Busse, F.H., (1978), Rept. Prog. Phys. *41*, 1929.

Busse, F.H. and N. Riahi, (1980), J. Fluid Mech. *96*, 243.

Coullet, P., S. Fauve and E. Tirapegui, (1985), J. Physique Lett. *46*, L-87.

Coullet, P. and S. Fauve, (1985), Phys. Rev. Lett. *55*, 2857.

Cross, M.C., (1982), Phys. Rev. A*25*, 1065.

Cross, M.C., (1983), Phys. Rev. A*27*, 490.

Cross, M.C., (1984), Phys. Rev. *A29*, 391.

Cross, M.C., (1988), Phys. Rev. A (in press). See also Phys. Rev. Lett *57*, 2935 (1986).

Cross, M.C. and A. Newell, (1984), Physica *10D*, 299.

Deissler, R.J., (1987), Physica *25D*, 233.

DiPrima, R.C. and H.L. Swinney in *Hydrodynamic Instabilities and the Transition to Turbulence*, H.L. Swinney and J.P. Gollub Eds., Topics in Applied Physics Vol. 45, Springer-Verlag, 1981.

Gertsberg, V.L. and G.I. Sivashinsky, (1981), Prog. Theor. Phys. *66*, 1219.

Gierer, A. and H. Meinhardt (1972) Kybernetik *12*, 30. See also H. Meinhardt *Models of Biological Pattern Formation*, Academic, London, 1982.

Hall, P., (1980), Proc. Roy. Soc. London *A372*, 547.

Hall, P., (1984), Phys. Rev. A*29*, 2921.

Hohenberg, P.C. and M.C. Cross, (1987), in *Fluctuations and Stochastic Phenomena in Condensed Matter Physics*, L. Garrido Ed., Lecture Notes in Physics Vol. 268 p.55, Springer (New York), 1981.

Howard, L.N. and N. Kopell, (1981), Stud. Appl. Math. *64*, 1.

Kramer, L., E. Ben-Jacob, H. Brand and M.C. Cross, (1982), Phys. Rev. Lett. *49*, 1891.

Kramer, L. and H. Riecke, (1985), Z. Phys. *B59*, 245.

Kuramoto, Y., (1978), Prog. Theor. Phys. (Kyoto) Supp. *64*, 346.

Kuramoto, Y., (1984), Prog. Theor. Phys. (Kyoto) *71*, 1182.

Newell, A.C., (1974), Lectures in Applied Mathematics, *15*, 157.

Newell, A.C. and J.A. Whitehead, (1969), J. Fluid Mech. *38*, 279.

Oster, G.F., (1987) *Proceedings of the Conference on Non-Linearity in Biology and Medicine*, Los Alamos (1987) to appear in Math. Biosci (1987).

Ohta, T. and K. Kawasaki, (1987), (preprint).

Platten, J.C. and J. Legros, *Convection in Liquids*, Springer-Verlag 1984, Ch. IX.

Pomeau, Y. and P. Manneville, (1981), J. Phys. (Paris) *42*, 1067.

Riecke, H. and H-G. Paap, (1986), Phys. Rev. *A33*, 547.

Riecke, H. and H-G. Paap, (1987), Phys. Rev. Lett. *59*, 2570.

Segel, L.A., (1969), J. Fluid Mech. *38*, 203.

Shraiman, B., (1986), Phys. Rev. Lett. *57*, 325.

Sivashinsky, G.I., (1979), Acta Astronautica *6*, 569.

Stuart, J.T. and R.C. DiPrima, (1978), Proc. R. Soc. Lond. *A362*, 27.

Swift, J. and P.C. Hohenberg, (1977), Phys. Rev. A*15*, 319.

Tam, W.Y., W. Horsthemke, Z. Noszticzius and H.L. Swinney, (1988), (preprint).

Turing, A.M., (1952), Phil. Trans. R. Soc. Lond. B*237*, 37.

Whitham, G.B., (1970), J. Fluid Mech. *44*, 373.

Winfree, A.T., (1984), Physica *12D*, 321.

TWO NONEQUILIBRIUM PHASE TRANSITIONS: STOCHASTIC HOPF BIFURCATION AND ONSET OF RELAXATION OSCILLATIONS IN THE DIFFUSIVE SINE-GORDON MODEL

Luis L. Bonilla

Física Teórica, Universidad de Sevilla
Sevilla, Spain

1. INTRODUCTION

The purpose of this lecture is presenting two examples of nonequilibrium phase transitions characterized by a time-periodic order parameter in the ordered phase. In both cases we can understand the phase transition as the self-synchronization of elementary nonlinear oscillators coupled by diffusion and in contact with a white-noise source. In our first example (the stochastic supercritical Hopf bifurcation) the collective oscillation appears as a soft excitation, [1]. In our second example (the diffusive sine-Gordon model in 2D), there is a hard excitation of a collective relaxation oscillation when an applied external field goes beyond some critical value, [2]. This transition is found at temperatures below the Kosterlitz-Thouless critical temperature.

In the past few years there has been some interest in systems displaying collective temporal oscillations, [3-6]. While deterministic systems described by an oscillatory order parameter have been extensively studied, [3], this is not the case if there are noise sources and the system has infinitely many degrees of freedom. Typical examples are

(a) the Brusselator, [3], with an additive white-noise forcing term, whose mean-field version was analyzed by Scheutzow, [4];

(b) the Poincaré oscillator with an additive white-noise forcing term, [5]:

$$\partial_t u_1 = \Lambda\, u_1 - \Omega u_2 + D\Delta u_1 + \sqrt{T}\, w_1(\underline{x},t),$$

$$\partial_t u_2 = \Lambda\, u_2 + \Omega u_1 + D\Delta u_2 + \sqrt{T}\, w_2(\underline{x},t), \qquad (1.1)$$

Here $\Lambda = -\alpha_1 - \lambda_1(u_1^2+u_2^2)$, $\Omega = -\alpha_2 - \lambda_2(u_1^2+u_2^2)$; $\langle w_i \rangle = 0$, $\langle w_i(\underline{x},t) w_j(\underline{x}',t') \rangle = \delta_{ij}\delta(\underline{x}-\underline{x}')\ \delta(t-t')$.

(c) The diffusive sine-Gordon model in the presence of an external field and an additive white noise term (this model is related to the overdamped Fukuyama-Lee model of sliding charge-density waves, [6]):

$$D^{-1}\partial_t u = \Delta u + E - (\alpha/\beta a^2)\sin\beta u + \sqrt{(2/D)}\ w(\underline{x},t),$$
$$\langle w \rangle = 0,\ \langle w(\underline{x},t)\ w(\underline{x}',t')\rangle = \delta(\underline{x}-\underline{x}')\ \delta(t-t'). \tag{1.2}$$

These examples are described by nonlinear stochastic reaction-diffusion equations over possibly unbounded supports in one, two, or three dimensions. If we "disconnect" noise and diffusion in (a) to (c), the resulting dynamical systems are capable of oscillatory behavior in time. Turning an additive white noise on, but not diffusion, yields a stochastic system of few degrees of freedom which is characterized by a unique stable stationary probability distribution. Thus no true phase transitions in the usual sense are possible, [7] : we would need a system with infinitely many degrees of freedom for a phase transition to appear. While this is achieved by adding the diffusion terms, it is interesting to see that phase transitions to oscillatory states are already present in the corresponding mean-field versions of models (a), (b) and (c). In the mean-field version, the laplacian is discretized and changed to

$$\Delta u(\underline{x},t) \quad \rightarrow \quad - N^{-1}\sum_{j=1}^{N} (u_i(t)-u_j(t)), \qquad N\rightarrow+\infty.$$

In the thermodynamic limit, $N^{-1}\ \Sigma u_j(t)$ becomes the average of u, and the resulting equations are easy to analyze. In all cases, the corresponding probability distributions bifurcate from stable stationary distributions to time-periodic ones as a control parameter crosses some threshold value. In cases (a) and (b), the one-particle probability distribution undergoes a supercritical Hopf bifurcation whereas in case (c) a time-periodic distribution is created by a saddle-node separatrix. Except for the ordered phase being characterized by a time-periodic order parameter (the mean value of u), these bifurcations are analogous to equilibrium phase transitions.

Conventional wisdom has it that: (i) the mean-field version of a model is equivalent to the short-range diffusive interaction for large space dimension; (ii) sometimes fluctuations destroy the mean-field phase transition in low enough space dimension. Thus, an analysis of the short-range models (a), (b) and (c) needs to be carried out before we may conclude that self-synchronization phenomena are present in them: extrapolation of the mean-field results might be unwarranted. We will now perform such an analysis for models (b) and (c). Near the critical point model (a) can be reduced to model (b) by projection methods, [3].

In Section 2 we derive an effective equation for the order parameter of the transition, i. e., for the mean value of the solution of the stochastic equation. As usual, this equation has a meaning only after renormalization of the original problem is performed. We do this for model (b) in Section 3, near 4 space dimensions. In Section 4 we analyze model (c) in two space dimensions. We find, [2] :

(i) There is a Kosterlitz-Thouless phase transition described by the Kosterlitz renormalization-group equations, just as in the equilibrium case.

(ii) Below the critical temperature, a reduced effective equation holds asymptotically as the infrared regulator (1/volume) goes to zero.

(iii) The self-synchronization transition takes place as the external field goes beyond a certain critical value.

2. MARTIN-SIGGIA-ROSE FORMALISM AND EFFECTIVE EQUATION

As in the case of the mean-field models, [4-6], the average value of $u(\underline{x},t)$ in (1.1) or (1.2) is the order parameter that characterizes the self-synchronization process. Near the transition point one may expect a slow time and space evolution of the order parameter toward its stable value (critical slowing-down). This slow evolution is described by a reduced equation which is asymptotically valid as the control parameter (temperature, external field, ...) approaches its critical value. We shall call <u>effective equation</u> to such a reduced equation. The effective equation plays the same role as the normal form in bifurcation theory, [8]. For constant external fields and constant order parameter, it is similar to the equation of state in equilibrium phase transitions, [9]. We shall now derive the effective equation for a generic model with additive white noise. Generalizations to colored and multiplicative noises are straightforward, [10].

Let us consider the equation

$$(D^{-1}\,\partial_t - \Delta + m^2)\, u = f(u) + \sqrt{(2/D)}\; w(x,t), \qquad t>0, \tag{2.1}$$

in d space dimensions. with $f(0)=0$, $f'(0)=0$. Here u, f, and w are n-component vector functions; w is a zero-mean, delta-correlated gaussian white noise. For the time being, let us consider the initial condition $u=0$ at $t=0$. There are several ways of deriving the effective equation for $\langle u(x,t)\rangle$ from (2.1). We shall use the path-integral version of the Martin-Siggia-Rose (MSR) formalism. [10-11].

Let us suppose that the stochastic equation (2.1) with zero initial condition has a unique solution. Then the path probability density for $u(x,t)$ is related to that of $w(x,t)$ by means of

$$\begin{aligned} P\{u(x,\tau),\ 0\leq\tau\leq t\} &= \det\{K-f'(u)\}\ P\{w(x,\tau),\ 0\leq\tau\leq t\}\ \delta(u(x,0)) \\ &= N(t)\ \delta(u(x,0))\ \exp\{-R(0)\textstyle\int_0^t\int f'(u)\ dxdt - \frac{1}{2}\int_0^t\int w^2 dxdt\}. \end{aligned} \tag{2.2}$$

Here $N(t)$ is a normalization constant, and we have used $\ln\det A = \mathrm{Tr}\ln A$. R is the gaussian response function:

$$K\ R(x-x',t-t') \equiv (D^{-1}\partial_t - \Delta + m^2)\ R(x-x',t-t') = \delta(x-x')\ \delta(t-t'), \tag{2.3}$$

$$\begin{aligned} R(x,t) &= \theta(t)\ (4\pi t)^{-d/2}\exp(-m^2Dt - x^2/4Dt) \\ &= \theta(t)\ D\ (2\pi)^{-d} \int \exp\{-Dt(m^2+k^2) - ik.x\}\ dk. \end{aligned} \tag{2.4}$$

$R(0)$ depends on the discrctization procedure used to give meaning to (2.2). With pre-point discretization, [10], $R(0)=0$, which simplifies the formulas that follow. In (2.2),

$$\sqrt{D/2}\ \{(D^{-1}\partial_t - \Delta + m^2)u - f(u)\},$$

is to be substituted for $w(x,t)$. By using a gaussian integration, we can write (2.2) as

$$\begin{aligned} P\{u(x,\tau),\ 0\leq\tau\leq t\} = \tilde{N}(t)\ \delta(u(x,0)) \int\exp\Big[\int_0^t\int\{&-\tilde{u}^2D^{-1} \\ &+ i\ \tilde{u}[Ku - f(u)]\}dxdt\Big]\ d\tilde{u}(x,\tau). \end{aligned} \tag{2.5}$$

To derive the effective equation. we will use the generating functional for moments of u and response functions. This functional is the average of $\exp[\int\{J_u(x,t)u(x,t) + i\ J_v(x,t)\tilde{u}(x,t)\}dx\ dt]$ by means of (2.5) [11], i.e., defining $v = i\tilde{u}$,

$$Z\{J_u,J_v\} = \int du\ dv\ \delta(u(x,0))\ \exp\Big[\int_0^t\int dxdt\{v^2/D + v\{Ku - f(u) + J_v\} + J_u u\}\Big] \tag{2.6}$$

We choose the normalization constant implicit in (2.6) so that $Z\{0,0\} = 1$, for example. Functional derivatives of $\ln Z$ with respect to J_u yield correlation functions, whereas derivatives with respect to J_v yield response functions (J_v in (2.6) is equivalent to adding an external field in (2.1)). The effective equation for $\langle u(x,t)\rangle$ can in principle be found as follows:

(i) Find the Legendre transform of $\ln Z\{J_u,J_v\}$,

$$\Gamma\{U,V\} = \int_0^t \int (J_u U + J_v V)\, dx\, dt - \ln Z\{J_u,J_v\}, \tag{2.7}$$

where J_u and J_v are functionals of $U(x,t)$ and $V(x,t)$ which we obtain by solving

$$U(x,t) = \delta \ln Z/\delta J_u(x,t), \quad V(x,t) = \delta \ln Z/\delta J_v(x,t). \tag{2.8}$$

(ii) As the Legendre transform is an involution, we have

$$J_u(x,t) = \delta\Gamma\{U,V\}/\delta U(x,t), \quad J_v(x,t) = \delta\Gamma\{U,V\}/\delta V(x,t). \tag{2.9}$$

When $J_u=0$, $J_v=0$, we have $U(x,t)= \langle u(x,t)\rangle$ and $V(x,t)=0$. Actually, it can be proven that $V=0$ if $J_u=0$, no matter what J_v and U are. This can be done either by invoking causality (Bausch et al., 11), or by using Ward identities corresponding to the supersymmetric field theory that results when $\det[K - f'(u)]$ is exponentiated by means of anticommuting fields, [12]. In any case, the effective equation is given by

$$0 = J_v(x,t) = \delta\Gamma\{\langle u(x,t)\rangle,0\}/\delta V(x,t). \tag{2.10}$$

Before we continue, let us say something about the initial conditions for (2.10). Clearly, the zero initial condition for (2.1) implies a zero initial condition for (2.10). The gaussian generating functional Z_o (equal to Z with $f(u)=0$) can be explicitly calculated:

$$\ln Z_o\{J_u,J_v\} = -\tfrac{1}{2} \int d1\, d2 \{J_u(1)\, C(1,2)\, J_u(2) + J_u(1) R(1-2) J_v(2) + J_v(1) R(2-1) J_u(2)\}, \tag{2.11}$$

where $1=(x_1,t_1)$, $1-2= (x_1-x_2,t_1-t_2)$, etc. $R(x,t)$ is the response function (2.4) and $C(1,2)$ is the correlation function

$$C(1,2) = C(2,1) = -(2\pi)^{-d} \int_{|t_1-t_2|}^{t_1+t_2} d\alpha \int dk\, \exp\{-\alpha D(k^2+m^2) - ik.x\}$$

$$= - (2\pi)^{-d} \int dk \frac{e^{-i\underline{k}.(\underline{x}_1 - \underline{x}_2)}}{k^2 + m^2} \{e^{-D|t_1-t_2|(k^2+m^2)} - e^{-D(t_1+t_2)(k^2+m^2)}\} \quad (2.12)$$

Notice that C=0 when $t_1= t_2=0$ because all moments are zero initially. From (2.6) we see that when the nonlinearity f(u) is present,

$$Z\{J_u,J_v\} = N \exp\{-\int_0^t \int dx\, dt\, \delta/\delta J_v(x,t)\, f(\delta/\delta J_u(x,t))\}\, Z_o\{J_u,J_v\}, \quad (2.13)$$

which will be used later to set up a perturbative theory.

Suppose now that $u(x,0)= u_o(x)$, a fixed function. Repeating the previous calculations, we now obtain (2.13) with the following Z_o instead of (2.11):

$$\ln Z_o\{J_u,J_v\} = - \tfrac{1}{2}\int d1\, d2\, \{J_u(1)C(1,2)J_u(2) + 2\, J_u(1)R(1-2) \{J_v(2) - u_o(x_2)\delta(t_2)\}\}, \quad (2.14)$$

where R(1-2) and C(1,2) are as in (2.4) and (2.12). In fact, (2.13) holds for a Z_o to be computed from the probability density:

$$P_0(u(x,t)) = \langle\delta(u(x,t) - \int_0^t\int dx'\, dt'\, R(x-x',t-t')\, [\sqrt{2/D}\, w(x',t') - J_v(x',t')] - \int dx'\, R(x-x',t)\, u_o(x'))\rangle_w.$$

Here $J_v(x,t)$ is the same external field as in (2.6). With this density we can compute

$$Z_o\{J_u,J_v\} = \int \exp\int dx\, dt\, J_u u\, P_o(u)\, du(x,t) = \langle\exp\{\int d1\, d2\, J_u(1) R(1-2)\, [\sqrt{2/D}\, w(2) - J_v(2) + u_o(x_2)\, \delta(t_2)]\}\rangle_w = (2.14),$$

once the average over the white noise is performed.

Thus we have proved that a nonzero initial condition is equivalent to adding an impulse $- u_o(x)\delta(t)$ to the external field $J_v(x,t)$ and keeping zero initial u. The effective equation at $J_v=0$ is therefore,

$$- u_o(x)\, \delta(t) = \delta\Gamma\{\langle u\rangle,0\}/\delta V(x,t), \quad t>0, \quad (2.15)$$

with zero initial $\langle u\rangle$. Except for corrections due to the nonlinearity, (2.15) is equivalent to solving (2.10) with the initial condition $\langle u\rangle = u_o$. Notice that a random initial condition is thus equivalent to adding a random impulse at t=0 to the effective equation. We may then obtain an equation for the double average of u (with respect to w and with respect to the initial conditions) by explicitly integrating over the initial probability distribution of $u_0(x)$ in (2.6). The price we pay is a more complicated correlation function C(1,2).

In the models here studied, the order parameter varies on a long time scale. We are therefore interested in the behavior of our systems for large times. The following can be proven by arguments already used in problems of stochastic quantization, [13]:

(i) For polynomial nonlinearities f(u), we may use the first term in (2.12) for C(1,2), which then becomes a function $C(1-2)= C(|x_1-x_2|,|t_1-t_2|)$. The difference with the theory resulting from (2.12) decays exponentially to zero as the time elapses. The isotropic correlation C(1-2) also results when the initial time is set equal to $-\infty$: this implies the upper limit of the α-integral in (2.12) to be infinity. Notice that equilibrium propagators in Schwinger's representation are also given by the α-integral in (2.12) with limits of integration 0 and infinity, [13].

(ii) The maximum external time in a diagram corresponding to a correlation function is a higher bound for the α-integrals associated with the diagram, [13].

(iii) When all external times are equal and tend to infinity, the diagrams corresponding to a correlation function add up to those corresponding to the equilibrium distribution (if such a distribution exists). This is true no matter what the initial condition for (2.1) is.

In what follows, we deal with an equation having a polynomial nonlinearity (the stochastic Hopf bifurcation), (1.1), and with an equation having a sine nonlinearity, (1.2). In both cases we take the initial time to be $t_0= -\infty$, and find the effective equation. This equation is valid for long enough times if $t_0=0$.

3. STOCHASTIC HOPF BIFURCATION

A. Unrenormalized effective equation.

It is convenient to use complex variables and parameters when dealing with (1.1):

$$u_o = u_1 + iu_2, \qquad w_o = w_1 + iw_2, \qquad D_o\,\alpha_o = \alpha_1 + i\,\alpha_2,$$
$$D_0\,\lambda_o/\,6 \;= \lambda_1 + i\lambda_2. \tag{3.1}$$

Then we have to consider the equation

$$\{\partial_t + D_o(\alpha_o - \Delta)\}u_o = -\tfrac{1}{6}\,D_o\lambda_o \quad |u_o|^2u_o + \sqrt{T_o}\;w_o(x,t), \qquad t>0$$
$$\langle w_o\rangle = 0, \qquad \langle \overline{w}_o(x,t)w_o(x',t')\rangle = 2\delta(x-x')\delta(t-t'),$$
$$\langle w_o(x,t)w_o(x',t')\rangle = 0. \tag{3.2}$$

Here the zero subscripts stand for unrenormalized quantities.

At zero temperature, $T_o=0$, it is inmediate to see that the zero solution of (3.2) is stable if $\mathrm{Re}D_o\alpha_o>0$, and unstable if $\mathrm{Re}D_o\alpha_o<0$. In the latter case, there is a homogeneous limit cycle solution

$$u_o(t) = (-6 \quad \mathrm{Re}D_o\alpha_o/\mathrm{Re}D_o\lambda_o)^{\frac{1}{2}}\ \exp(it\Omega_o),$$
$$\Omega_o = \mathrm{Im}D_o\alpha_o - \mathrm{Im}D_o\lambda_o\ \mathrm{Re}D_o\alpha_o/\mathrm{Re}D_o\lambda_o, \tag{3.3}$$

which is stable if

$$\kappa \equiv \mathrm{Re}D_o + \mathrm{Im}D_o\ \lambda_o\ \mathrm{Im}D_o/\mathrm{Re}D_o\lambda_o>0.$$

For $\kappa<0$, strongly turbulent solutions are possible at $d=1$, [14] (see also Dr. Coullet's lectures for phenomena in higher dimensions). Let us see now how this picture changes when we add the white noise.

The generating functional corresponding to (3.2) is

$$Z\{J_u,\overline{J_u},J_v,\overline{J_v}\} = \int du_o d\overline{u_o} dv_o d\overline{v_o}\ \exp\{-A\{u_o,\overline{u_o},v_o,\overline{v_o}\}$$
$$+ \mathrm{Re}\int(J_u\overline{u_o} + J_v\overline{v_o})\ dx\,dt\}, \tag{3.4}$$

$$-A = \int\left[\tfrac{1}{2}T_o|v_o|^2 + \mathrm{Re}\{\overline{v_o}\left[\partial_t u_o + D_o(\alpha_o + \lambda_o|u_o|^2u_o/6)\right]\}\right]\ dx\,dt. \tag{3.5}$$

When $T_o = 2\ \mathrm{Re}D_0$ and $\mathrm{Im}\lambda_o = 0$, we recover the critical dynamics of the standard ϕ^4 model, [11]. From (3.4) and (3.5), we find that the effective equation is now

$$2\ \delta\Gamma\{U_o,\overline{U_o},0,0\}/\delta\overline{V_o}(x,t) = J_v(x,t). \tag{3.6}$$

We need now an approximation scheme to calculate the left side of (3.6). Assuming that the noise strength T_o is small enough, the path integral (3.4) may be approximated by the steepest descent method, [9] Appendix 6.1, [2]. The result is an asymptotic expansion in powers of $T_o^{\frac{1}{2}}$. Such an expansion is used to solve (2.8) for J_u, $\overline{J_u}$, J_v, $\overline{J_v}$ in terms of U_0, $\overline{U_0}$, V_o, $\overline{V_o}$. Then the Legendre transform Γ can be obtained. In the resulting effective equation (3.6) there appear Fourier integrals containing $U_0(x,t)$ and $\overline{U_0}(x,t)$. These functions vary slowly with time and space so that they can be treated as constants. Keeping only the leading-order correction to the deterministic equation, we obtain

$$\begin{aligned}
&D_0^{-1}\partial_t U_0 - \Delta U_0 + U_0\Big[\alpha_o+\lambda_o|U_0|^2/6 + \tfrac{1}{3}T_o\, D_0\lambda_o(\mathrm{Re}D_0)^{-2}\, \Big\{-\tfrac{2}{\varepsilon}\Big(\frac{\mathrm{Re}D_0\phi}{\mathrm{Re}D_0} \\
&+ \tfrac{1}{12}\lambda_o|U_0|^2\Big) + \tfrac{1}{2}\big(1- \mathrm{Im}D_0\, \mathrm{Im}\phi/(\psi\mathrm{Re}D_0) + \tfrac{\lambda}{12\psi}\, |U_0|^2\big)(\mathrm{Re}\phi+\psi)\, \ln(\mathrm{Re}\phi+\psi) \\
&+ \tfrac{1}{2}\big(1+ \mathrm{Im}D_0\, \mathrm{Im}\phi/(\psi\mathrm{Re}D_0) - \tfrac{\lambda}{12\psi}\, |U_0|^2\big)(\mathrm{Re}\phi-\psi)\, \ln(\mathrm{Re}\phi-\psi) \\
&+ 2\, D_0\, \mathrm{Im}\phi\, (\overline{D_0}\mathrm{Im}\phi+ i\lambda_o\mathrm{Re}D_0|U_0|^2/12)\big[|D_0|^2(\mathrm{Im}\phi)^2-|\lambda_o U_0^2\, \mathrm{Re}D_0/6|^2\big]^{-1} \\
&\Big\{\frac{\mathrm{Re}D_0\phi}{\mathrm{Re}D_0}\, \ln\frac{\mathrm{Re}D_0\phi}{\mathrm{Re}D_0} + \frac{\mathrm{Re}D_0\overline{\phi}}{\mathrm{Re}D_0}\ln\frac{\mathrm{Re}D_0\overline{\phi}}{\mathrm{Re}D_0} -(\mathrm{Re}\phi+\psi)\ln(\mathrm{Re}\phi+\psi) - (\mathrm{Re}\phi-\psi)\ln(\mathrm{Re}\phi-\psi)\Big\}\Big\}\Big] \\
&= J_v, \qquad (3.7)
\end{aligned}$$

Here

$$\phi=\alpha_o +\lambda_o|U_0|^2/3, \quad \psi^2= |\lambda_o U_0^2|^2/36 - (\mathrm{Im}\phi)^2, \qquad d= 4-\varepsilon,\ \varepsilon\to 0+.$$

B. Renormalization.

The integrals in (3.6) have been dimensionally regularized, [9]. Then d= 4-ε and the space dimension is close to 4. One can show that the theory (3.4-5) is renormalizable by power counting if the following renormalized variables and fields are introduced, [1,2]:

$$\begin{aligned}
&\alpha_o= \mu^2\, \alpha Z_\alpha(1+i\gamma Z_\gamma), \qquad D_0 = Z_\delta\delta(1+i\nu Z_\nu), \qquad T_o = T\, Z_T; \\
&\lambda_o= \mu^\varepsilon(Z_1 + i\, Z_2)(\lambda_1+ i\lambda_2)K_d, \qquad V_0 = \mu^{3-\varepsilon/2}\, Z_v^{\frac{1}{2}}\, V, \\
&U_0 = \mu^{1-\varepsilon/2}\, Z_u^{\frac{1}{2}}\, U, \qquad K_d = 2^d\, \pi^{d/2}\, \Gamma(d/2). \qquad (3.8)
\end{aligned}$$

Here μ is an arbitrary parameter with dimensions of 1/(legth), The dimensionless constants Z_i (all of which are real except for Z_u and Z_v) are power series in the coupling constants λ_1, λ_2, ν with coefficients depending on γ and ε. If minimal substraction is adopted, insertion of (3.8) in (3.7) yields

$$Z_\alpha = 1 + 4\tilde{\lambda}_1/(3\varepsilon) + O(\ell^2), \quad Z_\gamma = 1 - 4(\tilde{\lambda}_1 - \tilde{\lambda}_2/\gamma)/(3\varepsilon) + O(\ell^2),$$
$$Z_1 = 1 + 10\tilde{\lambda}_1/(3\varepsilon) + O(\ell^2), \quad Z_2 = 2\tilde{\lambda}_2/(3\varepsilon) + O(\ell^2),$$
$$Z_\delta,\ Z_T, Z_\nu,\ Z_u,\ Z_V = 1 + O(\ell^2),$$
$$\tilde{\lambda}_i = \lambda_i T/(2\delta), \qquad \ell = \tilde{\lambda}_1,\ \tilde{\lambda}_2,\ \nu. \tag{3.9}$$

(See De Dominicis and Peliti, [11], for a similar renormalization in another model. Notice that I used the particular choice $T = 2\delta$ in [1], which is not necessary in general).

From (3.9) and the usual renormalization-group theory, [9], we can find how U, J_V, x and t scale with α as this parameter approaches its critical (bifurcation) value 0:

First, α, λ_i, ... change as follows when the scaling parameter is changed,

$$\mu d\tilde{\lambda}_1/d\mu = -\varepsilon\tilde{\lambda}_1 + \frac{10}{3}\tilde{\lambda}_1^2 - \frac{2}{3}\tilde{\lambda}_2^2, \quad \tilde{\lambda}_1(1) = \tilde{\lambda}_1,$$
$$\mu d\tilde{\lambda}_2/d\mu = -\varepsilon\tilde{\lambda}_2 + 4\tilde{\lambda}_1\tilde{\lambda}_2, \qquad \tilde{\lambda}_2(1) = \tilde{\lambda}_2,$$
$$\mu d\alpha/d\mu = -(2 - 4\tilde{\lambda}_1/3)\alpha, \qquad \alpha(1) = \alpha. \tag{3.10}$$

(The physical values correspond to $\mu = 1$).

Second, by imposing $\alpha(\mu) = 1$, we find α as a function of the physical value $\alpha = \alpha(1)$. Then the limit $\alpha \to 0$ corresponds to $\mu \to 0$. For $d < 4$ ($\varepsilon > 0$) the only stable (in fact, globally stable) fixed point of (3.10) is $\tilde{\lambda}_1^* = 3\varepsilon/10$, $\tilde{\lambda}_2^* = 0$ (a purely real coupling constant).

Third, from the equation for U and the fact that length and time scale as $1/\mu$ and μ^{-2}, respectively, we find that, as $\alpha \to 0-$,

$$U = \langle u \rangle \propto |\alpha|^{\frac{1}{2} - 3\varepsilon/20}, \quad \alpha\gamma \propto |\alpha|^{1+\varepsilon/5},$$
$$x \propto |\alpha|^{-\frac{1}{2} - \varepsilon/10}, \quad t \propto |\alpha|^{-1-\varepsilon/5}. \tag{3.11}$$

These scaling relations confirm that U is a slowly-varying order parameter which evolves on the scales

$$X = |\alpha|^{\frac{1}{2}+\varepsilon/10}x, \qquad T = |\alpha|^{1+\varepsilon/5}\,t, \qquad \alpha\to 0. \tag{3.12}$$

C. Renormalized effective equation.

We want to write the effective equation for rescaled variables $U(X,T)$, X, T, ..., which are of order 1 as $\alpha\to 0$. In principle, all we should do is substituting the rescaled variables in (3.7). The trouble is that consistency within our renormalization scheme requires expanding any function such as $f(y)^{a\varepsilon} \sim 1 + a\varepsilon \ln f(y)$, $\varepsilon\to 0+$, no matter what the size of $f(y)$. Thus we obtain several logarithmic terms in the final equation which are artifacts of the ε-expansion. To get rid of these terms, a complicated analysis similar to that by Lawrie, [15], of the equation of state of the ϕ^4 model, is necessary. We won't provide this analysis here. With this caveat, the effective equation is

$$\begin{aligned}\partial_T U - \Delta_X U = H - U\Big[\, & \chi + |U|^2 + \tfrac{\varepsilon}{20}\{\, 3(\chi + 3|U|^2)\ln(\chi + 3\,|U|^2) \\ & + (\chi+|U|^2)\ln(\chi+|U|^2) + 6\,\chi\ln 2 - 9(\chi+|U|^2)\ln 3\} \\ & + \tfrac{\ell_1}{6}\{3(\chi+3|U|^2)\ln(\chi+3|U|^2) + (\chi+|U|^2)\ln(\chi+|U|^2) \\ & + 6\chi\ln 2 - 9\,(\chi+|U|^2)\ln 3 - 10\,|U|^2\ln|U|^2\} \\ & - i\, g_2(\varepsilon)\,\ell_2\,|U|^2\Big] + o(\ell_i).\end{aligned} \tag{3.13}$$

Here we have redefined the field U and the variables X and T as follows

$$\begin{aligned} X &= g_x(\varepsilon,\ell_1)\,|\alpha|^{\frac{1}{2}+\varepsilon/10+\ell_1/3}x, \\ T &= g_t(\varepsilon,\ell_1)\,|\alpha|^{1+\varepsilon/5+2\ell_1/3}t\,D, \\ U(x,t) &\longrightarrow g_u(\varepsilon,\ell_1)\,|\alpha|^{\frac{1}{2}-3\varepsilon/20+\ell_1/3}\,U(X,T), \\ J_v(x,t) &= D\,g_v(\varepsilon,\ell_1)\,|\alpha|^{3/2+\varepsilon/20+\ell_1}\,H(X,T). \end{aligned} \tag{3.14}$$

χ is $\text{sign}\,\alpha$, with $\chi=0$ if $\alpha=0$. $\ell_1 = \check{\lambda}_1 - \tilde{\lambda}_1{}^*$, $\ell_2 = \tilde{\lambda}_2$ are the differences from the fixed point coupling constants and we assume

$$\varepsilon^2\ln\varepsilon \ll \ell_i \ll \varepsilon, \tag{3.15}$$

so that only the terms present in (3.7) need to be considered. The terms multiplying ℓ_i (3.13) are present if α is in the region

$$\varepsilon^{-1} \ll -\ln|\alpha| \ll (\varepsilon\ell_i)^{-\frac{1}{2}}. \tag{3.16}$$

If α is smaller, we go deeper into the critical region, and these correction-to-scaling terms should be dropped from (3.13). The functions g_i are chosen so as to make U=1 a solution of the effective equation for $\chi=-1$, and (3.13) to be compatible with the relation

$$H = U\,|U|^{2+\varepsilon}, \quad \text{at } \alpha=0, \text{ with } H \text{ and } U \text{ constant fields.}$$

We have also set $\mathrm{Im}\alpha = 0$ by multiplying all fields in the generating functional (3.4) by a convenient phase factor.

If we set $\ell_i = 0$ in (3.13) (α in the critical region), the resulting effective equation is universal, independent of the renormalization scheme chosen, [9]. The correction to the oscillation frequency is thus zero in the critical region. In the region (3.16), the oscillation frequency is non-universal and it scales as T does in (3.14). Let the external field be zero, but consider nonzero initial conditions. The scaling (3.14) together with (2.15) show that only small initial conditions, of order $|\alpha|^{\frac{1}{2}-3\varepsilon/20+\ell_1/3}$, evolve in the slow scales (3.14). Larger departures from the stable value of the order paramter presumably relax in a faster scale until they are of the indicated order of magnitude.

4. DIFFUSIVE SINE-GORDON EQUATION IN 2D

A. Unrenormalized theory.

For the sine-Gordon model (1.2), the MSR generating functional is given by (2.6) with

$$K = D_0^{-1}\partial_t - \Delta, \qquad f(u_0) = -\frac{\alpha_0}{\beta_0 a^2}\sin\beta_0 u_0 \,. \tag{4.1}$$

The external field E_0 may be included in J_v. The response function is (2.4) with d=2, $m_0=0$, whereas the correlation function is

$$C(x,t) = -(2\pi)^{-2}\int dk\, \exp(-D_0|t|k^2 - ik.x')\,\theta(k^2-m_0^2)/k^2. \tag{4.2}$$

C has been regularized by means of an infrared cutoff m_0 and an ultraviolet cutoff a such that $x'^2 = x^2 + a^2$. The effect of these cutoffs is analogous to having the theory defined on a square lattice of area $(2\pi/m_0)^2$ and lattice constant a. We have set $t_0 = -\infty$. As $m_0^2 D_0 t$ becomes large, C decays as

$$C(x,t) \sim -(4\pi m_0^2 D_0|t|)^{-1} J_0(m_0 x)\exp\{-D_0 m_0^2 |t|\} \to 0, \tag{4.3}$$

whereas as $t \to 0$, $m_0^2(x^2+a^2) \to 0$,

$$C(x,t) \sim C(x,0) + |t|^{3/4}(\ldots),\ C(x,0) \sim (4\pi)^{-1} \ln\{cm_0^2(x^2+a^2)\},$$
$$c = \exp\{-2\int_0^1 ds\, J_1(s) \ln s - 2\int_1^\infty ds J_0(s)/s\}. \tag{4.4}$$

To find $\Gamma\{U,V\}$ and then the effective equation, we have to evaluate Z from Z_0 given by (2.11) via some perturbative scheme. The following obvious formula and its generalization to averages of products of exponentials is helpful,

$$\begin{aligned}\langle e^{i\beta_0 u_0(x,t)}\rangle_g &= Z_0\{J_u + i\beta_0\delta(x-.)\,\delta(t-.),J_v\}/Z_0\{J_u,J_v\}\\ &= \exp\{\beta_0^2 C(0)/2 - i\,\beta_0\beta_1(x,t;\underline{J})\},\end{aligned} \tag{4.5}$$

$$\beta_1(x,t;\underline{J}) = -\,\delta \ln Z_0\{\underline{J}\}/\delta J_u(x,t),\quad \underline{J} = (J_u,J_v). \tag{4.6}$$

We denote by $\langle .\ \rangle_g$ the average over the gaussian path integral (2.6) with $f=0$. $C(0) = C(0,0) = (4\pi)^{-1}\ln(cm_0^2 a^2)$. By means of (4.5) and similar formulas, we find a perturbative series for Z in powers of α_0. This series enables us to solve (2.8) and to obtain the α_0-expansion of the Legendre transform Γ in (2.7). Notice that the α_0-expansion is very different from the steepest descent method of Section 3. Γ is

$$\begin{aligned}\Gamma\{U_0,V_0\} = &-\int\big[D_0^{-1}V_0^2 + V_0\{(D_0^{-1}\partial_t-\Delta)U_0 + A_0 \sin\beta_0 U_0\}\big]\, dx\, dt\\ &-\tfrac{1}{2}A_0^2 \int d1d2\ V_0(1)\big[V_0(2)\{(\mathrm{ch}\beta_0^2C-1) \sin\beta_0U_0(1) \sin\beta_0U_0(2)\\ &\quad - (\mathrm{sh}\beta_0^2C-\beta_0^2C)\cos\beta_0U_0(1)\cos\beta_0U_0(2)\}\\ &\quad -2\beta_0 R(1-2)\{(\mathrm{ch}\beta_0^2C-1) \cos\beta_0U_0(1) \sin\beta_0U_0(2)\\ &\quad +\mathrm{sh}\beta_0^2C\ \sin\beta_0U_0(1)\ \cos\beta_0U_0(2)\}\big]+O(A_0^3)\end{aligned} \tag{4.7}$$

$$\text{Here,}\quad A_0 = \alpha_0 \exp\{\beta_0^2 C(0)/2\}/(\beta_0 a^2),\ C = C(1-2). \tag{4.8}$$

The following formulas help us simplifiying (4.7) and extracting its singular part as $a\to 0+$:

$$\begin{aligned}\int d1d2\ f(1,2)\ R(1-2)\ C(1-2)^n = &\frac{1}{n+1}\int d1d2\ \theta(t_1-t_2)C(1-2)^{n+1}\ \partial_{t_2}f(1,2)\\ &-(n+1)^{-1}\int d1d2\ \delta(t_1-t_2)C(1-2)^{n+1}f(1,2),\end{aligned} \tag{4.9}$$

$$\beta_o^2 \int d1d2\ R(1-2)\ \{ch\beta_o^2C(1-2)-1\}\ f(1,2) =$$
$$\int d1d2\ \{sh\beta_o^2C(1-2)-\beta_o^2C(1-2)\}\ \{\theta(t_1-t_2)\partial_{t_2}f(1,2)-\delta(t_1-t_2)f(1,2)\} \tag{4.10}$$

$$\beta_o^2\int d1d2\ R(1-2)sh\beta_o^2C(1-2)\ f(1,2) =$$
$$\int d1d2\ \{ch\beta_o^2C(1-2)-1\}\ \{\theta(t_1-t_2)\partial_{t_2}f(1,2)-\delta(t_1-t_2)\ f(1,2)\}. \tag{4.11}$$

The result is

$$\begin{aligned}\Gamma\{U_o,V_o\} = &-\int[D_o^{-1}V_o^2 + V_o\{(D_o^{-1}\partial_t-\Delta)U_o + A_o\ \sin\beta_oU_o\}]dx\ dt\\ &-\tfrac{1}{2}A_o^2\int d1d2V_o(1)\left[V_o(2)+2\theta(t_1-t_2)\ \beta_0^{-1}\partial_tU_o(2)\right]\{(ch\beta_o^2C-1)\\ &\sin\beta_oU_o(1)\ \sin\beta_oU_o(2)-(sh\beta_o^2C-\beta_o^2C)\ \cos\beta_oU_o(1)\ \cos\beta_oU_o(2)\}\\ &-\beta_0^{-1}A_o^2\int d1\ dx_2\ \left[(sh\beta_o^2C-\beta_o^2C)\ \cos\beta_oU_o(1)\ \sin\beta_oU_o(2)\right.\\ &\left.+(ch\beta_o^2C-1)\ \sin\beta_oU_o(1)\ \cos\beta_oU_o(2)\right]\ V_o(1)\big|_{t_1=t_2}\\ &+O(A_o^3).\end{aligned} \tag{4.12}$$

<u>Proof of (4.9-11):</u>

By (2.4) and (4.2),

$$\int d1d2\ R(1-2)C(1-2)^nf(1,2) = \frac{(-1)^n}{n+1}D\int d1d2\{\Pi dk_s\}\ \ \theta(t_1-t_2)\ f(1,2)$$
$$\Sigma_s[\Pi_j'\ k_j^{-2}]\ \exp[-D(t_1-t_2)\Sigma k_j^2-i(x_1-x_2)\Sigma k_j]$$

Here Π' excludes precisely k_s^{-2} ; all sums are from 1 to n+1. In this expression the infrared and ultraviolet cutoffs have not been explicitly displayed. Obviously,

$$D\ e^{\cdots}\Sigma\ \Pi_j'(\ldots) = D\ \frac{\Sigma_s k_s^2}{\Pi_j k_j^2}\ \ e^{\cdots}\ \ = \Pi_j k_j^{-2}\ \partial_{t_2}e^{\cdots}\quad ,$$

which allows us integrating by parts in the t_2-integral. The result is (4.9) because at $t_2=-\infty$ the regularized correlation function C(1-2) is zero according to (4.3). The proofs of (4.10) and (4.11) are similar. Notice that we need the infrared regulator $m_o\neq0$ to obtain this result. We also need $m_o\neq0$ to be able to set $t_o=-\infty$ and forget about initial conditions. In fact, had we used the correlation function C(1,2) of (2.12) ($t_o=0$ with zero initial conditions), we would have written:

$$\int d1d2\ R(1-2)\ C(1,2)^n\ f(1,2) =$$

$$\frac{(-D)^{n+1}}{n+1} \int d1d2\ \Pi dk_s\ \theta(t_1-t_2) f(1,2) \left[\partial_{t_2} - \Sigma_j \delta(\alpha_j - t_1 - t_2)\right]$$

$$\int \Pi_s d\alpha_s\ \exp\{-D\Sigma \alpha_j k_j^2 - i(x_1-x_2)\Sigma k_j\}.$$

Here the α_j's are integrated from (t_1-t_2) to (t_1+t_2), and we have used

$$e^{-D(t_1-t_2)k_j^2} = \partial_{t_2}\{\int_{t_1-t_2}^{t_1+t_2} d\alpha_j e^{-D\alpha_j k_j^2}\} - e^{-D(t_1+t_2)k_j^2}$$

in the expression (2.4) for the response function. Integration by parts yields the result (4.9) plus a term

$$D \int d1d2\ f(1,2) C(1,2)^n \int dk\ \exp\{-D(t_1+t_2)k^2 - ik(x_1-x_2)\}\theta(t_1-t_2).$$

When computing the effective equation or vertex functions, functional differentiation destroys at least one of the integrals over 1,2. In the resulting expression this extra term becomes exponentially small in comparison with the terms kept in (4.9) (take the limit $t_1, t_2 \to +\infty$ with $|t_1-t_2|$ fixed). It is clear that there is one extra term similar to this one in each of the equations (4.10) and (4.11). For the same reason, we can ignore it.

We have therefore proved that a zero initial condition in (1.2) doesn't influence the vertex functions, correlations or effective equation after large times. This is so even though the sine nonlinearity is not polynomial and the proofs developed for stochastic quantization are not applicable. A nonzero initial condition is dealt with as in Section 2, and it doesn't have an effect at large times either. Notice that the exponential decay of $C(x,t)$ takes place in the limit $D_o m_o^2 t \to +\infty$. As $m_o^2 \sim 1/L^2$ (L^2= area of the system), our effective equation involve long enough time scales (see below).

B. Renormalization.

In Ref. [16] it is argued that the static sine-Gordon theory is renormalizable near $\beta_o^2 = 8\pi$. We shall prove that the ultraviolet singularities in (4.12) coincide with those of the static theory (to the order considered in this paper). From (4.4),

$$A_o = \frac{\alpha_o \exp \beta_o^2 C(0)/2}{\beta_o a^2} = \frac{\alpha_o (cm_o^2 a^2)^{\beta_o^2/8\pi}}{\beta_o\ a^2} = \alpha_o (cm_o^2)^{1+\delta_o} a^{2\delta_o}\ \beta_o^{-1},$$

if $\beta_o^2/8\pi \equiv 1+\delta_o$. Except for a β_o factor due to the definition of potential energy, the singularities that appear when this expression is expanded in powers of δ_o are present in the equilibrium theory, [16]. The only ultraviolet singularities of $O(\alpha_o^2)$ in (4.12) appear in the integral

$$\int d1\, dx_2\, V_o(1) \exp\{-\beta_o^2 C(1-2)\} \sin\beta_o(U_o(2)-U_o(1)) \Big|_{t_2=t_1} \qquad (4.13)$$

To extract these singularities we use a multipole (gradient) expansion

$$U_o(x_2,t_1)-U_o(x_1,t_1) = (x_2-x_1).\nabla U_o(1) + \tfrac{1}{2}(x_2-x_1)^2:\nabla\nabla U_o(1) + \dots$$

Insertion of this in (4.13) yields

$$\tfrac{1}{4}\int d1\, V_o(1)\quad U_o(1) \int_{|x|<B} dx\, x^2 \exp\{-\beta_o^2 C(x,0)\} + \dots \qquad (4.14)$$

Here B is a cutoff such that $a \ll B$, and that $m_o B =$(fixed number)$\ll 1$. By using (4.4) we see that the integral (4.14) diverges logarithmically as $a \to 0+$. The terms not explicitly represented in (4.14) do not contain short-distance singularities. Comparison with [16] shows the term (4.14) to contain <u>exactly</u> the same singularity as in the static (equilibrium) case.

To the order considered in (4.12) (and one may argue as in [16] to extend this), only two constants are needed to renormalize the generating functional:

$$\alpha_o = Z_\alpha \alpha,\quad \beta_o = Z_\phi^{-\frac{1}{2}}\beta,\quad \delta_o = Z_\phi^{-1}(1+\delta)-1,\quad D_o = Z_\phi D,\quad E_o = Z_\phi^{-\frac{1}{2}}E,$$

$$m_o = m\mu,\qquad U_o = Z_\phi^{\frac{1}{2}}\, U,\qquad V_o = Z_\phi^{\frac{1}{2}}\, V. \qquad (4.15)$$

The constants Z_α, Z_ϕ are power series in α and δ with coefficients depending on μa. We choose these coefficients so that only terms which are singular as $a \to 0+$ are cancelled (minimal substraction). The renormalized action integral is

$$-A = \int\left[v^2/D + v\{D^{-1}\partial_t u - Z_\phi \Delta u + Z_\alpha Z_\phi \frac{\alpha}{\beta a^2}\sin\beta u - E\}\right]dxdt. \qquad (4.16)$$

Notice that no renormalization of the infrared cutoff is necessary (see below). Here μ is a parameter with dimensions of inverse length. With this choice of units and renormalization, the Z's are exactly the same as in the static case:

$$Z_\alpha = 1 - \delta \ln\mu^2 a^2,$$
$$Z_\phi = 1 + \frac{\alpha^2}{64} \ln\mu^2 a^2. \qquad (4.17)$$

The unrenormalized vertex functional doesn't depend on μ.This and the relation between bare and renormalized functionals yields the following renormalization-group equation (RGE) for the renormalized vertex functional, [9],

$$\left[\mu\partial_\mu + \beta_\alpha\partial_\alpha + \beta_\delta\partial_\delta - m\partial_m - \gamma_\phi D\partial_D - \tfrac{1}{2}\gamma_\phi \int dxdt\{U\frac{\delta}{\delta U} + V\frac{\delta}{\delta V}\}\right]\Gamma = 0,$$
$$\Gamma = \Gamma\{U(x,t),V(x,t);\alpha,\delta,m,D;\mu\}. \qquad (4.18)$$

$$\beta_\alpha = \mu\partial_\mu|_o\alpha = 2\alpha\delta, \quad \text{(subscript o means fixed bare quantities)}$$
$$\beta_\delta = \mu\partial_\mu|_o\delta = \alpha^2/32,$$
$$\gamma_\phi = \mu\partial_\mu|_0 \ln Z_\phi = \alpha^2/32. \qquad (4.19)$$

The β's and γ_ϕ in (4.19) have been calculated to lowest order in the coupling constants α,δ. In (4.18) all parameters and fields are dimensionless except for $V(x,t)$ and μ. It is convenient to nondimensionalize them too, so we'll use the dimensionless variables μx, $\mu^2 t$ and field $\mu^2\, V(\mu x,\mu^2 t)$. In this new units, Γ doesn't explicitly depend on μ. The solution of (4.18) is then

$$\Gamma\{U(\mu x,\mu^2 t),\mu^2 V(\mu x,\mu^2 t);x,y,m,D\} =$$
$$\Gamma\{U(\mu(s)x,\mu^2(s)t)E(s),\mu^2(s)V(\mu(s)x,\mu^2(s)t)\; E(s);x(s),y(s),m(s),D(s)\}$$
$$E(s) = \exp\{-\tfrac{1}{2}\int_0^s \gamma_\phi(r)dr\}. \qquad (4.20)$$

Here the functions of s are solutions of the characteristic equations corresponding to (4.18), i.e.,

$$\mu(s) = \mu e^s, \quad m(s) = m\, e^{-s}, \quad D(s) = D\, e^{-\{x(s)-x\}/2}, \qquad (4.21)$$

$$dx/ds = y^2, \quad x(0) = x; \quad x \equiv 2\delta,$$
$$dy/ds = xy, \quad y(0) = y; \quad y \equiv \alpha/4. \qquad (4.22)$$

We shall use the right side of (4.20) plus the characteristic equations (4.21-22) to find the left side of (4.20). The equations for $x(s)$ and $y(s)$ are the well-known Kosterlitz's RGE, [17]. Their phase plane is represented in Fig. 1. There the arrows correspond to $\mu(s)\to 0+$, $s\to-\infty$ (infrared). The trajectories in this Figure are given by

$$x^2 - y^2 = q \text{ (const.)}.$$

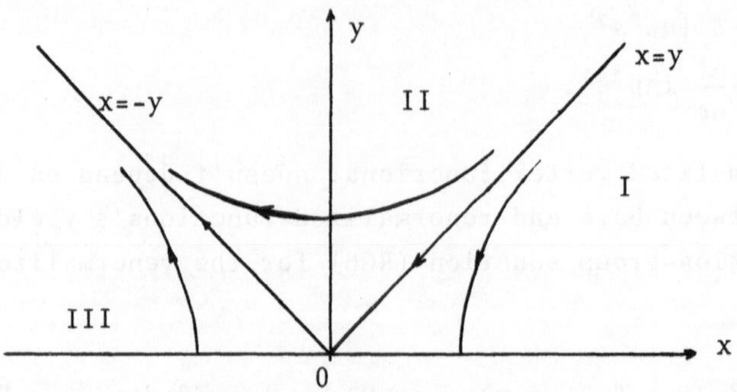

Figure 1

The separatrices $x=\pm y$ divide the upper half plane in three regions

I $(q>0,\ x>0)$; II $(q<0)$; III $(q>0,\ x<0)$.

To reach the critical region we need the relevant variable $m(s)$ to be fixed at some finite value, 1 for example, as s varies. Then $e^s = m$ and $s\to-\infty$ means letting the infrared cutoff (inverse of the size of the system) go to zero. In Regions II and III this limit causes $y(s)=\frac{1}{4}\alpha(s)$ to become large. Then the sine term in the action (4.16) becomes dominant, and the generating functional Z will be well approximated by a gaussian path integral obtained by linearizing about one of the zeroes of the sine term. The resulting theory has a finite correlation length and it will not be of interest to us. After crossing the separatrix $x=y$, we enter Region I where all initial conditions evolve towards the critical line $y=0$, $x\geq 0$ as $m\to 0+$ $(s\to-\infty)$. At $m=0$, the correlation length is infinite. It is this change of behavior (from $q<0$ and finite correlation length to $q>0$ and infinite correlation length) which characterizes the Kosterlitz-Thouless phase transition, [16-18].

Before we continue, let us find the physical meaning of q, the first integral of (4.22). For the equilibrium distribution corresponding to (1.2) to coincide with the usual one, [16], we need $\beta^2 D = 1$. The critical temperature for the K-T transition is reached on the separatrix $x=y$, i. e., $2(\beta_c^2/8\pi-1) = \alpha/4$, or $D_c = 1/[\pi(\alpha+8)] \leq 1/8\pi$. $q\to 0$ implies $D\to D_c$, so that $q \sim \alpha(D_c-D)/(8\pi D_c^2)$ and Regions I $(q>0)$ and II $(q<0)$ correspond to $D<D_c$ and $D>D_c$, respectively. q is thus the distance to the critical temperature.

We want to find the effective equation corresponding to (1.2) below the K-T critical temperature D_c (Region I in Fig.1). To do so, we'll use (4.20) together with the solutions of (4.22), which are:

$$y(m) = \frac{2\sqrt{q}\,y\,(\sqrt{q}+x)}{(\sqrt{q}+x)^2 - y^2 m^{2\sqrt{q}}} m^{\sqrt{q}} \sim \frac{2\sqrt{q}\,y}{\sqrt{q}+x} m^{\sqrt{q}},$$

$$x(m) \sim \sqrt{q} + y m^{2\sqrt{q}}/(\sqrt{q}+x); \qquad m \to 0+,\ q>0. \tag{4.23}$$

$$y(m) = x(m) = 1/|\ln m|, \qquad q = 0. \tag{4.24}$$

Here $y(m) = y(s=\ln m)$, and $x=(q+y^2)^{\frac{1}{2}}$ and y are the physical values of the coupling constants, i.e., their values at s=0.

C. Renormalized effective equation.

As before, we want to write an effective equation for rescaled variables and fields which are of order 1 as $m \to 0+$ (thermodynamic limit) This can be achieved by rescaling as indicated in (4.20). Consider the right side of this equation, which we want to approximate by means of (4.12).

(i) The infrared cutoff in the correlation function C(1-2) is equal to 1 in the new units. In fact, after nondimensionalization the momenta k become k/μ, and therefore the infrared cutoff is $m_0/\mu = m$. In the right side of (4.20) we have $m(s) = 1$ instead of m, which proves our assertion. Similarly the ultraviolet cutoff is $\mu m a$ in the right side of (4.20).

(ii) The slow scales are $(m=e^s)$:

$$X = \mu(s)x = \mu m x, \quad T = \mu^2 m^2 t. \tag{4.25}$$

(iii) $\int J_V V\, dx\, dt$ becomes $\{E(s)/\mu^2(s)\} \int J_V V\, dX dT$, so that

$$E \text{ becomes } E(\mu m)^{-2} \exp\{x - x(m)/4\} \equiv \mathcal{E}. \tag{4.26}$$

$\mathcal{E} = O(1)$ is the reduced external field.

Let us insert (4.15) in (4.12) and then substitute the result in (4.20). After (4.23-6) are taken into account, we obtain

$$\begin{aligned}
\Gamma\{U,\mu^2V;x,y,m,D\} = & - \int \Big[D^{-1}V^2 + V\{D^{-1}\partial_T U - E^2(m)\big(1+\tfrac{y^2(m)}{4}\ln(\mu m a)^2\big)\,\Delta U \\
& + \sqrt{2/\pi}\, Y(m)\big(1+\tfrac{1}{2}x(m)\ln c\big)\sin\beta_\gamma U\}\Big]\, dX dT \\
& - \tfrac{1}{\pi} Y^2(m) \int d1d2\, V(1)\{V(2) + 2\theta(T_1-T_2)\beta^{-1}\partial_T U(2)\}\{(-1 \\
& + \mathrm{ch}\beta^2 C)\sin\beta_\gamma U(1)\sin\beta_\gamma U(2) - (\mathrm{sh}\beta^2 C - \beta^2 C)\cos\beta_\gamma U(1) \\
& \cos\beta_\gamma U(2)\} - Y^2(m)/(E(m)\pi\sqrt{2\pi}) \int d1 dX_2\, V(1) \\
& \{(\mathrm{ch}\beta^2 C - 1)\sin\beta_\gamma U(1)\cos\beta_\gamma U(2) + (\mathrm{sh}\beta^2 C - \beta^2 C) \\
& \cos\beta_\gamma U(1)\sin\beta_\gamma U(2)\}\big|_{T_2=T_1} + \ldots
\end{aligned} \tag{4.27}$$

Here we have used the abbreviations

$$\beta^2 \equiv 8\pi(1+ x(m)/2),$$
$$C \equiv C(1-2) \quad \text{with } m_o=1 \text{ and scales (4.25)},$$
$$1 \equiv (X_1,T_1),\ 2 \equiv (X_2,T_2),$$
$$\beta_\gamma \equiv \beta E(m),$$
$$Y(m) \equiv y(m)cE(m) = y(m)\ c\ \exp\left[\tfrac{1}{4}(x-x(m))\right]. \tag{4.28}$$

The effective equation is therefore,

$$D^{-1}\partial_T U = E^2(m)\ (1+\frac{Y^2(m)}{4}\ln(\mu m a)^2)\ \Delta U + E-\sqrt{2/\pi}\ Y(m)\ (1+x(m)\tfrac{1}{2}\ln c)\sin\beta_\gamma U$$
$$- 2Y^2(m)/(\pi\beta)\int dX'dT'\{(\mathrm{ch}\beta^2 C-1)\ \sin\beta_\gamma U\ \sin\beta_\gamma U' - (\mathrm{sh}\beta^2 C-\beta^2 C)$$
$$\cos\beta_\gamma U\ \cos\beta_\gamma U'\}\ \partial_T U' - \frac{Y^2(m)}{\pi\sqrt{2\pi}\ E(m)}\int dX'\{(\mathrm{ch}\beta^2 C-1)\ \sin\beta_\gamma U$$
$$\cos\beta_\gamma U' + (\mathrm{sh}\beta^2 C-\beta^2 C)\ \cos\beta_\gamma U\ \sin\beta_\gamma U'\}\big|_{T'=T}\ , \tag{4.29}$$

plus cubic terms in the coupling constants. This integrodifferential equation describes the evolution of the order parameter U= <u> towards its stable values as explained in Section 2. The time integral in (4.29) goes from $T'=-\infty$ to $T'=T$. U and U' mean U(X,T) and U(X',T') respectively. C is C(X-X',T-T'). The ultraviolet singularity of the last integral in (4.29) is compensated by the $\ln\mu ma\ \Delta U$ term.

D. Onset of the relaxation oscillation.

There is a redeeming feature in the complicated-looking equation (4.29): Y(m) becomes small with m. This allows for a perturbative analysis similar to that carried out for the mean-field version of (1.2), [19]. The idea is to consider the second-order terms in the coupling constants Y(m), x(m), as small disturbances or "imperfections" of the rest of the equation. Ignoring them for a moment, we have a deterministic diffusive sine-Gordon equation which can be explicitly integrated at least for stationary solutions. Let us call this equation "unperturbed". The unperturbed equation has stable stationary solutions

$$E = \sqrt{2/\pi}\ Y(m)\ \sin\beta_\gamma U_s, \qquad 0 \le \beta_\gamma U_s \le \pi/2 \quad (\mathrm{mod}\ 2\pi) \tag{4.30}$$

if $E \le \sqrt{2/\pi}\ Y(m)$. It also has "soliton" solutions (traveling waves connecting stationary solutions of type 4.30), but let me ignore them here. If $E \ge \sqrt{2/\pi}\ Y(m)$, there is a linearly stable solution of the unperturbed equation which is a relaxation oscillation for $\partial_T U$ (or for U restricted to an interval of length 2π). This space-independent solution appears as the "creation of a limit cycle from a homoclinic orbit containing a saddle-node", in the cylindrical phase plane of axes $U\epsilon[0,2\pi)$, $\partial_T U$, [19]. The relaxation oscillation is a closed loop that encircles the phase cylinder at a height proportional to E.

Let us go back to (4.29). What we have is an "imperfect bifurcation" of the type above described: small terms modify the creation of the limit cycle. Far from the critical field $E_c = \sqrt{2/\pi}\,Y(m)$, the imperfection has little effect, whereas for $|E-E_c| \sim Y^2(m)$ further analysis is needed. As in the mean-field case, [19], we can show that the "imperfection" just increases the critical field a little bit, without destroying the oscillatory solution. The critical exponents for some appropriately defined order parameter, $||U-U_c||$, are the classical mean-field ones ($||U-U_c|| \sim |E-E_c|^{\frac{1}{2}}$, for example).

Let us summarize what we have obtained so far, leaving a more detailed account for a future publication:

(i) We have obtained an approximate vertex functional for the diffusive sine-Gordon model valid for time scales of order L^2/D. As in the equilibrium case, there is a Kosterlitz-Thouless phase transition at a critical temperature.

(ii) Below the K-T temperature, an approximate effective equation holds asymptotically as the infrared regulator (inverse size of the system) goes to zero.

(iii) There is a critical external field (of order $L^{-2-b\sqrt{D_c-D}}$, or $L^{-2}/\ln L$ at $D=D_c$ in the original units) so that: for $E<E_c$ the order parameter evolves toward a fixed constant value or toward traveling wave solutions linking different constant values (the "solitons"); for $E>E_c$ the order parameter evolves toward a stable time-periodic, space-independent solution (which is a closed loop encircling the $U,\partial_T U$ - cylinder phase plane). One-parameter families of periodic traveling waves are also possible in this regime, depending on the initial values of the order parameter, [20].

We therefore have a self-synchronization transition with classical critical exponents at $E=E_c$ below the Kosterlitz-Thouless critical temperature.

ACKNOWLEDGEMENTS

I thank Prof. Garrido for his invitation to this Conference and Prof. Brey for fruitful conversations.

REFERENCES

1. L. L. Bonilla, Phys. Rev. Lett. 60, 1398 (1988).
2. L. L. Bonilla, unpublished.
3. G. Nicolis and I. Prigogine, Self-Organization in Nonequilibrium Systems. Wiley, N.Y. 1977; H. Haken, Synergetics, Springer, N.Y. 1977.
4. M. Scheutzow, Probab. Th. Rel. Fields, 72, 425 (1986).
5. M. Shiino, Phys. Lett. 111A, 396 (1985).
6. L. L. Bonilla, Phys. Rev. B35, 3637 (1987) and references here cited.
7. It is possible for the unique stationary distribution to change its shape when a control parameter changes, e.g., to develop several maxima. Such changes are known as "noise-induced transitions". See W. Horsthemke and R. Lefever, Noise-Induced Transitions. Springer, Berlin 1984.
8. V. I. Arnol'd, Geometric Methods in the Theory of Ordinary Differential Equations. Springer, N. Y. 1983.
9. D. J. Amit, Field Theory, the Renormalization Group and Critical Phenomena, 2nd ed. World Sci., Singapore 1984.
10. R. Phythian, J. Phys. A10, 777 (1977); A. Förster and A. S. Mikhailov in Self-Organization by Nonlinear Irreversible Processes (W. Ebeling & H. Ulbricht, eds). Springer, Berlin 1986. P. 89.
11. C. De Dominicis and L. Peliti, Phys. Rev. B18, 353 (1978); R. Bausch, H-K Janssen & H. Wagner, Z. Phys. B24, 113 (1976).
12. L. L. Bonilla in p. 15 of Random Media (G. Papanicolaou, ed.) IMA vol. 7. Springer, N.Y. 1987.
13. A. Gonzalez-Arroyo, Lect. Notes in Phys. 216, 171 (1985).
14. Y. Kuramoto, Suppl. Prog. Theor. Phys. 64, 346 (1978).
15. I. D. Lawrie, J. Phys. A14, 2489 (1981).
16. D. J. Amit, Y. Y. Goldschmidt & G. Grinstein, J. Phys.A13,585 (1980)
17. J. M. Kosterlitz, J. Phys. C7, 1046 (1974).
18. P. Minnhagen, Rev. Mod. Phys. 59, 1001 (1987).
19. L. L. Bonilla, J. Stat. Phys. 46, 659 (1987).
20. N. J. Kopell and L.N. Howard, Stud. Appl. Math. 52, 291 (1973); P. S. Hagan, SIAM J. Math. Anal. 13, 717 (1982).

Exactly Solvable Multistable Fokker-Planck Models with Arbitrarily Prescribed N Lowest Eigenvalues

H.R. Jauslin, Dept. Physique Théorique, Université de Genève,
CH-1211 Genève 4, Switzerland

Abstract: We present a method to construct Fokker-Planck models with some prescribed features, for which all eigenfunctions and the time-dependent transition probability can be explicitly calculated. The prescribed features can be formulated by choosing arbitrarily the N lowest eigenvalues. Alternatively one can prescribe some qualitative behavior for the potential, like the number and relative depth of wells and barriers.
The method is based on ideas of supersymmetric quantum mechanics and the theory of solitons, that can be traced back to the work of Darboux and Crum.

1. Introduction

Many properties of fluctuating systems depend essentially only on the shape of the potential at low energies, or on the lowest eigenvalues and eigenfunctions. In general it is not possible, even in one dimensional models, to determine the transition probabilities and eigenfunctions for a given problem. It is thus of interest to be able to construct potentials that have a prescribed set of N lowest eigenvalues, for which these quantities can be calculated exactly. The problem can also be formulated in terms of the construction of a potential having some prescribed qualitative features, like e.g. the number and depth of wells, or even some approximate quantitative behavior at low energies.

We consider the one-dimensional Fokker-Planck equation

$$\frac{\partial}{\partial t}P = -\frac{\partial}{\partial x}(K\ P) + \frac{\partial^2}{\partial x^2}P \tag{1.1}$$

where the drift K is taken to derive from a potential W:

$$K = -W' \tag{1.2}$$

where primes denote derivatives with respect to x. It is well known that it can be transformed into an equation that has the form of a Schrödinger equation with imaginary time: If P_S is a solution of the stationary Fokker-Planck equation and we define a new function $\psi(x,t)$ by

$$P(x,t) \doteq \sqrt{P_S(x)}\psi(x,t) \tag{1.3}$$

then Eq.(1.1) is transformed into

$$-\frac{\partial}{\partial t}\psi = -\frac{\partial^2}{\partial x^2}\psi + V\psi \tag{1.4}$$

with

$$V = (W')^2 - W'' \tag{1.5}$$

The groundstate φ_0 of this Schrödinger operator is given by

$$\varphi_0 = \sqrt{P_S} \tag{1.6}$$

and the corresponding eigenvalue is $E_0 = 0$. The transition probability density is related to the time-dependent propagator G of the Euclidean Schrödinger equation (1.4) by

$$P(x,t|x_0) = \frac{\varphi_0(x)}{\varphi_0(x_0)} \; G_N(x, x_0, t) \; e^{E_0 t}. \tag{1.7}$$

The present approach starts with a potential V_0 for which the corresponding transition probability and eigenfunctions are explicitly known. V_0 is then modified in such a way as to add N arbitrary eigenvalues at the bottom of the spectrum. This step involves the knowledge of the solutions of the original stationary Schrödinger equation at energies that are smaller than the bottom of the spectrum. The transition probability and eigenfunctions corresponding to the new potential V_N can be expressed in terms of the original ones, involving only derivatives and integrals.

The basic ideas of this method can be traced back to the work of Darboux [1] and of Crum [2]. These ideas have found applications in supersymmetric quantum mechanics [3] and in the study of solitons and inverse scattering theory [4,5]. Some special cases of this method have been treated in the literature: Eigenfunctions have been calculated in Refs.[6-9], and time-dependent solutions and propagators were obtained in Refs.[10,11]. In Refs.[12,13] expressions for transition probabilities in bistable Fokker-Planck models were obtained for particular initial conditions. An application to the construction of creation and annihilation operators for a whole class of Hamiltonians is discussed in Ref.[14].

2. Darboux-Crum Transformation

We consider a Schrödinger operator

$$H_- = -\frac{d^2}{dx^2} + V_-(x) \tag{2.1}$$

that has a ground state φ_0 with energy E_0, which we write as

$$\varphi_0 = e^{-W} \tag{2.2}$$

Then H_- admits the following representation

$$(H_- - E_0) = A^\dagger A \tag{2.3}$$

with

$$\begin{aligned} A &= \quad \frac{d}{dx} + W' \equiv \; \varphi_0 \frac{d}{dx} \varphi_0^{-1} \\ A^\dagger &= -\frac{d}{dx} + W' \equiv - \; \varphi_0^{-1} \frac{d}{dx} \varphi_0 \end{aligned} \tag{2.4}$$

We define a new operator

$$\begin{aligned} H_+ &\doteq AA^\dagger + E_0 \\ &= -\frac{d^2}{dx^2} + V_+ \end{aligned} \tag{2.5}$$

where

$$V_+ = V_- + 2W'' \equiv -V_- + 2E_0 + 2W'^2 \tag{2.6}$$

Remark that $V_\pm$ can also be written as

$$\begin{aligned} V_- &= W'^2 - W'' + E_0 \\ V_+ &= W'^2 + W'' + E_0 \end{aligned} \tag{2.7}$$

Comparison with Eq.(1.5) shows that transforming H_- into H_+ amounts to reversing the sign of the drift of the Fokker-Planck equation [15,16]. H_+ is called the supersymmetric partner of H_-. Their spectral properties are related as follows:

1) The spectrum of H_+ is equal to the one of H_- but without the eigenvalue E_0.

2)If f(x) satisfies the equation

$$H_- f = \Gamma f \tag{2.8}$$

where Γ can be either

a) a real number λ

or b) the operator $-\frac{\mathrm{d}}{\mathrm{dt}}$, (2.9)

then $g \doteq Af$ satisfies

$$H_+ g = \Gamma g \tag{2.10}$$

Conversely, if g satisfies (2.10) then $h \doteq A^\dagger g$ satisfies (2.8).

This properties are easily proven by direct insertion and the use of the polar decomposition of A and $A^\dagger$ (see Refs.[4,17]).

The normalized eigenfunctions $\varphi_{(+),n}$ and $\varphi_{(-),n}$ (for $n \geq 1$) corresponding to the common eigenvalue E_n of H_- and H_+ are given by:

$$\varphi_{(-),n} = (E_n - E_0)^{-1/2} \ A^\dagger \varphi_{(+),n} \tag{2.16}$$

$$\varphi_{(+),n} = (E_n - E_0)^{-1/2} \ A \varphi_{(-),n} \tag{2.17}$$

Property 2) allows to express the time-dependent solutions associated with H_- in terms of time dependent solutions associated with H_+, and vice versa.

3. Adding One Eigenvalue

We start with a Schrödinger operator

$$H_0 = -\frac{d^2}{dx^2} + V_0(x) \tag{3.1}$$

with a potential that at $\pm\infty$ behaves as

$$V_0 \sim c_\pm x^{\eta_\pm}, \qquad c_\pm, \eta_\pm \geq 0 \tag{3.2}$$

This includes potentials with purely discrete spectrum as well as e.g. the free particle. We will denote by λ_0 the lowest point of the spectrum.

We will construct a new Hamiltonian H_1 that has the same spectrum as H_0 but with one supplementary eigenvalue $\lambda_1 < \lambda_0$. The idea is to identify $(H_0 - \lambda_1)$ with the $(H_+ - E_0)$ of last section; the corresponding H_- is then identified as the new H_1: We want to represent H_0 as

$$(H_0 - \lambda_1) = A_1 A_1^\dagger \tag{3.3}$$

with

$$A_1 = \varphi_{(1),0} \frac{d}{dx} (\varphi_{(1),0})^{-1} = \frac{d}{dx} + W_1' \tag{3.4}$$

Then

$$H_1 \equiv -\frac{d^2}{dx^2} + V_1(x) \doteq A_1^\dagger A_1 + \lambda_1 \tag{3.5}$$

will have the desired properties.

Equation (3.3) together with eq. (3.4) is equivalent to

$$\begin{aligned} V_0 - \lambda_1 &= (W_1')^2 + W_1'' \\ &\equiv \varphi_{(1),0} \frac{d^2}{dx^2} \varphi_{(1),0}^{-1} \end{aligned} \tag{3.6}$$

or further

$$\left(-\frac{d^2}{dx^2} + V_0\right) \varphi_{(1),0}^{-1} = \lambda_1 \varphi_{(1),0}^{-1} \tag{3.7}$$

Thus, all that is needed in order to add one eigenvalue is to find a positive solution of the original Schrödinger equation (corresponding to the Hamiltonian (3.1)) for a $\lambda_1 < \lambda_0$, such that its inverse is normalizable. As we will see, for each λ_1 there is a one-parameter family of such solutions. The new potential is given by

$$\begin{aligned} V_1 &= V_0 - 2W_1'' \\ &= -V_0 + 2\lambda_1 + 2(\varphi_{(1),0} \frac{d}{dx} \varphi_{(1),0}^{-1})^2 \end{aligned} \tag{3.8}$$

The set of all solutions of (3.7) can be completely characterised by the following asymptotic properties (a more complete discussion is given in the Appendix of Ref.[17]):

1)There are two linearly independent positive solutions of (3.7) that are uniquely defined by the asymptotic conditions

$$h_1(x) \underset{x \to -\infty}{\sim\!\succ} (V_0 - \lambda_1)^{-1/4} \, e^{\int^x dy \, (V_0 - \lambda_1)^{1/2}} \to 0 \tag{3.9}$$

$$h_3(x) \underset{x \to +\infty}{\sim\!\succ} (V_0 - \lambda_1)^{-1/4} \, e^{-\int^x dy \, (V_0 - \lambda_1)^{1/2}} \to 0 \ . \tag{3.10}$$

h_1 and h_3 go exponentially to ∞ in the opposite limits.

2)With these two functions we can construct a one-parameter familly of solutions

$$g_0(x;\lambda_1,\alpha_1) \doteq \alpha_1\; h_1(x) + \; h_3(x) \tag{3.11}$$

that, for $\alpha_1 > 0$, are everywhere positive and their inverse is square-integrable. We can thus use g_0 for the addition of an eigenvalue. This familly, together with h_1 and h_3, are all the positive solutions of (3.7)(up to multiplication with a constant), because h_1 and h_3 are linearly independent, and a negative α_1 would lead to negative values for $x \to \infty$.

Remark that if V_0 is symmetric and we set $\alpha_1 = 1$ then g_0 is the (unique) even solution, whereas $\alpha_1 = -1$ gives the odd solution.

3) The new potential V_1 has the same asymptotic behavior (3.2) as V_0.

4. Adding Two Eigenvalues

One could now iterate the procedure of Section 3 and add eigenvalues $\lambda_1, \lambda_2, ..., \lambda_N$ succesively. This involves the determination of a positive solution of a new Schrödinger equation at each step. Since the potentials become more and more complicated this may look like a hopeless task. However, it turns out that it is sufficient to know the solutions $h_1(x;\lambda)$ and $h_3(x;\lambda)$ of the original H_0 for $\lambda = \lambda_1, \lambda_2, ..., \lambda_N$, and that one can construct H_N in one step.

In order to explain the procedure, we consider first the addition of a second eigenvalue λ_2.

Let $g_0(x;\lambda_1,\alpha_1)$ be a positive solution for $\lambda = \lambda_1$ of

$$H_0 f = \lambda f \tag{4.1}$$

as defined in eq. (3.11). Define

$$V_1 = V_0 - 2\,\frac{d^2}{dx^2}\,\ln g_0 \tag{4.2}$$

We know from Section 2., that if $f_0(x;\lambda)$ is a solution of eq.(4.1) for an arbitrary λ, then a solution of

$$H_1 f = \lambda f \tag{4.3}$$

is given by

$$f_1(x;\lambda) \doteq cA_1^\dagger\;\; f_0(x;\lambda) \tag{4.4}$$

$$\equiv -c\; {g_0}^{-1}(\lambda_1,\alpha_1)\det\begin{pmatrix} g_0(\lambda_1,\alpha_1) & f_0(\lambda) \\ g_0'(\lambda_1,\alpha_1) & f_0'(\lambda) \end{pmatrix} \tag{4.5}$$

$$\equiv -c\,\frac{\Omega_2(\;g_0(\lambda_1,\alpha_1)\;,\;f_0(\lambda)\;)}{\Omega_1(g_0(\lambda_1,\alpha_1))} \tag{4.6}$$

where c is a constant and

$$A_1^\dagger = -\;g_0(\lambda_1,\alpha_1)\,\frac{d}{dx}\,{g_0}^{-1}(\lambda_1,\alpha_1) \tag{4.7}$$

In eq. (4.6) we have introduced the notation

$$\Omega_n(f_1, f_2, ..., f_n) \doteq \det \begin{pmatrix} f_1 & \cdots & f_n \\ f_1' & \cdots & f_n' \\ \vdots & \ddots & \vdots \\ f_1^{(n-1)} & \cdots & f_n^{(n-1)} \end{pmatrix} \tag{4.8}$$

and $\Omega_0 \doteq 1$. In fact, all the solutions can be obtained this way. The general positive solution g_1 is given by

$$g_1 = \frac{\Omega_2(\, g_0(\lambda_1, \alpha_1), g_0(\lambda_2, -\alpha_2)\,)}{\Omega_1(g_0(\lambda_1, \alpha_1))} \tag{4.9}$$

where $\alpha_2 > 0$ is an arbitrary constant. The negative sign in $-\alpha_2$ has the effect that g_1 is positive and its inverse is normalizable [17].

Thus we can construct the new potential with two added eigenvalues as

$$\begin{aligned} V_2 &= V_1 - 2\,\frac{d^2}{dx^2}\,\ln g_1 \\ &= V_0 - 2\,\frac{d^2}{dx^2}\,(\,\ln g_1 + \ln g_0\,) \\ &= V_0 - 2\,\frac{d^2}{dx^2}\,\ln\,\frac{\Omega_2(\, g_0(\lambda_1, \alpha_1), g_0(\lambda_2, -\alpha_2)\,)}{\Omega_1(g_0(\lambda_1, \alpha_1))}\,\Omega_1(g_0(\lambda_1, \alpha_1)) \\ &= V_0 - 2\,\frac{d^2}{dx^2}\,\ln \Omega_2\,(g_0(\lambda_1, \alpha_1), g_0(\lambda_2, -\alpha_2)) \end{aligned} \tag{4.11}$$

Its ground-state and the eigenfunction corresponding to λ_1 are

$$\varphi_{(2),\lambda_2} = c_{2,2}\; g_1^{-1} \tag{4.12}$$

$$\varphi_{(2),\lambda_1} = c_{2,1}\; A_2^\dagger\; g_0^{-1} \tag{4.13}$$

where $c_{2,2}$ and $c_{2,1}$ are normalization constants. The normalized eigenfunctions for the eigenvalues $E_n > \lambda_1$ are given by

$$\varphi_{(2),n} = (E_n - \lambda_2)^{-1/2}\;\; A_2^\dagger \varphi_{(1),n} \tag{4.14}$$

$$= (E_n - \lambda_2)^{-1/2} (E_n - \lambda_1)^{-1/2}\;\; A_2^\dagger\;\; A_1^\dagger \varphi_{(0),n} \tag{4.15}$$

$$= Z_{2,n}\, B_2^\dagger\;\varphi_{(0),n} \tag{4.16}$$

In the last equality we have introduced the notation

$$B_N^\dagger \doteq A_N^\dagger A_{N-1}^\dagger ... A_1^\dagger \tag{4.17}$$

with

$$A_i^\dagger = -\; g_{i-1}\;\frac{d}{dx}\; g_{i-1}^{-1} \tag{4.18}$$

and for the normalization

$$Z_{N,n} \doteq \prod_{i=1}^{N} (E_n - \lambda_i)^{-1/2} \tag{4.19}$$

Notice that the operator $B_2^\dagger$ acting on a function f can be represented as

$$B_2^\dagger f = \frac{\Omega_3(\, g_0(\lambda_1, \alpha_1)\;,\; g_0(\lambda_2, -\alpha_2)\;,\; f\,)}{\Omega_2(\, g_0(\lambda_1, \alpha_1)\;,\; g_0(\lambda_2, -\alpha_2)\,)} \tag{4.20}$$

5. Adding N Eigenvalues

The results of last section are easily extended by induction to the simultaneous addition of N eigenvalues:

Given N arbitary values $\lambda_0 > \lambda_1 > \lambda_2 > ... > \lambda_N$ and N positive constants $\alpha_1, ..., \alpha_N$, we consruct the potential

$$V_N = V_0 - 2\frac{d^2}{dx^2}\ln\Omega_N(\ g_0(\lambda_1, +\alpha_1),\ g_0(\lambda_2,\ -\alpha_2),\ g_0(\lambda_3, +\alpha_3), ..., g_0(\lambda_N, (-1)^{N+1}\alpha_N))$$

(5.1)

The corresponding Hamiltonian H_N has the same spectrum as H_0 plus the eigenvalues $\lambda_1, \lambda_2, ..., \lambda_N$. Its ground-state is

$$\varphi_{(N),\lambda_N} = c_{N,N}\ g_{N-1}^{-1} \tag{5.2}$$

where g_{N-1} is the general positive solution of $H_{N-1}g_{N-1} = \lambda_N g_{N-1}$, which is given by (5.6). The eigenfunctions corresponding to $\lambda_1, \lambda_2, ..., \lambda_{N-1}$ are

$$\varphi_{(N),\lambda_i} = c_{N,i}\ A_N^\dagger ... A_{i+1}^\dagger g_{i-1}^{-1} \tag{5.3}$$

The normalized eigenfunctions for the eigenvalues $E_n > \lambda_1$ are given by

$$\varphi_{(N),n} = Z_{N,n}\ B_N^\dagger\ \varphi_{(0),n} \tag{5.4}$$

(with the notation introduuced in (4.17), (4.19)). The operator $B_N^\dagger$ can be represented by

$$B_N^\dagger f = (-1)^N \frac{\Omega_{N+1}(\ g_0(\lambda_1, \alpha_1), ..., g_0(\lambda_N, (-1)^{N+1}\ \alpha_N)\ ,\ f\ )}{\Omega_N(\ g_0(\lambda_1, \alpha_1), ..., g_0(\lambda_N, (-1)^{N+1}\ \alpha_N)\ )} \tag{5.5}$$

The relations (5.1)-(5.4) follow immediately from

$$g_{N-1} = \frac{\Omega_N(\ g_0(\lambda_1, \alpha_1), ..., g_0(\lambda_N, (-1)^{N+1}\ \alpha_N)\ )}{\Omega_{N-1}(\ g_0(\lambda_1, \alpha_1), ..., g_0(\lambda_{N-1}, (-1)^N\ \alpha_{N-1})\ )} \tag{5.6}$$

which is obtained by induction from the results of Section 4. (This is done e.g. in Ref.[5],p.176 for the case $V_0 = 0$, but the algebra is the same in the general case). Therewith we obtain eq.(5.1) by

$$\begin{aligned} V_N &= V_{N-1} - 2\ \frac{d^2}{dx^2}\ \ln\frac{\Omega_N}{\Omega_{N-1}} \\ &= V_{N-2} - 2\ \frac{d^2}{dx^2}\ \ln\frac{\Omega_N}{\Omega_{N-1}}\frac{\Omega_{N-1}}{\Omega_{N-2}} \\ &\ \vdots \\ &= V_0 - 2\ \frac{d^2}{dx^2}\ \ln\ \Omega_N \end{aligned} \tag{5.7}$$

The rest is an immediate consequence of the results of the previous sections.

6. Propagators

In this section we show that the time-dependent propagators G_N and G_{N-1} corresponding to the Hamiltonians H_N and H_{N-1} are related by the two following equivalent equations:

$$G_N(x,x_0;t) = -\, g_{N-1}^{-1}(x_0)\, A_N^{\dagger}(x) \int_{x_0}^{\infty} dz G_{N-1}(x,z,t)\, g_{N-1}(z) \tag{6.1}$$

$$\begin{aligned} G_N(x,x_0;t) =& \theta(t)\, e^{-\lambda_N t}\, \varphi_{(N),\lambda_N}(x)\, \varphi_{(N),\lambda_N}(x_0) \\ & \theta(t)\, A_N^{\dagger}(x) A_N^{\dagger}(x_0) \int_t^{\infty} ds G_{N-1}(x,x_0,s)\ e^{-\lambda_N(t-s)} \end{aligned} \tag{6.2}$$

By iteration we can express G_N in terms of the known G_0. The result is given in Eqs. (6.11) and (6.12). The validity of (6.1) and (6.2) can be checked directly by insertion. They were constructed as follows:

1) In Section 2 we saw that the time-dependent solutions of

$$-\psi_N(x,t) = H_N \psi_N(x,t) \tag{6.3}$$

can be expressed in terms of solutions of

$$-\psi_{N-1}(x,t) = H_{N-1} \psi_{N-1}(x,t) \tag{6.4}$$

by

$$\psi_N(x,t) = A_N^{\dagger} \psi_{N-1}(x,t) \tag{6.5}$$

Notice that in (6.5) we do not ask that the ψ are in the Hilbert space. Thus, in order to relate the propagators G_N and G_{N-1} we need to determine the initial condition $\psi_{N-1}(x,0)$ that gives

$$A_N^{\dagger} \psi_{N-1}(x,0) \ = \ \psi_N(x,0) \ = \ \delta(x-x_0) \tag{6.6}$$

The inversion of (6.6) gives

$$\begin{aligned} \psi_{N-1}(x,0) &= -g_{N-1}(x) \int_{-\infty}^{x} dz\ \ g_{N-1}^{-1}(z)\ \delta(z-x_0) \\ &= -g_{N-1}(x)\ g_{N-1}^{-1}(x_0)\ (1-\theta(x-x_0)) \end{aligned} \tag{6.7}$$

($\psi_{N-1}(x,0)$ is clearly not in L_2.) The propagator is given by

$$\begin{aligned} G_N(x,x_0,t) &= \theta(t)\ A_N^{\dagger}(x)\ \psi_{N-1}(x,t) \\ &= A_N^{\dagger}(x) \int_{-\infty}^{\infty} dz\ G_{N-1}(x,z,t)\ \psi_{N-1}(z,0) \end{aligned} \tag{6.8}$$

Inserting (6.7) into (6.8) we obtain (6.1).

2) The second expression (6.2) can be obtained easely in the case of a purely discrete spectrum, using the representation

$$G(x,x_0,t) = \theta(t)\sum_{n=0}^{\infty}\varphi_n(x)\varphi_n(x_0)\ \mathrm{e}^{-E_n t}, \tag{6.9}$$

and applying the transformation (2.16) to $\varphi_n(x)$ and $\varphi_n(x_0)$ and the identity

$$\int_t^{\infty} ds\ G_{N-1}(x,x_0,s)\ \mathrm{e}^{-\lambda(t-s)} = \theta(t)\sum_{n=0}^{\infty}\varphi_n(x)\varphi_n(x_0)\ (E_n-\lambda)^{-1}\ \mathrm{e}^{-E_n t}. \tag{6.10}$$

In the general case (spectrum with a continuous part) the same expression (6.2) can be obtained using Hilbert space methods, as described in Ref.[17].

3) We remark finally that after performing the iteration of (6.1) the result can be written compactly as

$$G_N(x,x_0;t) = (-1)^N B_N^{\dagger}(x)\int_{x_0}^{\infty} dz_N\int_{z_N}^{\infty} dz_{N-1}\ldots\int_{z_2}^{\infty} dz_1 G_0(x,z_1,t)\prod_{i=1}^{N}\left(g_{i-1}^{-1}(z_{i+1})\ g_{i-1}(z_i)\right) \tag{6.11}$$

with $z_{N+1}\equiv x_0$. The iteration of (6.2) gives (with $s_{N+1}\equiv t$)

$$G_N(x,x_0;t) = \theta(t)\ \sum_{i=1}^{N}\mathrm{e}^{-\lambda_N t}\ \varphi_{(i),\lambda_i}(x)\ \varphi_{(i),\lambda_i}(x_0) + \tag{6.12}$$

$$\theta(t)B_N^{\dagger}(x)B_N^{\dagger}(x_0)\int_t^{\infty} ds_N\int_{s_N}^{\infty} ds_{N-1}\ldots\int_{s_2}^{\infty} ds_1 G_0(x,x_0,s_1)\ \exp\left(-\sum_{i=1}^{N}(s_{i+1}-s_i)\lambda_i\right).$$

7. Conclusion

We have thus a method to construct families of exactly solvable models depending on 2N free parameters. One can use this freedom to construct potentials having some qualitative features, or that mimic models for a particular physical problems.

The asymptotic behavior is the same within a family. Thus the modifications occur in relatively localized regions. For instance, one observes that if two eigenvalues are put close to each other the potential develops a barrier between two wells, as one may expect from the theory of tunnel-splitting.

The case $V_0 = 0$ is related to soliton solutions of the Korteweg-deVries equation [18]. Since the number of solitons is equal to the number of eigenvalues, an appropriate choice of the parameters can give one potential-well per eigenvalue.

Thus, by appropriately choosing the eigenvalues λ_i and the constants α_i, one can construct solvable models that cover a wide range of qualitative behavior.

Appendix

As the simplest illustration of the method we consider the potential V_1 constructed by adding one eigenvalue $\lambda_1 = -\mu^2$ to the free particle ($V_0 = 0$): The general positive solution of (3.7) is

$$g_0(x; -\mu^2, \alpha) = \mathrm{ch}(\mu x; \alpha) \doteq \frac{1}{2}(\alpha\ e^{\mu x} + e^{-\mu x}). \tag{A.1}$$

The new potential is (by (3.8)):

$$V_1 = \frac{\alpha\mu}{\mathrm{ch}(\mu x; \alpha)} \tag{A.2}$$

and the normalized groundstate

$$\varphi_{1,\lambda_1}(x) = \frac{\sqrt{2\mu\alpha}}{\mathrm{ch}(\mu x; \alpha)} \tag{A.3}$$

The propagator can be easily calculated using (6.1):

$$G(x, x_0; t) = \frac{1}{\sqrt{4\pi t}}\ e^{-\frac{(x-x_0)^2}{4t}} + \frac{\mu\ e^{\mu^2 t}}{4\ \mathrm{ch}(\mu x_0; \alpha)\ \mathrm{ch}(\mu x; \alpha)}\left[\mathrm{Erf}\left(\frac{2\mu t + (x - x_0)}{\sqrt{4t}}\right) + \mathrm{Erf}\left(\frac{2\mu t - (x - x_0)}{\sqrt{4t}}\right)\right] \tag{A.4}$$

where Erf denotes the error function.
The corresponding Fokker-Planck equation has a drift

$$K = -2\mu\ \mathrm{th}(\mu x; \alpha) \doteq -2\mu\frac{\alpha\ e^{\mu x} - e^{-\mu x}}{\alpha\ e^{\mu x} + e^{-\mu x}}. \tag{A.4}$$

Its stationary state is

$$P_S = \varphi^2_{1,\lambda_1}(x) = \frac{2\mu\alpha}{\mathrm{ch}^2(\mu x; \alpha)} \tag{A.5}$$

and the transition probability density is given by

$$P(x, t|x_0) = \frac{\mathrm{ch}(\mu x_0; \alpha)}{\mathrm{ch}(\mu x; \alpha)}\frac{1}{\sqrt{4\pi t}}\ e^{-\frac{(x-x_0)^2}{4t}}\ e^{-\mu^2 t} + \frac{\mu\alpha}{4\ \mathrm{ch}^2(\mu x; \alpha)}\left[\mathrm{Erf}\left(\frac{2\mu t + (x - x_0)}{\sqrt{4t}}\right) + \mathrm{Erf}\left(\frac{2\mu t - (x - x_0)}{\sqrt{4t}}\right)\right]. \tag{A.6}$$

Acknowledgements: I would like to thank L.S. Schulman, F. Monti and M.-O. Hongler for very useful discussions. This work has been partially supported by the Fonds National Suisse de la Recherche Scientifique.

References

[1]- G. Darboux, C.R. Acad. Sci. Paris **94** (1882) 1456; and *Théorie des surfaces*, Vol. 2, Gauthier-Villars, Paris 1894.

[2]- M.M. Crum, Quart. J. Math. Oxford (2), **6** (1955) 121.

[3]- E. Witten, Nucl. Phys. **B 188** (1981) 513.

[4]- P.A. Deift, Duke Math. J. **45** (1978) 267.

[5]- P. Deift, E. Trubowitz, Comm. Pure Appl. Math. **29** (1979) 121.

[6]- W.M. Zheng, J. Math. Phys.**25** (1984) 88.

[7]- M. Razavy, Am. J. Phys. **48** (1980) 285.

[8]- B. Mielnik, J. Math. Phys. **25** (1984) 3387.

[9]- C.V.Sukumar, J. Phys. **A 18** (1985) 2917 and 2937, J. Phys. **A 19** (1986) 2297 , and J. Phys.**A 20** (1987) 2461 .

[10]- B. Gaveau, L.S. Schulman, J. Phys. **A 19** (1986) 1833.

[11]- M.J. Englefield, J. Phys.**A 20** (1987) 593.

[12]- M.-O. Hongler, W.M. Zheng, J. Stat. Phys. **29** (1982) 317.

[13]- M.-O. Hongler, W.M. Zheng, J. Math. Phys. **24** (1983) 336.

[14]- H.R. Jauslin, Helv. Phys. Acta **61** (1988) 901.

[15]- H. Risken, *The Fokker-Planck Equation*, Springer Verlag, 1984.

[16]- Th. Leiber, F. Marchesoni, H. Risken, Phys. Rev. Lett. **59** (1987) 1381.

[17]- H.R. Jauslin, J. Phys. **A 21** (1988) 2337 .

[18]- G.L. Lamb, Jr., *Elements of Soliton Theory*, J. Wiley, New York 1980.

PHASE AND FREQUENCY DYNAMICS IN LASER INSTABILITIES

N.B. Abraham
Institute for Scientific Interchange, Villa Guallino,
10129 Torino, ITALY
and
Department of Physics, Bryn Mawr College, Bryn Mawr, PA 19010 USA

INTRODUCTION

It would seem natural that studies of laser dynamics would contain many examples of the dynamical changes in the optical phase or its derivative, the optical frequency, in addition to the numerous reports of amplitude (or intensity) pulsations. Surprisingly, (though the early discoverers of laser instabilities noted both amplitude and phase contributions) such dynamics are rarely reported, either because they do not exist in the systems under study or because they were not observed or measured.

One reason for the relative absence of references to phase dynamics is that most optical signals are detected by measurements of the intensity of the electric field. Such measurements are insensitive to the phase of the electric field or to the optical carrier frequency. This "natural" form of experimental measurement has influenced the way that theoreticians view their equations and the way that numerical solutions of the equations are presented.

Another reason that the phase is often ignored is that in a number of simple problems the phase of an optical system is decoupled from the amplitude and it does not enter the dynamics of the variables in a crucial way. In these cases the phase is not of interest in studies of the dynamical instability of steady state solutions.

One strongly held view is that optical instabilities most often arise from "population pulsations" under a mechanism wherein weak optical sidebands added to a strong optical field cause a pulsating intensity which modulates the population inversion of the medium.

When the modulation frequency matches a natural oscillation frequency of the population inversion the latent sidebands can parametrically draw energy of the strong field. However, if the modulation of the population inversion is a real sinusoidal modulation at the frequency separation of the sidebands from the strong oscillating frequency, the population modulation causes the generation of a pair of equally spaced and in-phase optical sidebands which correspond to an amplitude modulation of the electric field. This is a simple description of an "amplitude instability" which frequently occurs in optical systems. The "frequency modulated" case in which the sidebands are out of phase and thus cause intensity and population inversion modulation at the second harmonic of their spacing from the strong mode is usually formally considered (though in some critical cases it has been ignored). Because in many cases it is less likely to cause instabilities, it is often not discussed or not carefully examined.

Instabilities in only the phase of the electric field at first have no effect on the intensity. Thus, they have no initial effect on the population inversion. One may think of these as purely dispersive instabilities. When phase instabilities grow stronger they frequently lead to changes in the amplitude as they eventually couple to the amplitude variables, particularly when strong phase pulsations are involved.

It is fair to say that both kinds of instabilities and also their linear combinations can arise. When such instabilities occur they involve coupling among the several variables and cannot be uniquely assigned as arising from pulsations in only one of the variables. Indeed, when the pulsations arise from FM instabilities, the pulsations of the population inversion are often minor consequences of the process and not the origin. Unfortunately, FM instabilities when noted are often then monitored for the ensuing intensity pulsations, so that much of the crucial dynamical evolution of the phase is ignored.

It is important to identify the particular physics of phase and amplitude decompositions in optical physics. Most optical dynamics problems are solved in a slowly-varying amplitude approximation in which one assumes a carrier wave oscillating at an optical frequency with an amplitude that varies more slowly.

$$E(r,t) = E(r,t)\exp(i(kz-\omega t)) + c.c. \quad (1)$$

Such an approximation is generally valid when the decay rates of the relevant atomic levels are much less that the optical frequency corresponding to the energy difference between the coupled levels and when the bandwidth of the illuminating or emitted fields is relatively narrow.

The slowly varying nature of the amplitude with respect to k and ω permits a reduction of the second-order Maxwell's inhomogeneous wave equation

$$\nabla^2 E - (1/c^2)\ddot{E} = -\mu\ddot{P} \quad (2)$$

to a first-order equation

$$[(1/ik)\nabla_t^2 + (\partial/\partial z) + (1/c)(\partial/\partial t)]E = (\omega^2/ik)P. \quad (3)$$

A single longitudinal mode approximation (or better, a longitudinally uniform field approximation) leads to elimination of the z-derivative and this is reasonable when the longitudinal boundaries are close enough together to give a mode spacing which exceeds the bandwidth of the slowly varying amplitude and the bandwidth of the material response. The imperfect reflection of the mirrors is then expressed as a damping factor for the field and the resonance condition gives this a complex form. After these approximations one is left with a "wave equation" reduced to the form

$$\dot{E} = -i\nabla_t^2 E - \kappa E - i\delta E + \tilde{\mu}P. \quad (4)$$

The resulting equation is similar to diffusion problems in fluids, chemical reactions and heat transport as the transverse Laplacian appears in the optical case from the effects of diffraction. The difference is that the "diffusion coefficient" in the optics case is pure imaginary.

The plane wave approximation eliminates the transverse derivative as well with the further simplification of the "wave equation" to the form

$$\dot{E} = -\kappa E - i\delta E + \tilde{\mu}P. \qquad (5)$$

Further equations are then needed to describe the response of the material polarization P to the driving by the electric field. Even when the dynamical response is fast, P is typically a nonlinear complex function which can be written

$$P = \chi(E)E \qquad (6)$$

where $\chi(E)$, the complex susceptibility, depends on the amplitude, phase, and frequency of the electric field.

This sequence of approximations is usually well justified except for the plane wave approximation which has little experimental validity. However often the boundary conditions and the spatial extent of the medium provide such strong transverse mode selection that the transverse effects make negligible contributions to the dynamics.

It is the slowly varying amplitude $E(r,t)$ which is separated into amplitude and phase given by

$$E(r,t) = \rho(r,t)\exp(i\Theta(r,t)). \qquad (7)$$

The phase may be free or locked in steady state solutions. Usually it is only locked when there is an external field with a definite phase which breaks the phase symmetry of the free-running problem. Instabilities may come in the amplitude alone, in coupled amplitude and phase dynamics, or purely in the phase. Pure phase instabilities typically signal a frequency shift from the frequency of the reference frame (that is, the frequency of the carrier wave) selected for the problem. These may correspond to the growth of a single sideband or to the loss of stability of the assumed steady state solution in favor of a steady state with a different optical frequency. In systems with an external phase reference, phase instabilities signal frequency modulation. For systems of ordinary differential equations in time, pure phase instabilities require two phase rather than just one such as the phase of the electric field and the atomic polarization. In contrast, if one keeps transverse spatial dependence of the field, using partical differential equations in space and time, one may find pure phase instabilities in space in the style of the Kuramoto-Sivashinski equation.

In many other cases the phase does not play a singular role, but it plays an important role as a necessary participant in the dynamics of a coupled system of amplitude and phase variables. For example, the detuning of a laser changes the relative phase of the electric field and the material polarization. Typically this phase difference also then enters the dynamics of the laser system. Intensity pulsations are then coupled to pulsations in the phase difference, and thereby, typically, to pulsations in the phase and frequency of the electric field.

A brief list will help to reveal the breadth of phase instabilities or examples where the phase enters the dynamics of lasers in important ways.

Single mode lasers with detuning have important temporal oscillations in the phase when they develop spontaneous instabilities. Both inhomogeneously broadened lasers and homogeneously broadened lasers have been shown to exhibit these effects /1-7/. The full extent of the phase and frequency modulation has only recently begun to be appreciated in these cases. One important consequence of the entry of the phase into the dynamics is a reduction in the chaotic nature of the pulsations and an overall increase in the stability of the laser. More periodic pulsations are observed instead of chaotic pulsations and the thresholds for instabilities are increased.

In addition, the transition of single mode lasers to multimode oscillations sometimes involves pure amplitude modulation, signalling the onset of pairs of sidemodes and at other times involves "phase instabilities" which signal transitions to two-mode operation or to the switching of stability from one mode to another mode. The phase instabilities are found in homogeneously broadened lasers with detuning /8,9/ and are found in inhomogeneously broadened lasers even without detuning /10/.

Perhaps the earliest case where the phase and frequency dynamics were known to be important is that of a laser with an injected signal from another laser. Instabilities arise when the injected signal is not quite resonant with the second laser and the instead of phase and frequency locking there may result intensity and phase pulsations of very complicated forms /11-13/. These often take the form of transient locking of the two frequencies and then a slip in the optical phase by 2π radians before nearly locked conditions resume. This form of phase dynamics is typical of coupled but incompletely locked oscillators and it is not surprising that it also appears for optical oscillators.

Phase and frequency dynamics have also been found to play an important role in the interaction of counter-propagating modes in a bidirectional ring laser. Early studies found that detuning was essential to achieve spontaneous switching between the two modes /14/. Only more recently has it been observed (both experimentally and theoretically) that this instability includes alternating shifts in the phases and the frequencies of both modes /15,16/. Recognizing these shifts of phase and frequency helps us to interpret the origin of the instabilities and the nature of the resulting dynamical evolution.

Another multimode case of interest involves the interaction of a relatively large number of modes. Recent studies /17/ have shown that the modes exhibit not only amplitude fluctuations from their interactions, but that they also undergo shifts in phase and frequency as part of time-dependent interactions.

It should also be noted that phase dynamics is often discussed when one considers the response of a laser to noise. Phase noise and frequency noise contribute to the linewidth of the laser and are often traced to the intrinsic spontaneous emission which provides interference. The Schawlow-Townes linewidth formula applies to lasers where the phase dynamics are simple and decoupled from the amplitude dynamics. In semiconductor lasers, where there is a strong intensity dependent dispersion, the AM and FM noise spectra are coupled and correlated. This also provides a reminder that semiconductor lasers in various applications are good candidates to show important phase dynamics as well, but there have been few investigations of these phenomena /18/.

Phase dynamics have been found in lasers with transverse spatial variations as the transition from plane wave pattterns to stationary transverse spatial variations arises from spatial phase instabilities /19/. Phase dynamics are also known in other optical systems including one photon optical bistability /20/ (where detuning is essential to the achievement of instabilities, indicating phase and amplitude coupling), in subharmoninc generation in a cavity /21/ and in two-photon optical bistability /22/ (where detuning is not essential and where in resonance the instabilities are purely in the phase). The richness and variety of this list make it all the more surprising that phase instabilities are often not followed with attention to the phase and frequency modulation or variations that result in the final solutions.

OUTLINE

In these notes, details of several of these cases will be reviewed to present some examples and characteristics of phase and frequency evolution both in causing the instabilities and in sustaining the pulsations. Specific choices are drawn from recent studies of single mode ring lasers (both inhomogeneously broadened and homogeneously broadened) and of bidirectional ring lasers.

Perhaps it is worth a comment as to why it seems useful to refer to both "phase" and "frequency" dynamics when the frequency is defined to be the derivative of the phase. Of course there is no precise distinction except in a few specialized cases. The use of both terms serves as a reminder that some kinds of evolution are better described in terms of the time dependence of the phase while others are better described in terms of the time dependence of the derivative of the phase. For example, a system may maintain a nearly constant phase and then abruptly (or rapidly) change to a new nearly constant phase. In this case one is interested to know the physical meaning of the particular values of the phase that the system selects or the particular values of the phase jumps that appear. The sudden frequency shift during such a rapid phase change serves only to record the change. In contrast, there may be periods of relatively constant frequency followed by rapid transitions to a new frequency. In this case one may search for physical insight regarding the process of selecting the different frequencies or the frequency changes. Of course, more generally, there may be complicated and continuous evolution of both the phase and its derivative making the distinctions of less relevance in interpreting the results.

THE ENTRANCE OF THE PHASE INTO THE DYNAMICS OF SINGLE MODE LASERS

The phase can be shown to enter the dynamics of the laser in two simple ways. First, we consider the effect of detuning of the laser cavity from resonance on the steady state solutions of the laser. Second, we consider the effect of detuning on the stability of the steady state solutions.

The Steady State Frequency of the Laser

The equations of the single mode laser are given by /13/

$$\dot{E} = -\kappa E - i\delta_{AC}E - \kappa A \int P_\delta g(\delta)\,d\delta \tag{8a}$$

$$\dot{P}_\delta = -\gamma_\perp(1+i\tilde{\delta})P - \gamma_\perp E D_\delta \tag{8b}$$

$$\dot{D}_\delta = -\gamma_{||}(D_\delta - 1) - (\gamma_{||}/2)(P_\delta^*E + P_\delta E^*) \tag{8c}$$

where $g(\delta)$ is an inhomogeneous lineshape factor which can be reduced to a delta function to recover the homogeneously broadened case. A is the pump parameter, and δ_{AC} is the detuning of the laser cavity from the atomic resonance and δ is the detuning of each subgroup of atoms from the center of the atomic line. E, P_δ and D_δ are the slowly varying amplitudes of the field in the cavity, the atomic polarization and the atomic population inversion, respectively, with corresponding decay rates κ, $\gamma_\perp$ and $\gamma_{||}$, and the tilde (~) denotes division by $\gamma_\perp$. E and P are complex variables, while D is a real variable. The equations are written in the reference frame of the resonant frequency of the atomic line center so that the slowly varying amplitudes multiply carrier waves at a much higher optical frequency.

The homogeneously broadened laser is described by five equations while the inhomogeneously broadened laser requires integro-differential equations or the integral can be approximated by a sum over a large number of subgroups. The homogeneously broadened case can be reduced to three real variables if the detuning is set to zero, while the inhomogeneously broadened case retains many complex variables even with resonant tuning.

The nontrivial steady states of the laser are found by setting the the intensity and the population inversion to constants and seeking solutions which correspond to a particular frequency of the electric field of the form

$$E(t) = E_{ss} \exp(-i\Delta\gamma_\perp t). \tag{9}$$

With a detuning δ_{AC}, the steady state frequency Δ of the laser changes. Hence the frequency of the laser is a function of the parameters.

For the homogeneously broadened laser the relation is relatively trivial:

$$\Delta = -\tilde{\delta}_{AC}/(1+\tilde{\kappa}). \tag{10}$$

The frequency is given by the detuning of the laser cavity from resonance and by the relative decay rates of the field in the cavity and of the polarization of the medium. Hence it is independent of the laser intensity (and the laser excitation parameter) as shown in Figure 1.

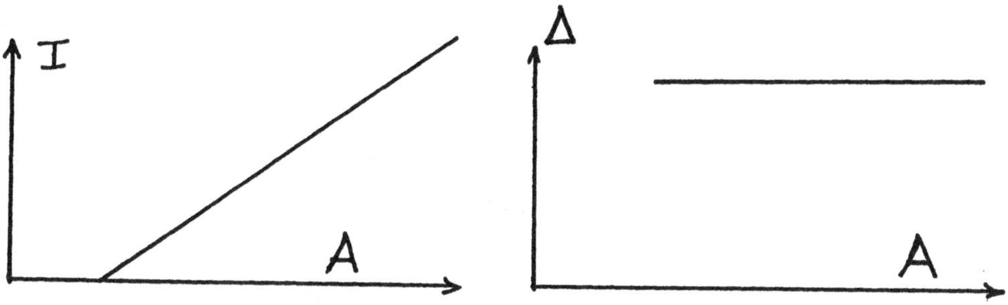

Figure 1. Schematic plots of intensity and laser frequency versus excitation for a homogeneously broadened laser.

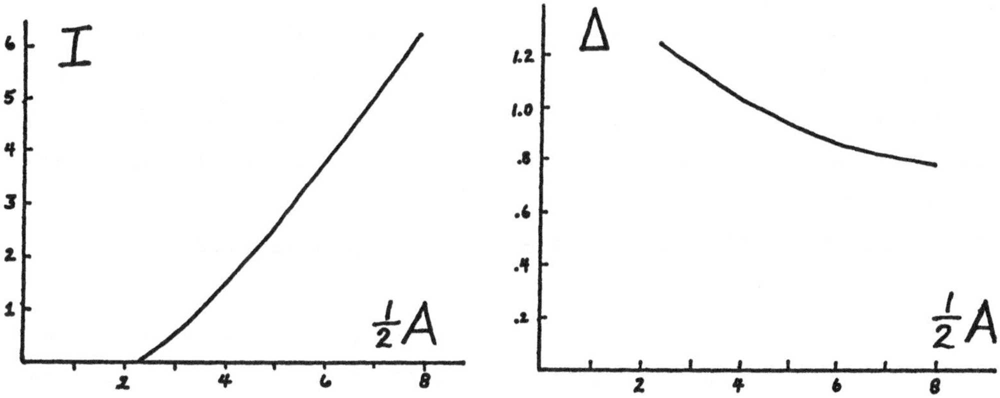

Figure 2. Intensity and laser frequency versus excitation for an inhomogeneously broadened laser with detuning of the cavity from the atomic resonance, selected from ref. 1. Parameter values are $\tilde{\sigma}_D = 4.93$ (inhomogeneously broadened linewidth of Gaussian profile); $\tilde{\kappa} = 6.19$; $\tilde{\gamma} = 0.243$ $(\gamma_\parallel/\gamma_\perp)$.

In contrast, the inhomogeneously broadened laser with detuning has a laser frequency that depends on the excitation parameter (and thus also on the intensity of the laser). An example for the solution of the equations for such a laser selected from reference 1 is shown in Figure 2.

In a naive view of the dynamics of these two lasers one would predict a difference in the correlation between frequency and intensity pulsations. In the inhomogeneously broadened case one might suppose that the frequency of the steady state is determined by the laser intensity and thus if the laser intensity changes in time it is reasonable to expect the laser frequency to vary in time. In contrast, it would seem at least possible that there could be intensity variations without frequency variations in the homogeneously broadened case. These arguments are made with a view which might be called the "adiabatic theory" of laser frequencies. However, the laser frequency is determined by the dispersive properties of the medium. In steady state equilibrium these are set by the degree of excitation of the medium and the degree of saturation of the medium by the intensity. In contrast, if the intensity is changing in time, the material variables are not in equilibrium and the dispersive properties of the medium are not generally able to adiabatically follow the intensity. In particular, the single mode laser instabilities arise precisely when the atomic variables are unable to adiabatically follow the fluctuations in the field and so the naive prediction turns out to be overly simplified.

The Stability of the Steady State Solutions of the Laser

The stability analysis of the steady state solution of the homogeneously broadened laser in resonance results in a secular equation for the eigenvalues which can be written as /5,6/:

$$\lambda(\lambda+\kappa+\gamma_\perp)\{\lambda^3+a_2\lambda^2+a_1\lambda+a_0\} = 0 \tag{11}$$

where

$$a_2 = \gamma_{||}+\gamma_\perp+\kappa$$

$$a_1 = \gamma_{||}(\gamma_\perp+A+\kappa)$$

$$a_0 = 2\gamma_{||}\gamma_\perp\kappa(A-1).$$

The cubic term governs the stability of the amplitudes of the three variables for the equations. The $\lambda = 0$ eigenvalue governs the sum of the phases of the field and polarization, indicating the arbitrariness of the absolute phase of the system. The $\lambda = -(\kappa+\gamma_\perp)$ eigenvalue governs the difference in the phases of the field and polarization, indicating that this phase difference is stabilized. The factorizing of the secular equation indicates that the amplitude dynamics (at least as they relate to the stability of the steady state solution) are decoupled from the phase dynamics.

In the inhomogeneously broadened case (for Gaussian or Lorentzian broadening lineshapes), the secular equation can be written in the following form /5/:

$$\lambda\, F(\lambda) \;=\; 0, \tag{12}$$

where F is a complicated function involving λ and the relaxation rates, the excitation parameter A, and the lineshape function. Rather surprisingly $F(\lambda)=0$ has a finite number of roots which can be found by rather tedious algebra in the case of the Lorentzian lineshape or by numerical solutions for the case of the Gaussian lineshape.

The $\lambda = 0$ eigenvalue which remains separated again corresponds to the absolute phase of the system. In contrast with the homogeneously broadened case, the relative phase(s) enters into the dynamics even around the resonant steady state solution.

With detuning, both secular equations become more complicated, but the important complication occurs in the homogeneously broadened case where the phase dynamics were decoupled in resonance. [The inhomogeneously broadened laser with detuning has a secular equation similar to that of the resonant inhomogeneously broadened laser.] In the homogeneously broadened case the secular equation for the detuned laser can be written as /13,14/:

$$\lambda\{\lambda^4 + b_3\lambda^3 + b_2\lambda^2 + b_1\lambda + b_0\} = 0 \tag{13}$$

where

$$b_3 = 2\gamma_\perp + 2\kappa + \gamma_{||}$$

$$b_2 = 2\gamma_{||}(\gamma_\perp+\kappa) + (\gamma_\perp+\kappa)^2 + (A-1-\Delta^2)\gamma_\perp\gamma_{||} + (\kappa-\gamma_\perp)^2\Delta^2$$

$$b_1 = [(\gamma_\perp+\kappa)^2 + \gamma_{||}(3\kappa+\gamma_\perp)(A-1-\Delta^2) + (\kappa-\gamma_\perp)^2\Delta^2]\gamma_{||}$$

$$b_0 = 2\kappa\gamma_{||}\gamma_\perp(\gamma_\perp+\kappa)(A-1-\Delta^2)\,.$$

The only eigenvalue remaining as a separate factor is that for the absolute phase. Detuning brings the relative phase into the dynamics. The involvement of the relative phase from the start of the instability is illustrated in Figure 3. Here we have slightly switched the gain from below to above the instability point of the

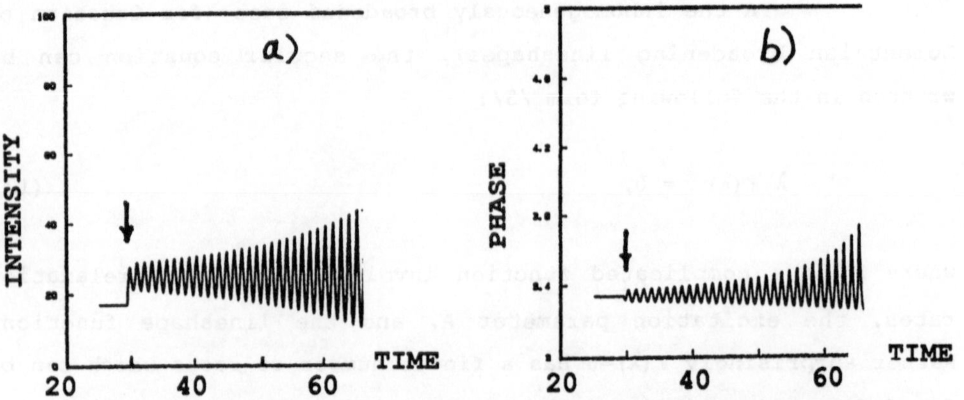

Figure 3. (results of T. Mello). a) Intensity evolution in the homogeneously broadened laser model for parameters $\tilde{\gamma} = 1.0$; $\tilde{\kappa} = 3.0$; $\tilde{\delta}_{AC} = 0.8$; after a switch in the excitation parameter from A = 17.9 to 25.9; b) Corresponding evolution of the phase difference.

laser. The system responds with growing intensity oscillations shown in Figure 3a. Figure 3b shows the evolution of the difference between the field and polarization phases corresponding to this instability. We see that from the very start the phase of the system has entered the dynamics. It turns out that in any, rotating reference frame we choose, the phase of the field also begins to oscillate, indicating that there is frequency modulation as well as amplitude modulation which can be observed and measured.

A close examination of the evolution of the phases of the two complex variables directly shows that both of them oscillate while their sum remains constant. Thus the involvement of the relative phase in the pulsations is reflected in phase and frequency modulation of the laser output field.

The detuning which brings the relative phase into the dynamics also has other important effects on the stability of the laser and on the nature of the instabilities. The most striking effect is that the threshold for laser instabilities is increased. This is shown in Figure 4. We see that in all cases the instability threshold is much higher with detuning than in resonance. Whether it is the involvement of the relative phase that contributes to this stabilization of the laser is difficult to separate. It also appears that as the detuning increases strongly that the instability threshold may be going to infinity. In particular, in the case of

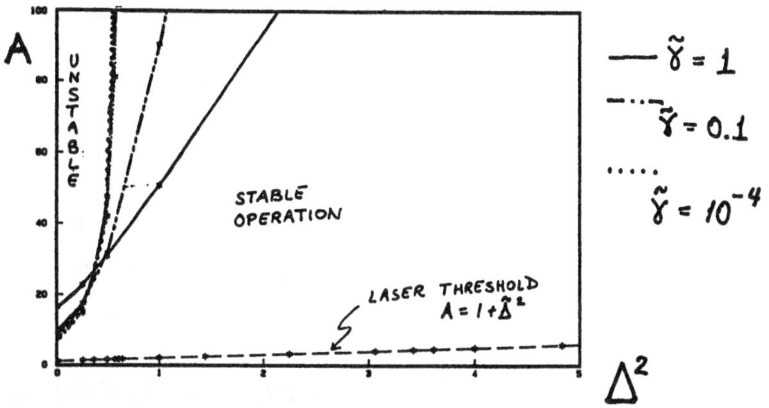

Figure 4. Instability boundary of the steady state laser solutions plotted for A vs. Δ^2 for several different values of $\tilde{\gamma}$. Parameter value: $\tilde{\kappa}$ = 3.0. (Results of Tina Mello.)

small γ it appears that the instability boundary reaches a vertical asymptote.

In addition to the stabilizing effect of the detuning, it is also found that with sufficient detuning the instabilities at the laser threshold become subcritical rather than supercritical (meaningthat at the instability threshold the pulsations are small amplitude modulations around the steady state solution instead of large pulses) /6/. This also means that with detuning the region of coexistence of large amplitude pulsing solutions and stable steady states (which exists on resonance below the instability boundary) diappears.

Another effect is that in the instability domain, the detuning tends to make the pulsations periodic rather than the chaotic solutions that are found on resonance /4a,6/.

TIME DEPENDENT PHASE AND FREQUENCY WITH MODERATE DETUNING

Several examples of time-dependent intensity and phase for the inhomogeneously broadened laser with detuning are shown in Figures 5, 6 and 7. With increasing excitation for fixed detuning the pulsingpattern changes from simply periodic (Fig. 5a) to alternating ("period-doubled") pulses (Figs. 5b,c). The phase of the electric

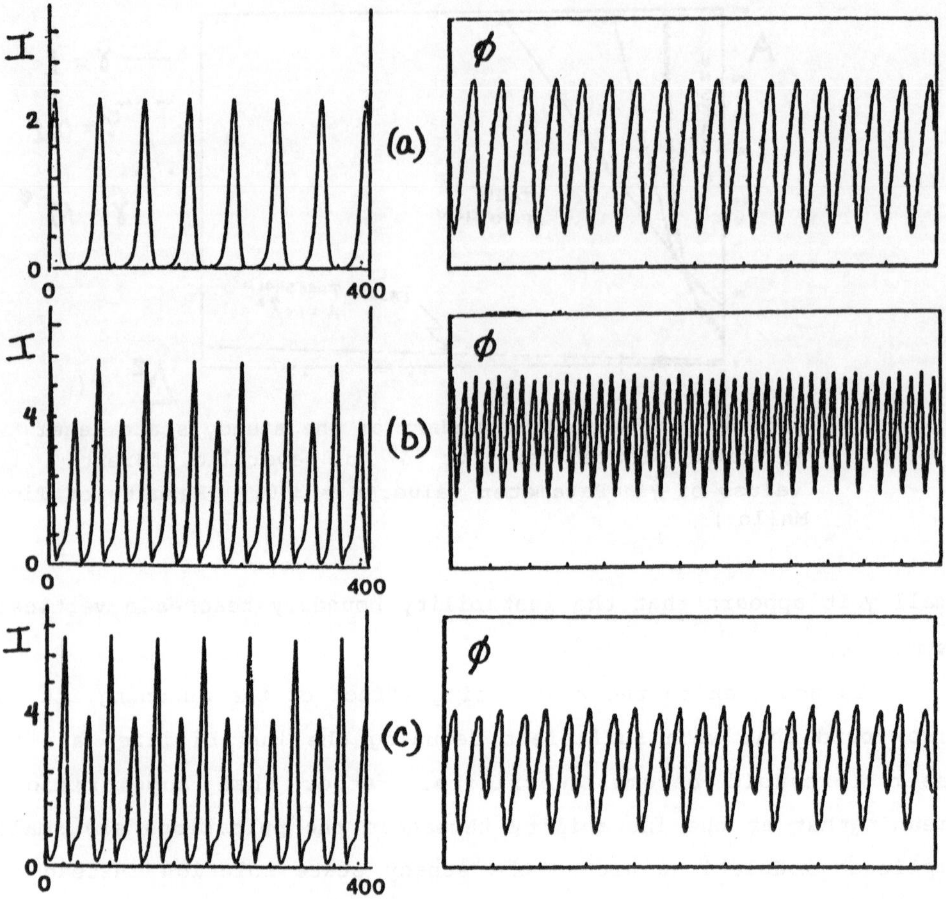

Figure 5. Intensity and phase (relative to the average phase drift) versus time for an inhomogeneously broadened laser with intial detuning of the cavity for parameters equal to those used in Fig. 2 and excitation values given by a) A=6.74, b) 7.92, c) 8.14. (Results of M.F.H. Tarroja, Ph.D. Thesis, Bryn Mawr College,1988.)

field amplitude in each case in Figure 5 is plotted after the average linear drift in the phase has been subtracted away. The solutions were calculated in the rotating reference frame giving a constant electric field amplitude for the steady state solution. The average drift that had to be subtracted away represents a dynamically induced frequency shift.

In Figures 6 and 7 we present two other ways to illustrate the solutions. The phase of the solution of Fig. 5c is plotted in Figure

6 in the rotating reference frame of the steady state solution to show that there is an average drift as well as modulation of the phase. The drift is a shift of the frequency of the pulsing solution from the frequency of the steady state solution. In Figure 7 the solution in that rotating frame is shown in a plot of Re E vs. Im E giving a quasiperiodic representation of the solution with the amplitude pulsations superimposed on the rotation in the complex plane due to the dynamically induced frequency shift.

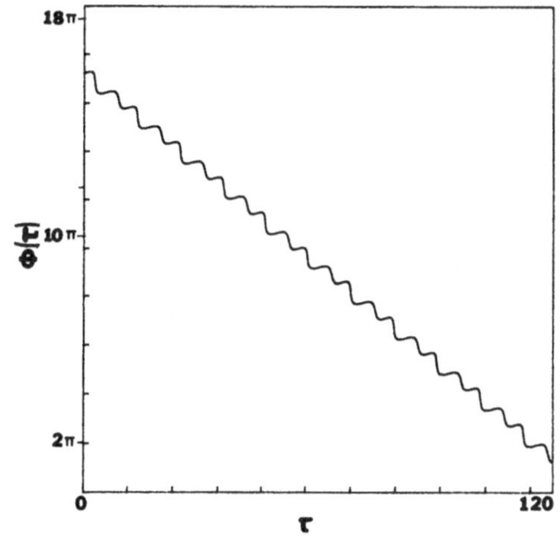

Figure 6. Phase versus time for Figure 5c showing the average phase drift and its modulation. From Ref. 1.

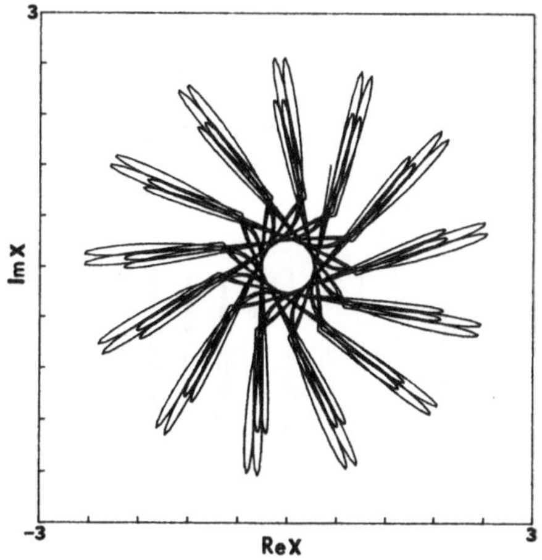

Figure 7. Representation of the solution in Figure 5c in the (Re E, Im E) plane. From Ref. 1.

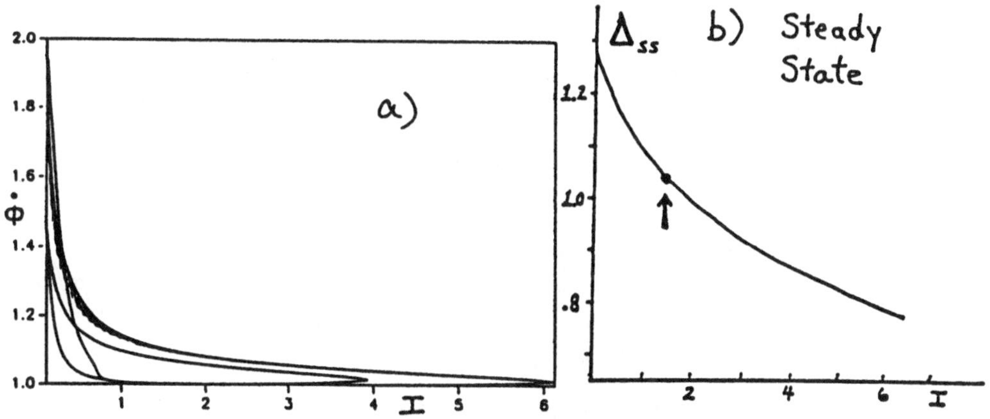

Figure 8. Frequency versus intensity for a) steady state frequency versus steady state intensity for solutions of Fig. 2 and b) instantaneous frequency versus intensity for the time dependent solution of Fig. 5c for A = 8.14 (I_{ss} = 1.56; Δ_{ss} = 1.03). (From M.F.H. Tarroja, Ph.D. Thesis.)

A further way to visualize this solution is to plot the frequency versus the intensity. Figure 8 shows these plots for both the steady state solutions of Figure 2 and for the time dependent solution of Figure 5c. These two examples have all of the same parameters except that in the first case the parameter A is varied to draw the curve while in the second case the parameter A is held fixed and the time dependence of the solution "draws" the curve. We see that the dynamical pulsations bear a qualitative similarity to motion along the steady state curve but there is more complex modulation. We also see a lag between the frequency and intensity depending on the slope of the intensity. Much larger frequencies are found near

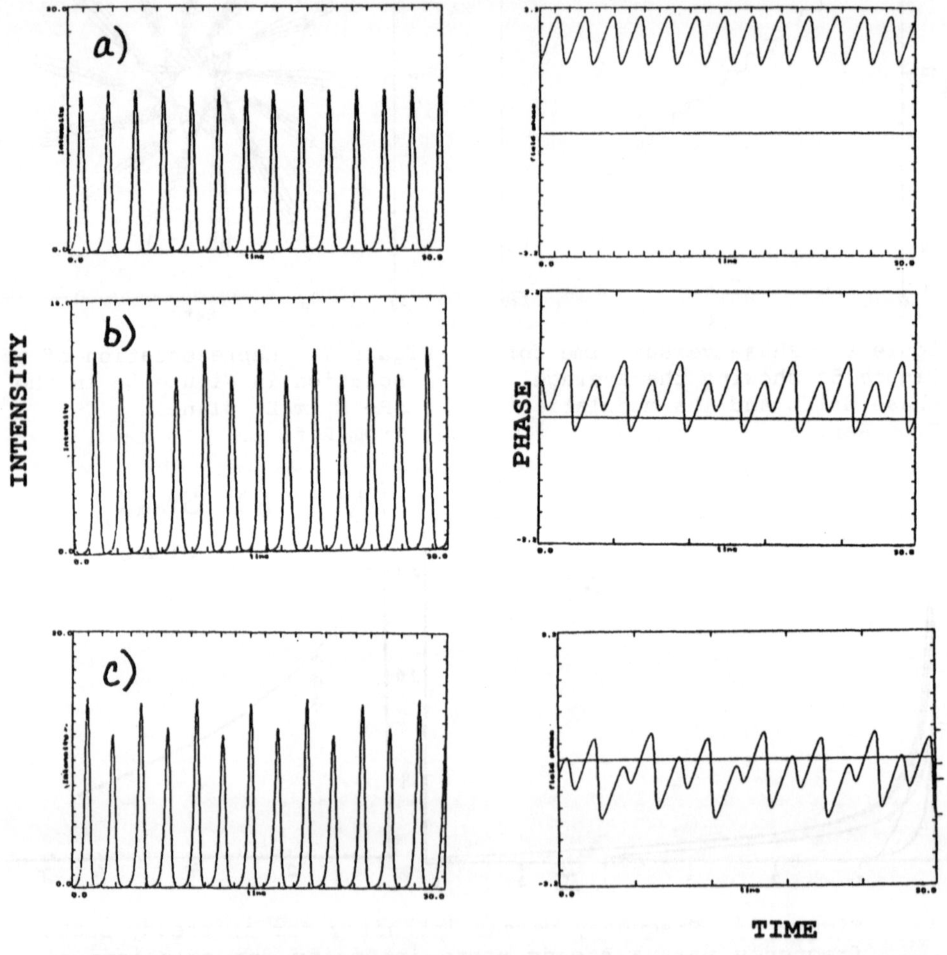

Figure 9. Intensity versus time and phase (relative to average phase drift) versus time for three different detunings of the homogeneously broadened laser with parameters given by A=15; $\tilde{\gamma} = 0.25$; $\tilde{\kappa} = 2.0$ and detunings a) $\tilde{\delta}_{AC} = 0.66$; b) $\tilde{\delta}_{AC} = 0.75$; and c) $\tilde{\delta}_{AC} = 0.9$. (Results of H. Zeghlache.)

zero intensity than are found for the steady state solution. These larger frequencies correspond to relatively rapid phase changes inbetween the intensity pulsations.

Similar behavior is seen in the homogeneously broadened case. Figure 9 shows samples of the intensity pulsations and of the phase for three different detunings which give period four, period two and period one pulsations, respectively.

For comparison, we take samples of the representation of Figure 9b for Figures 10 and 11 to show the behavior of the phase in the reference frame of the steady state solution and the behavior of the attractor in that reference frame when displayed in a plot of the real versus imaginary parts of the field amplitude.

The true structure of the attractor is better revealed by displaying the real versus imaginary parts of the slowly varying amplitude after a change of reference frame to that corresponding to the average slope of the phase as shown in Figure 12. In this reference frame the attractor is represented as a simple two-cycle, just as would have been expected from the intensity pulsations.

In both Figs. 6 and 10 we see that the phase is nearly constant during each intensity pulse and then slips by an amount of about .5π during the nearly zero intensity between pulses. In fact, the frequency during the pulse changes slightly and is, on average, somewhat different from the steady state frequency. In general it is even closer to the atomic resonance frequency than was the steady state frequency (which is already "pulled" away from the cavity resonance frequency and toward the atomic frequency). By contrast,

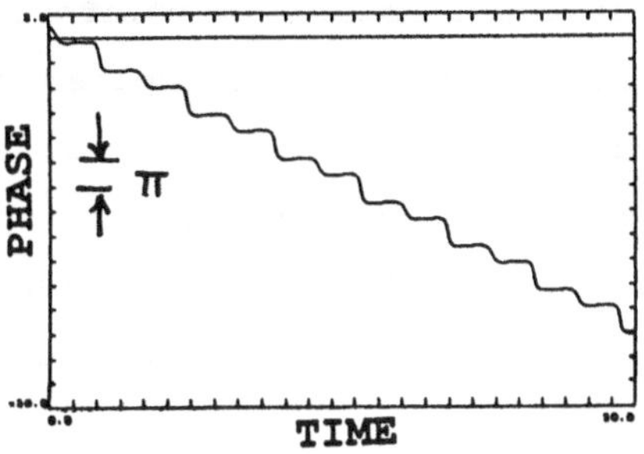

Figure 10. Phase versus time.

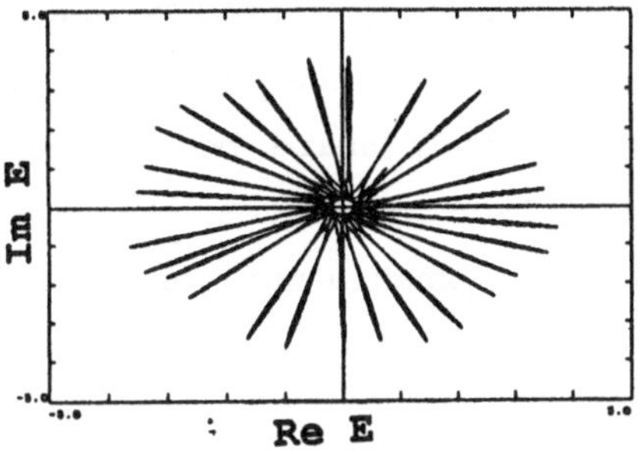

Figure 11. Re E vs. Im E.

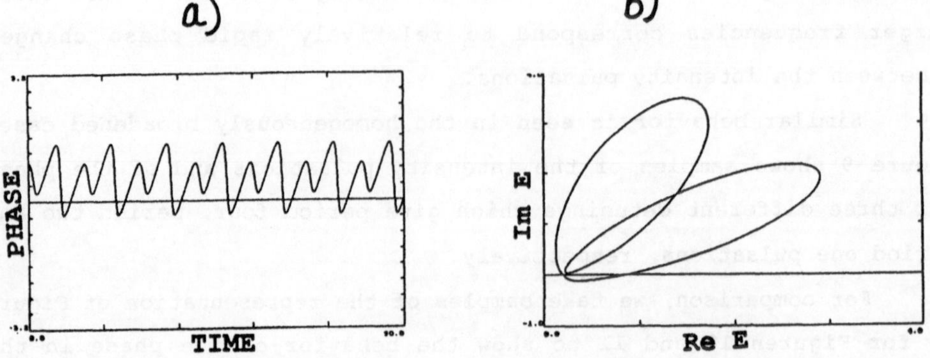

Figure 12. Representation of the solution from Fig. 9b in the rotating reference frame with frequency given by the average slope of the plot in Fig. 10. a) phase vs. time; b) Re E vs. Im E.

the average slope of the phase corresponds to a reference frame with frequency nearly equal to the laser cavity. Thus the strong pulling of the laser frequency in steady state operation is disrupted by the pulsations which "favor" large frequency modulation with a proper reference frame quite close to the cavity frequency. Perhaps this solution takes its reference frequency to be that of a nearly empty cavity because of bleaching of the medium by the intense pulses. The detailed physics of these frequencies and frequency shifts is now a matter open for further discussion and interpretation which may lead to improved understanding of the nature of AM and FM pulsations in single mode detuned lasers.

The similarity between the homogeneously broadened and inhomogeneously broadened cases is truly remarkable. It reinforces the view that the dynamics are much more similar than implied by their steady state behaviors in response to parameter variations. This also reinforces the view that what was once viewed as the "special experimental case" of the inhomogeneously broadened lasers is only a simple extension of the behavior of the laser-Lorenz model.

BEHAVIOR OF THE PHASE IN SINGLE MODE LASERS AT AND NEAR RESONANCE

In resonance the equations for the homogeneously broadened laser reduce to three real equations (isomorphic to the Lorenz equations /23/). Even in this case the phase has meaning. If we decompose a variable X or Y which can be positive or negative into amplitude and phase, the phase represents the sign of the variable. Changes of sign in the variable lead to jumps by pi radians in this simplified "phase". For the Lorenz case the solutions are often chaotic leading to irregular jumps. Aizawa /24/ has shown that the sign changes in the chaotic case have the form of a shot noise process.

Examples of the solutions in resonance for the homogeneously broadened and inhomogeneously broadened cases are shown in Figure 13.

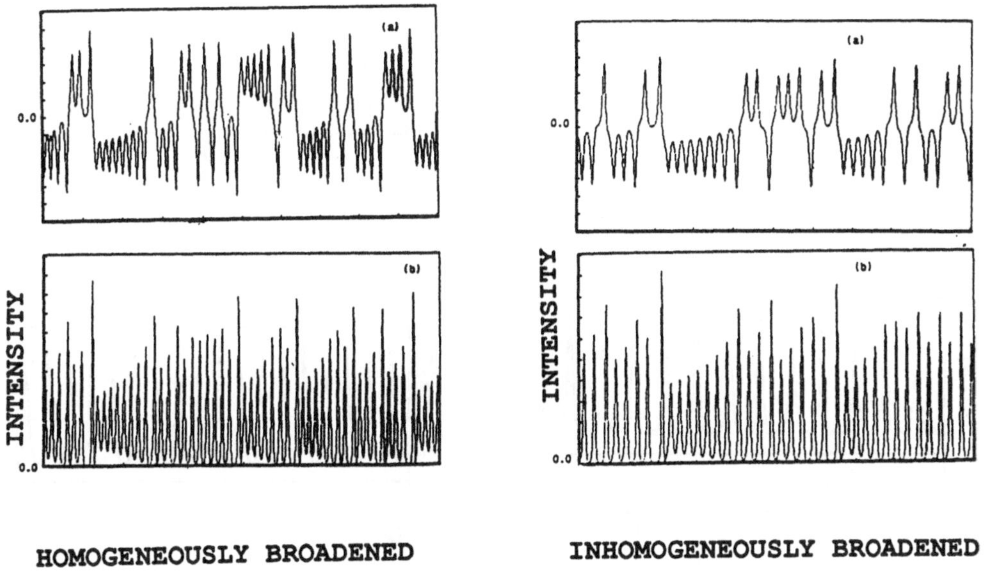

Figure 13. Intensity and electric field amplitude for the homogeneously broadened and inhomogeneously broadened laser models in resonance. (Solutions from M.F.H. Tarroja.)

It has recently been shown /4a/ that for very small detunings the structure of these pulsations is not visibly altered. Interestingly enough, the phase changes dramatically. In the resonant case there are alternations by π radians in the phase. With small detunings the phase still makes transitions of nearly π, but although they are rapid they are no longer abrubt. Instead of alternating as in the resonant case these transitions are always in the same direction. The fact that the transitions are less than π means that the amplitude never passes through zero in the detuned case, however, the fact that they are nearly π indicates that the

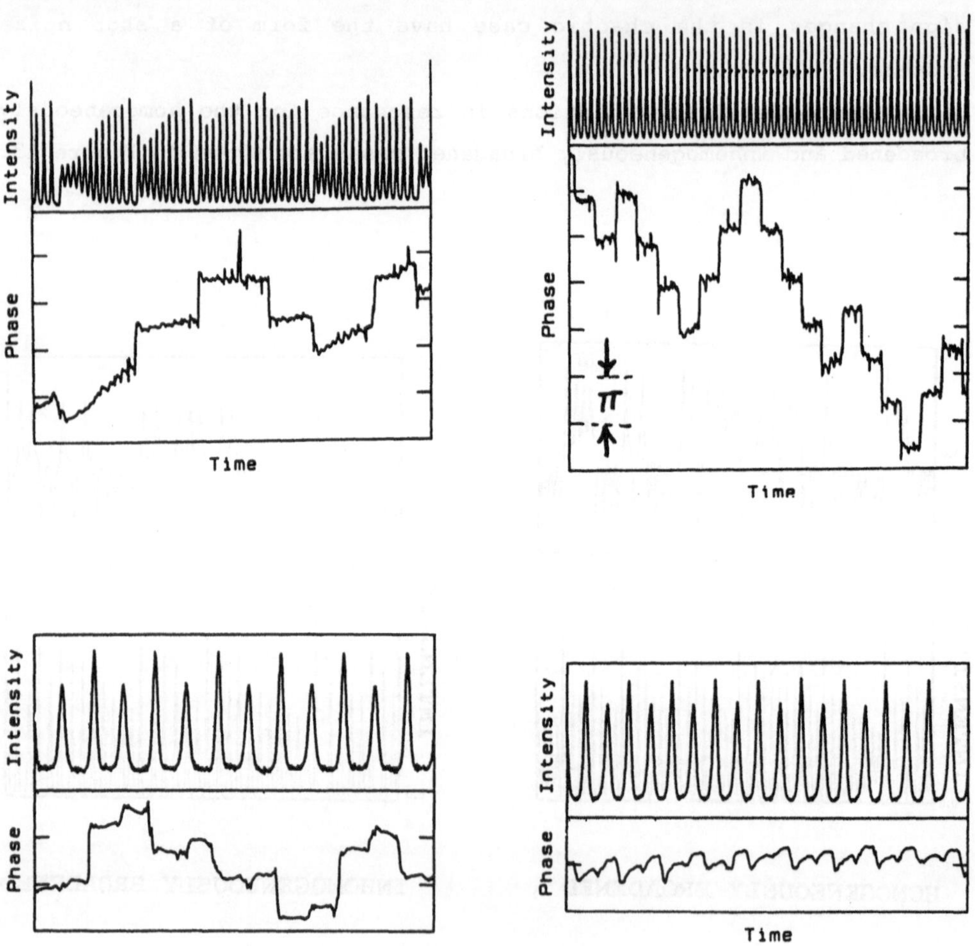

Figure 14. Examples of intensity pulsations and corresponding phase pulsations measured for a single-mode FIR laser (courtesy of C.O. Weiss, see Ref. 4b).

population inversion and the complex amplitude of the longitudinal modulation of the population inversion at a spatial periodicity of $\lambda/2$, where λ is the wavelength. κ_1, κ_2, and $\gamma_{||}$ are the loss rates of the two field variables and the population variables, respectively. Δ is the detuning in units of $\gamma_{||}$, A is the excitation parameter and tilde denotes division by $1+\Delta^2$.

The nontrivial steady states are two unidirectional solutions, which are stable in resonance, and an unstable bidirectional solution. The unidirectional solutions lose stability with sufficient detuning. If $\gamma_{||} \ll \kappa$, the amount of detuning required for instability is quite small. Just beyond the instability boundary, the pulsations correspond to nearly regular alternation of the two modes in a kind of alternate square-wave pulsing. During these pulsations we have found that a combination of the phases (in this case there are three complex variables) remains nearly constant except for relatively abrupt jumps by π whenever one of the two modes passes near zero.

Careful analysis of these transitions has shown that the attracting solution makes a jump from the stable manifold near a unidirectional solution to the unstable manifold near that solution. On these two manifolds the phase of the weak field changes sign.

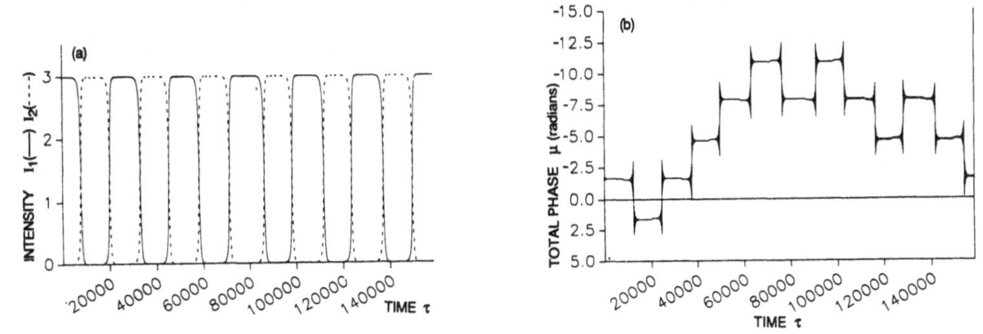

Figure 15. a) Intensities of the two modes in a solution of the bidirectional ring laser model using $\kappa_1=1.4\times10^7$; $\kappa_2=1.401\times10^7$; $\gamma_{||} = 2.5\times10^3$; $A = 4.0$, $\Delta = 0.1$. b) Combined phase of the two fields and the population grating ($\phi_1-\phi_2+\psi$) versus time for the solution shown in a).

attractor with detuning is very similar to that in resonance. One can imagine that the attractor is nearly the same as in the resonant case but that now it has a small rotation in the (Re E, Im E) plane. However, no simple expression for this rotation has yet been found and, indeed, it appears that it may be chaotic.

Recently it became possible to measure such phase dynamics during laser pulsations /4b/. Experimentally this requires a stable heterodyne reference system and the ability to extract both the in-phase and in-quadrature components of the laser signal with respect to the frequency reference. Several examples are shown in Figure 14 where we see both phase jumps of π at the end of spiral type intensity pulsations, spiral pulsations without phase jumps, jumps of less than π corresponding to periodic pulsations, and more complicated patterns. With these measurements much more detailed assignments of the character of instabilities are possible than can be made based on the intensity pulsation patterns alone.

PHASE AND FREQUENCY DYNAMICS IN A BIDIRECTIONAL RING LASER

A simple model of the interaction of counterpropagating modes of abidirectional ring laser shows similarly instructive phase and frequency dynamics. The model corresponds to two modes interacting with homogeneously broadened atoms. The equations governing this interaction are /14,15/:

$$\dot{E}_1 = -\kappa_1 E_1 + \kappa_1 (1+i\Delta)\tilde{A}(E_1 D_0 + E_2 D_1^*) \tag{14a}$$

$$\dot{E}_2 = -\kappa_2 E_2 + \kappa_1 (1+i\Delta)\tilde{A}(E_2 D_0 + E_1 D_1) \tag{14b}$$

$$\dot{D}_0 = -\gamma_{||}(D_0 - 1) - \tilde{\gamma}_{||} D_0(|E_1|^2 + |E_2|^2) - \tilde{\gamma}_{||}(E_1 E_2^* D_1 + E_1^* E_2 D_1^*) \tag{14c}$$

$$\dot{D}_1 = -\gamma_{||} D_1 - \tilde{\gamma}_{||} D_1(|E_1|^2 + |E_2|^2) - \tilde{\gamma}_{||} E_1^* E_2 D_0, \tag{14d}$$

where E_1, E_2, D_0 and D_1 are the slowly varying complex amplitudes of the two modes, the real amplitude of the longitundinal average

Physically this corresponds to the weak mode switching from receiving destructive interference from the scattered strong mode to receiving constructive interference from the scattered signal.

Further examination of the frequencies of the two modes reveals that each mode makes transitions among three different optical frequencies as shown in Figure 16a. When one mode dominates the other (the nearly unidirectional condition), the dominant mode assumes the frequency of the undirectional steady state solution(s). The weaker mode, when it is decaying in intensity, has been shifted by a dynamical dispersion to an optical frequency more detuned from the atomic resonance. After the phase jump at the weakest point of the intensity the frequency of the weak mode shifts to be more resonant with the atoms.

Thus one can physically interpret the changes in the growth and decay of a mode as corresponding to more or less favorable resonance with the atoms and to constructive versus destructive interference with the scattered field from the strong mode. Furthermore, the fact that the two modes have different frequencies means that the population grating they form is not stationary but moves longitudinally in time, alternating direction. This additional (and somewhat unexpected degree of freedom for the solution) helps to sustain the pulsations.

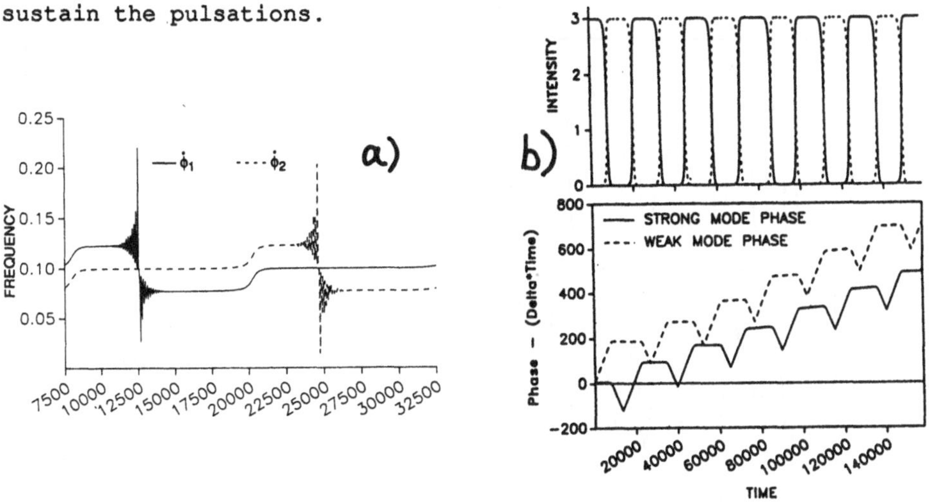

Figure 16. Frequencies and phases of the two modes for the solution shown in Fig. 15. a) frequencies of the two modes; b) phase of the two modes in the rotating reference frame of the steady state solutions.

Another way to represent these frequencies is to plot the phase in the rotating reference frame of the steady state solutions as shown in Figure 16b. The difference in the average drifts in the phases of the two modes is due to the slight different in cavity loss for the two modes. The three frequencies are here represented by three slopes. The constant phase regions are when one mode dominates and assumes the frequency of the unidirectional steady state solutions.

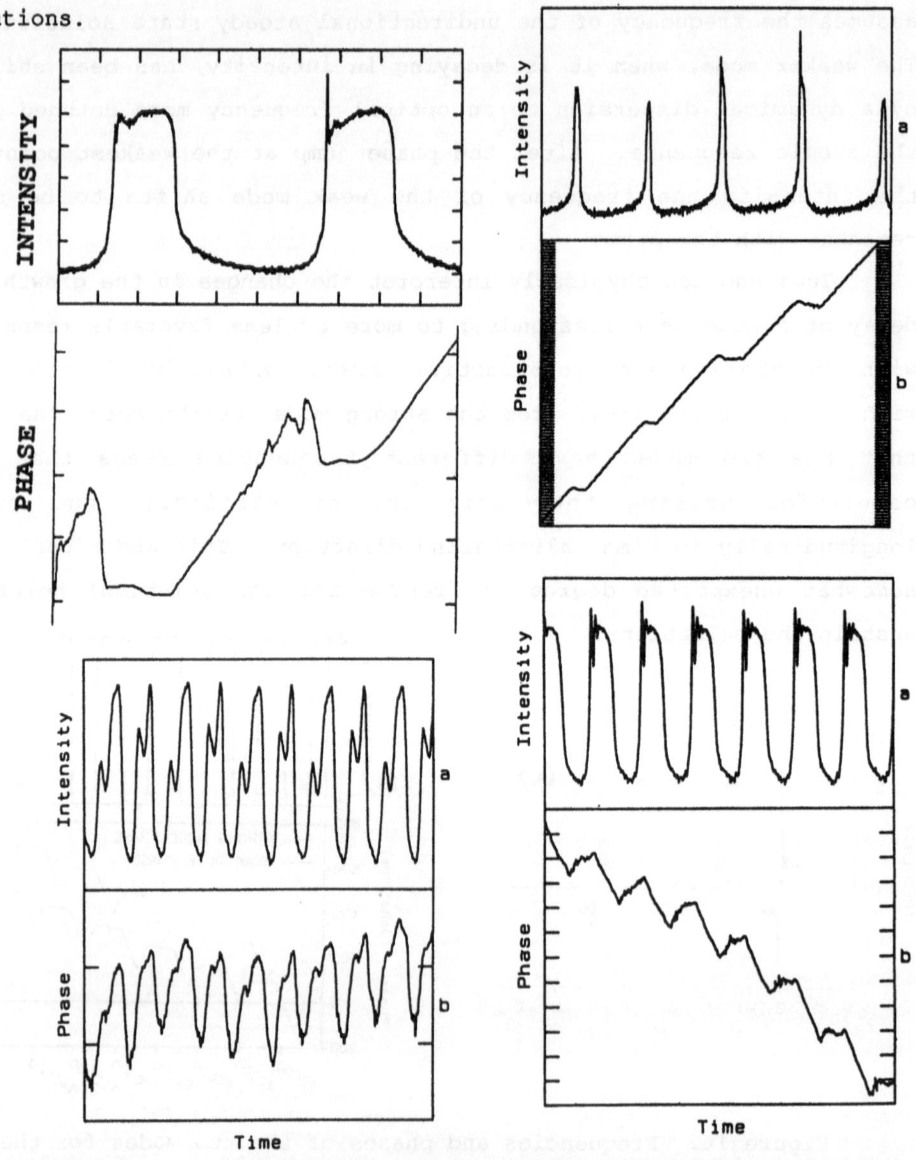

Figure 17. Intensities and phases versus time for various types of pulsations measured for one mode of a bidirectional laser (From Ref. 17). Phase divisions are π radians.

18 for the corresponding field and intensity pulsations shown in Figure 13.

Phase dynamics is eliminated in converting the complex amplitude to the intensity, so the qualitative differences in the spectra help to point out the hidden features of the FM part of the spectrum. For example, in the case of the spiral chaos shown in Figure 18, significant amounts of the low frequency modulation are purely FM with the rather remarkable consequence that the power spectrum of the electric field is broadband while the intensity spectrum has a peak some 20-30 dB above the broadband part of the spectrum.

The mixture of AM and FM contributions in the electric field power spectrum is even greater in the detuned laser case, all the more reason to advance this study rather than concentrating only on the intensity.

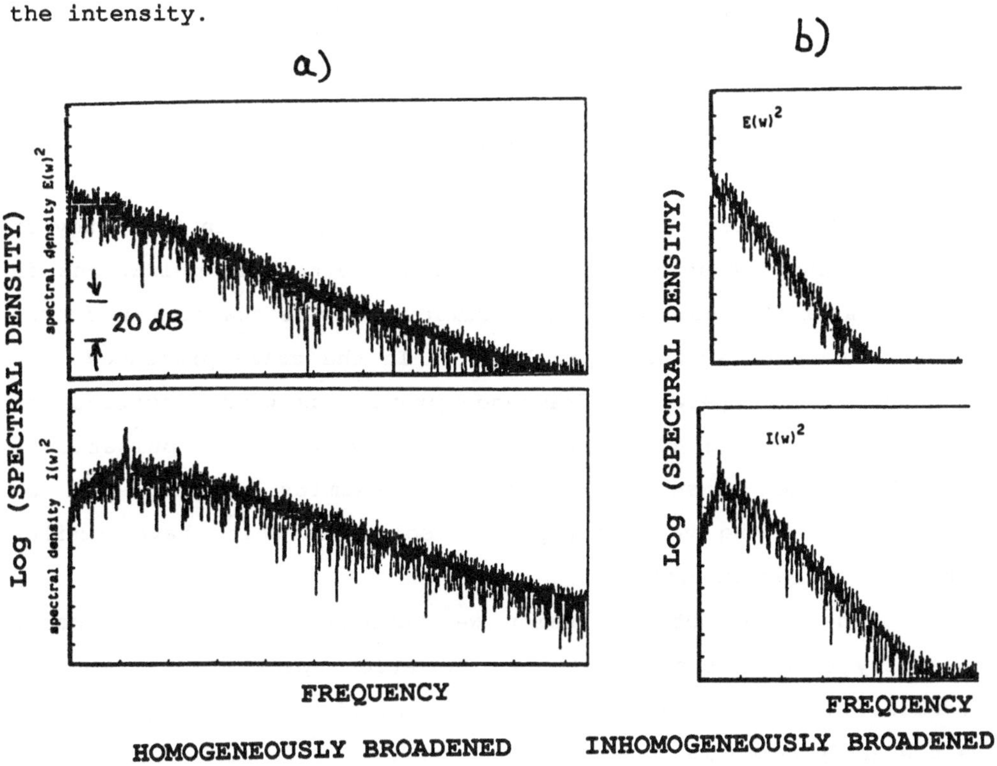

Figure 18. Intensity and electric field power spectra for the spiral type pulsations of resonantly tuned lasers (time dependent signals shown in Figure 13) a) homogeneously broadened case, b) inhomogeneously broadened case.

Hence we see that study of the dynamics of the phases and frequencies provides both new information and new insight into the physical processes at work.

Experimental measurements of the optical phase of a bidirectional FIR ring laser /16/ as shown in Figure 17 show many of the phenomena seen in the numerical modelling. The FIR ring laser is only slightly different in pressure and other operating conditions from those used to obtain results similar to the solutions of the laser-Lorenz model. Thus one would expect the proper model to also include equations for the dynamical evolution of the polarization. Instead, the numerical results for the model with adiabatic elimination of the polarization seem to do a good job of modelling what is observed experimentally. This indicates that the critical interactions causing the dynamical frequency and phase shifts in the experiment are preserved in the simplified model.

CONCLUSIONS, DISCUSSIONS, SPECULATIONS

Rich dynamical information has been uncovered in the study of the phase and frequency variations during dynamical pulsations in several laser systems. Similar information may be awaiting us in the study of many other laser problems. In the cases presented, the dynamics of the phase and frequency provide us with important new insights regarding the physical processes of the pulsations. Recording the phase and frequency dynamics permits one to discriminate among various types of solutions which have similar intensity pulsation patterns.

One important conclusion we can draw is that one should carefully examine both amplitude modulation spectra and frequency modulation spectra in the cases for dynamical evolution, just as it is presently commonly done in noise studies. This will require that spectra be calculated for the complex amplitude variables.

In other recent articles the difference between the spectra of the intensity and the spectra of the complex field amplitude have been noted /1,25,26/. Examples of these spectra are shown in Figure

Experimentally one cannot directly extract the electric field amplitude or even its power spectrum from the detected photocurrent which is proportional to the intensity. Instead it is necessary to heterodyne a stable reference laser with the laser whose dynamics are to be measured. If the two lasers are detuned from each other by more than the bandwidth of the spectrum of either, the intensity pulsation spectrum is separated from the heterodyne spectrum which carries the information about the amplitude and phase modulation. The combined signals can be written as:

$$E = E_L(t)\exp(i\omega_L t) + E_R\exp(i\omega_R t) \qquad (15a)$$

$$I(t) = I_L(t) + I_R + 2\mathrm{Re}[E_L(t)+E_R]\cos(\omega_R-\omega_L)t. \qquad (15b)$$

The spectrum of E_L is now transferred to a modulation of a detectable frequency $\omega_R-\omega_L$ which can be separated from 0. Several such separations for an inhomogeneously broadened laser are shown in Figure 18.

Similar intensity power spectra with broadband features and sharp peaks can have distinctly different heterodyne spectra. Immediately obvious are symmetries of the spectra relative to the center frequency. If there is a peak at $\omega_R-\omega_L$ this means $\langle E_L\rangle=0$ and the attractor is symmetric. If there is no peak at $\omega_R-\omega_L$ then $\langle E_L\rangle\neq 0$ and the attractor is asymmetric. In this way the heterodyne spectrum alone is sufficient to distinguish symmetry broken attractors from truly period-doubled attractors. A sharply double peaked heterodyne spectrum indicates nearly periodic, 200% amplitude modulation, while a broadband spectrum corresponds to the irregular switching of the positive and negative values of E characteristic of the classic Lorenz chaos spirals.

Monitoring of the phase is also essential in determining symmetries of various solutions /1,18/. For example, the resonant laser-Lorenz equations have an inversion symmetry of $E \rightarrow -E$, $P \rightarrow -P$, $D \rightarrow D$. With this symmetry, solutions can have even parity under the inversion or solutions can have complementary solutions under inversion. The former are symmetric and the latter are asymmetric.

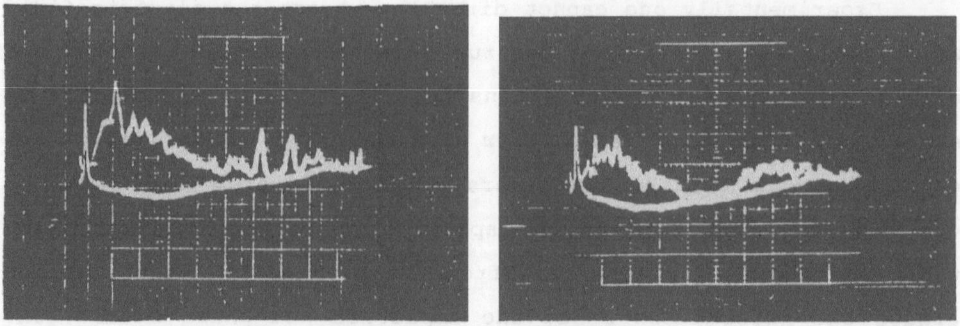

Figure 18. Sample homodyne (flush left) and corresponding heterodyne spectra from measurements of M.F.H. Tarroja on a He-Xe 3.51 micron single mode ring laser. Frequency scale is 100 MHz full scale and vertical scale is 10 dB per division.

The asymmetric solutions which include a change of sign of the electric field amplitudehave two different intensity pulses for a single cycle around the attractor while the symmetric solutions have two equal intensity pulses per cycle. The proper periodicity of the attractor cannot be determined from the intensity alone. In fact, more complex pulsations of symmetric or asymmetric types are possible as noted by Sparrow /27/. Calling positive pulses x and negative pulses y, the Lorenz equations are found to have various complex attractors described by sequences such as xy, x^2y^2, x^2y, x^3y^3, or xyx^2y^2 before repeating. Only phase or field measurements can distinguish among subsets of these sequences and thereby aid in proper naming of the attractors and of the transitions among attractors /28/.

A further topic of interest, is that the five equations of the laser-Lorenz model do not specify the phase of the electric field in steady state. Hence, in the five-variable space, there is a family of possible nontrivial steady-state solutions which corresponds to a circle of constant radius in the complex (Re E, Im E) plane. The arbitrary absolute phase means that the solution can be any one of these points. When the pulsations begin, there is still an arbitrary absolute phase, but the phase is no longer constant in any particular reference frame. The periodic and the chaotic solutions have a frequency shift which means that there is deterministic motion as if the solution spirals in a helix around the circle of steady state points in the (Re E, Im E) plane. Of course, the rotation around the circle merely represents the frequency shift of the "proper" reference frame for the pulsing solution from that of the steady state solution. In the proper frame the solution is just a small amplitude limit cycle about a single (stationary) fixed point.

For certain detunings and excitations, the instability is a very small amplitude pulsation with a small shift in the optical frequency. In this case one would see a simple helical motion around the circle of fixed points.

Further surprises and insights from these considerations sure await continuing investigations.

ACKNOWLEDGEMENTS

The results presented here are the compilation of fruitful collaborations with M.F.H. Tarroja, D.K. Bandy and L.M. Narducci on the inhomogeneously broadened laser, with C.O. Weiss on the measurement of FIR laser instabilities, with T. Mello, G.L. Lippi, H. Zeghlache and Paul Mandel on the study of the homogeneously broadened laser model, and with Zeghlache, Mandel, Lippi, Mello and L.M. Hoffer on the study of the bidirectional ring laser. I am grateful to them for stimulating collaborations, discussions and interpretations. This formulation and interpretation of these results was supported in part by a fellowship from the Alexander von Humboldt Stiftung, a grant from the Institute for Scientific Interchange in Torino, Italy, and the hospitality of L.A. Lugiato and the Dipartimento di Fisica del'Politecnico in Torino.

REFERENCES:

1. M.F.H. Tarroja, N.B. Abraham, D.K. Bandy, and L.M. Narducci, Phys. Rev. A **34**, 3148 (1986).
2. A.V. Uspenskiy, Sov. Rad. Eng. Electron. Phys. **9**, 60 (1964).
3. N.B. Abraham, M.F.H. Tarroja, L.M. Hoffer, G.L. Lippi, P. Mandel and H. Zeghlache, in *Solitons and Optical Chaos*, eds., H. Morris and D. Heffernan (Plenum, NY, to be published).
4. a) H. Zeghlache, P. Mandel, N.B. Abraham, and C.O. Weiss, submitted to Phys. Rev. A, June 1988; b) C.O. Weiss, N.B. Abraham, and U. Hübner, sub. Phys. Rev. Lett., May 1988.
5. N.B. Abraham, P. Mandel and L.M. Narducci, in *Progress in Optics XXV*, ed. E. Wolf (North Holland, Amsterdam, 1988) pages 1-190.

6. P. Mandel and H. Zeghlache, Opt. Commun. **47**, 146 (1983); H. Zeghlache and P. Mandel, J. Opt. Soc. Am. B **2**, 18 (1985).
7. S.T. Hendow and M. Sargent, III, J. Opt. Soc. Am. B **2**, 84 (1985).
8. L.M. Narducci, J.R. Tredicce, L.A. Lugiato, N.B. Abraham, and D.K. Bandy, Phys. Rev. A **33**, 1842 (1986).
9. L.M. Hillman, R.W. Boyd, and C.R. Stroud, Jr., Opt. Lett. **7**, 426 (1982).
10. P. Mandel, Opt. Commun. **53**, 249 (1985); in *Optical Instabilities*, eds., R.W. Boyd, M.G. Raymer, and L.M. Narducci (Cambridge U. Press, Cambridge, 1986) p. 262; and in *Frontiers in Quantum Optics*, eds., R. Pike and S. Sarkar (Hilger, Bristol, 1986) p. 430; also D.K. Bandy and L.M. Narducci, private communication. See also ref. 5.
11. D.K. Bandy, L.M. Narducci and L.A. Lugiato, J. Opt. Soc. Am. B **2**, 202 (1985).
12. J.R. Tredicce, F.T. Arecchi, G.L. Lippi, G.P. Puccioni, J. Opt. Soc. Am. B **2**, 173 (1985).
13. F. Morgensen, G. Jacobsen, A. Olesen, Opt. Quantum Electron. **16**, 183 (1984).
14. H. Zeghlache, P. Mandel, N.B. Abraham, L.M. Hoffer, G.L. Lippi and T. Mello, Phys. Rev. A **37**, 470 (1987).
15. L.M. Hoffer, G.L. Lippi, N.B. Abraham, and P. Mandel, Opt. Commun. **69**, 219 (1988).
16. N.B. Abraham and C.O. Weiss, Opt. Commun., submitted June 1988.
17. W. Brunner, R. Fisher, H. Paul, J. Opt. Soc. Am. B **2**, 202 (1985); T. Ogawa, E. Hanamura, Opt. Commun. **61**, 49 (1987) and Appl. Phys. B **43**, 139 (1987); R. Müller and P. Glas, J. Opt. Soc. Am. B **2**, 184 (1985).
18. C.H. Henry, IEEE J. Quantum Electron., **QE-19**, 1391 (1983).
19. R. Lefever, L.A. Lugiato, K. Wang, N.B. Abraham and P. Mandel, Phys. Lett. submitted August 1988. (See Y. Kuramoto, *Chemical Oscillations, Waves and Turbulence*, (Springer-Verlag, Berlin 1984). Even in such a case, as described by P. Coullet in this conference, the "wimpy" K-S turbulence in the phase couples strongly to amplitude dynamics when the gradient of the phase is large leading to "macho" turbulence of coupled amplitude and phase dynamics, particularly in a single space dimension.
20. L.A. Lugiato, in *Progress in Optics XXI*, ed. E. Wolf (North Holland, Amsterdam, 1984) pp. 69-216; M. LeBerre, E. Ressayre, A. Tallet and H.M. Gibbs, Phys. Rev. Lett. **56**, 274 (1986); M. LeBerre, E. Ressayre, A. Tallet, H.M. Gibbs, D.L. Kaplan and M.H. Rose, Phys. Rev. A **35**, 4020 (1987).
21. C.M. Savage and D.F. Walls, Optica Acta **30**, 557 (1983).
22. L.A. Lugiato, P. Galatola, M. Vadacchino, N.B. Abraham and L.M. Narducci, manuscript in preparation, July 1988.
23. E.N. Lorenz, J. Atmos. Sci. **20**, 130 (1963); H. Haken, Phys. Lett. **53A**, 77 (1975).
24. Y. Aizawa, Prog. Th. Phys. **68**, 64 (1982).
25. N.B. Abraham, in *Optical Instabilities*, eds., R.W. Boyd, M.G. Raymer and L.M. Narducci (Cambridge U. Press, Cambridge, 1986) page 46.
26. N.B. Abraham, A.M. Albano, B. Das and M.F.H. Tarroja, in *Fundamentals of Quantum Optics II*, ed. F. Ehlotzky, (Springer, Heidelberg, 1987) page 32.
27. C. Sparrow, *The Lorenz Equations: Bifurcations, Chaos and Strange Attractors* (Springer, Heidelberg, 1982).
28. A. Arneodo, P. Coullet and C. Tresser, Phys. Lett. **81A**, 197 (1981); Y. Kuramoto and S. Koga, Phys. Lett. **92A**, 1 (1982).

FLUCTUATIONS AND CRITICAL PHENOMENA IN REACTION-DIFFUSION SYSTEMS

Alexandre S. Mikhailov

Department of Physics, Moscow State University
117234 Moscow USSR

Studies of nonequilibrium phase transitions in reaction-diffusion systems constitute an important branch of modern self-organization theory. It is well known that in many situations we find here very close analogies with the second order phase transitions at thermal equilibrium (see, e.g., [1]). In a present paper I want, however, to draw attention to a less common situation found in systems with autocatalytic (or chain) chemical reactions or in biological systems with reproduction processes. As I show below (for a detailed review, see [2,3]) this situation is characterized by the mathematical models with a positively defined real order parameter, which leads to profound differences in the fluctuation behavior from the standard second-order phase transitions.

1. Noise-induced phase transition

In 1976 W.Horsthemke and M.Malek-Mansour [4] investigated a model given by the stochastic Verhulst equation

$$n = \lambda n - n^2 + f(t)\, n \qquad (1)$$

with a white noise random force $f(t)$. This model refers, for instance, to ecological systems, where it describes a saturated logistic growth of a population in presence of random fluctuations in the birth and death rates. It can be viewed, as well, as an effective model for an auto-catalytic chemical reaction.

It was found that Eq.(1) leads to a complicated fluctuation behavior. Namely, a steady state probability distribution $p_s(n)$ has two qualitatively different forms, depending on the value of the noise intensity S . If $S > \lambda$ and the Stratonovich interpretation of Eq.(1) is used, then probability $p_s(n)$ has a maximum at $n = S - \lambda$ and vanishes at $n = 0$. On the other hand, if $S < \lambda$ this probability distribution diverges at $n \rightarrow 0$ (but remains normalizable) and has no maxima. To describe this qualitative change in the fluctuation behavior, observed with increasing the noise intensity , W. Horsthemke and M. Malek-Mansour introduced a concept of noise-induced transitions. Many other examples of noise-induced transitions, including noise-induced bistability [5] or noise-induced oscillatory behavior [6], were found later. These examples and various applications are reviewed in [7].

However, in all these cases we have only point-like (not distributed) systems and, therefore, as it is well known from the theory of phase transitions, we cannot expect here any genuine singularities which are typical for phase transitions in systems of nonzero dimensionality. Hence, it seems interesting to ask whether such noise-induced transitions are possible, as well, in distributed systems and, if so, what are their distictive properties.

An example of a noise-induced phase transition in a distributed system with diffusion was investigated in [8-10].

Suppose that we have the following set of chemical reactions, that includes two competing autocatalitic reactions,

$$X + Z \rightarrow 2X, \quad Y + Z \rightarrow 2Y, \qquad X \rightarrow 0, \quad Y \rightarrow 0, \quad Z \rightarrow 0, \quad W \rightarrow Z \tag{2}$$

If we denote [X] = N, [Y] = n, [Z] = M and assume that the concentration of the substrate W is kept constant, then, under ideal mixing, this set of chemical reactions will be described by a system of three differential equations

$$\dot{N} = (B M - A) N, \qquad \dot{n} = (b M - a) n, \qquad \dot{M} = Q - (\gamma + B N + b n) M. \tag{3}$$

A slightly different system of differential equations describes an *ecological* model. Suppose that we have two biological species with population densities N and n , correspondingly, which compete for the same renewable resource - namely, for the food M that grows at a rate Q. If we assume that the birth rates of both species are proportional to the amount of food M available, this ecosystem will obey the equations

$$\dot{N} = (B M - A) N, \qquad \dot{n} = (b M - a) n, \qquad \dot{M} = Q - \gamma M - C N - c n. \tag{4}$$

Since the processes characterized by Eqs.(3) and Eqs.(4) are very similar (see [10]), I discuss below only the ecological model. When two inequalities

$$A/B < a/b, \quad A/B < Q/\gamma \tag{5}$$

are satisfied, species (N) are strong and species (n) are weak. In the long run the strong species survive and the weak species extinct, so that asymptotically with $t \to \infty$ a steady state

$$N = N_1 \, , \, M = M_1 \, , \, n = 0 \qquad (6)$$

is reached, where

$$M_1 = A/B \, , \, N_1 = (Q - \gamma M_1)/C \, . \qquad (7)$$

This steady state corresponds to the attractive focus of the differential Eqs.(4).

Now suppose that the food growth rate Q fluctuates in time, $Q \to Q + f(t)$, so that $\langle f \rangle = 0$. Would that change the result of a competition and, specifically, would that make coexistence possible ? The answer is *no*.

Indeed, the first two of Eqs.(4) can be rewritten as

$$(d/dt) \ln N = B M - A \, ,$$
$$(d/dt) \ln n = b M - a \, . \qquad (8)$$

After exclusion of M and integration we obtain from Eqs.(8) the relationship

$$n^{B/b} = \text{const} \cdot \exp(- B p_c t) \cdot \langle N \rangle \, . \qquad (9)$$

If a statitistical steady state exists in which $\langle N \rangle$ does not depend on t, then in this steady state in the limit $t \to \infty$ we find from (9) that $\langle n^{B/b} \rangle = 0$ and, therefore, $\langle n \rangle = 0$ because $n \geq 0$.

Next suppose that the weak species are mobile, but the food growth rate is constant:

$$\dot{n} = (b M - a) n + D \Delta n \, , \, Q = \text{const}. \qquad (10)$$

Would that give to the weak species some evolutionary advantages ?

If all the regions of the medium were initially populated by the immobile strong species (N) the answer is again *negative*. It can be easily shown that then only the homogeneous attracting steady state exists and it corresponds to extinction of the weak species.

Sometimes the above-mentioned general results are expressed in the form of an "ecological theorem": *steady coexistence of two species relying upon the same renewable resource is impossible*. Below I show that this "theorem" needs an important correction. Namely, *if the food growth rate fluctuates in space and time and the weaker species are mobile, while the stronger ones are immobile, then steady statistical coexistence of the two species becomes possible*.

Such coexistence settles down only when the intensity of food growth rate fluctuations exceeds some threshold level and, hence, this effect represents a genuine *noise-induced* phase transition in a distributed system with diffusion.

Now I want to outline the scheme of derivation (for detailes of it, see [10]).

Our system is described by equations

$$\dot{N} = (B\,M - A)\,N\,,$$
$$\dot{n} = (b\,M - a)\,n + D\,\Delta n\,, \qquad (11)$$
$$\dot{M} = Q - \gamma\,M - C\,N - c\,n + f(\mathbf{r},t)\,.$$

Here $f(\mathbf{r},t)$ is a Gaussian noise with a zero average ($\langle f\rangle = 0$) and a pair correlation function

$$\langle f(\mathbf{r},t)f(\mathbf{r}_1,t_1)\rangle = 2\gamma\theta\,\exp(-\kappa_f|\mathbf{r} - \mathbf{r}_1|)\,\delta(t - t_1)\,. \qquad (12)$$

Hence, $r_f = 1/\kappa_f$ is the typical spatial size of an individual fluctuation and θ characterizes the intensity of fluctuations. In absence of both species (when $N = n = 0$) we have $\langle \delta M^2\rangle_0 = \theta$.

By introducing the new variables

$$p = M - M_1,\; q = \ln(N/N_1) \qquad (13)$$

we obtain from Eqs. (11)

$$\dot{q} = B\,p\,,$$
$$\dot{p} = -\gamma\,p - (\omega_0{}^2/B)(e^q - 1) - c\,n + f(\mathbf{r},t), \qquad (14)$$
$$\dot{n} = b(p - p_c)n + D\,\Delta n,$$

where we have introduced the notations $\omega_0{}^2 \equiv BCN_1$ and $p_c \equiv a/b - A/B$.

If the intensity θ of noise $f(\mathbf{r},t)$ is sufficiently small we have $\langle q^2\rangle \ll 1$. Then after linearization of Eqs.(14) in respect to q we obtain the equations

$$\dot{q} = B\,p$$
$$\dot{p} = -\gamma\,p - (\omega_0{}^2/B)\,q + f(\mathbf{r},t) - c\,n \qquad (15)$$
$$\dot{n} = b(p - p_c)n + D\,\Delta n\,.$$

Hence, our ecological system effectively reduces to a set of identical damped harmonical oscillators that are positioned in every space point and interact through the field n . In absence of oscillators this field relaxes to $n = 0$. Oscillators are permanently excited by an external random force $f(\mathbf{r},t)$. Their excitation leads to creation of the *breeding regions* for the field n where $p > p_c$.

In the mean-field approximation we find from Eqs.(15) (see [10]) that would be violated). We continue the discussion of the fluctuation properties of such phase transitions in Section 6.

In conclusion of the present Section I want to remark that the above results have an important ecological interpretation. In a fluctuating environment, mere mobility gives an evolutionary advantage that makes coexistence of weak and strong species possible. The individuals of the mobile weak species survive since they more effectively utilize the food growth rate fluctuations ! In slightly different words this means that the *variability* of the environment can be considered as an "additional resource", utilization of which makes coexistence of the two species possible.

2. Decay and reproduction in fluctuating media

Another important example of a nonequilibrium phase transition with a positive order parameter is provided by the generalized Verhulst equation:

$$\dot{n} = -\alpha n + f(\mathbf{r},t)\, n - \beta n^2 + D\, \Delta n \,. \tag{26}$$

This equation describes reproduction, decay and diffusion of a certain substance (depending on a particular application, this might a population of bacteria, chemical radicals, etc.). We assume that the decay (or death) rate α is constant and that there is some nonlinear mechanism, which puts a limit to the unbounded growth. If such limitation occurs at sufficiently small values of the population density n , it is always possible to perform decomposition in powers of n and to keep only the first nonlinear damping term, i.e. the term $-\beta n^2$.

Below we consider (see also [11,12,3]) a situation when reproduction (or breeding) of a substance happens only within certain *reproduction centers* that are created randomly and independently by some external agent, i.e.

$$f(\mathbf{r},t) = \sum_j g(\mathbf{r} - \mathbf{r}_j,\, t - t_j), \tag{27}$$

where

$$g(\mathbf{r},t) = J\, \chi(r)\, \sigma(t) \tag{28}$$

and $\sigma(t) = 1$ for $0<t<\tau_0$, $\sigma(t) = 0$ othervise. We assume that $\chi(r) \geq 0$, $\chi(0) = 1$ and that $\chi(r)$ vanishes for $r \gg r_0$. Hence, all reproduction centers are identical; they have intensity J , life-time τ_0 , spatial size r_0 and the form, given by the function $\chi(r)$. An average number of reproduction centers per unit volume per unit time

$$\langle n \rangle = \begin{cases} 0 \ , & \theta < \theta_c \\ (\nu/\beta)(\theta/\theta_c - 1) \ , & \theta > \theta_c \end{cases} \qquad (16)$$

where for

$$Dk_f^2 \gg \omega_o \gg \gamma \gg bp_c \qquad (17)$$

in three dimensions (d = 3) we have

$$\theta_c = (p_c/b)Dk_f^2 \ , \qquad (18)$$

$$\nu = bp_c \ , \ \beta = (3\sqrt{2}/4)(b^2cp_c/\gamma)(\omega_oDk_f^2)^{-1/2}. \qquad (19)$$

The mean-field approximation is valid when the inequality

$$(\langle \delta n^2 \rangle)^{1/2} \ll \langle n \rangle \qquad (20)$$

holds. (Note that in contrast to the standard theory of the second order phase transitions there is no divergency in $\langle \delta n^2 \rangle$ here since all integrals at small r 's are effectively cut by r_f). Direct estimation gives

$$(\langle \delta n^2 \rangle)^{1/2}/\langle n \rangle = 2^{1/4} \ (bp_c/\omega_o)^{1/2}(\omega_o/Dk_f^2)^{1/4} \ll 1 \ . \qquad (21)$$

Hence, in three dimensions the mean-field approximation for this particular noise-induced transition turns out to be valid whatever close to the point of transition, provided that inequalities (17) are satisfied.

It is possible to derive the effective Ginzburg-Landau-like equation for such nonequilibrium phase transition (see [10]). It has the form

$$\dot{\eta} = \nu \ (\theta/\theta_c - 1) \ \eta - \beta \ \eta^2 + D \ \Delta n + \varphi(\mathbf{r},t) \ n \ . \qquad (22)$$

The "order parameter" η is defined as a slow component of n , i.e.

$$\eta = \sum_{|\omega|<\Omega} n(\omega) \ e^{-i\omega t} \ , \qquad (23)$$

where $\Omega \ll bp_c, \gamma, \omega_o, Dk_f^2$. Note that the noise $\varphi(\mathbf{r},t)$ enters *multiplicati-vely* in Eq.(22); it is defined as

$$\varphi = (1/\omega_o^2)\partial/\partial t \ \tilde{f}(\mathbf{r},t) \ , \qquad (24)$$

where $\tilde{f}$ is a slow component of the original random force $f(\mathbf{r},t)$,

$$\tilde{f} = \sum_{|\omega|<\Omega} f(\omega) \ e^{-i\omega t} \ . \qquad (25)$$

Eq.(22) is typical for nonequilibrium phase transitions with a positively defined order parameter: it includes only *quadratic* nonlinearity (which is sufficient when η is nonnegative) and *multiplicative* noise (othervise the condition of nonnegativity of η

is taken equal to m. It is convenient to introduce, as well, a dimensionless concentration of reproduction centers c as $c = m r_0^d \tau_0$, where d is the medium dimension (d = 1,2,3). Note that in the limit c >> 1 our random field $f(\mathbf{r},t)$ becomes Gaussian.

There are two principal problems related to Eq.(26): what is the enpopulation (or explosion) threshold m_{cr} and what are the fluctuation properties of the enpopulated state established for $m > m_{cr}$?

Note that the explosion threshold is determined by the condition that in absence of the nonlinear damping term, the volume average $\langle n(\mathbf{r},t)\rangle \to \infty$ as $t \to \infty$ for $m > m_{cr}$.

Let us estimate the total production of the reproducing substance on a single reproduction center in absence of nonlinear damping. This is described by an equation

$$-\dot{n} = -D\,\Delta n - J\,\chi(r)\,n\,,\quad 0 \le t \le \tau_0\,. \qquad (29)$$

We multiplied its both sides by -1 in order to make more clear the analogy with the Schrödinger equation. Indeed, Eq.(29) can be interpreted as a Schrödinger equation with imaginary time and the potential $U(r) = -J\,\chi(r)$.

Since $\chi(r) \ge 0$, every reproduction center corresponds to some potential well. Energy levels E of the bounded states in such potential well are negative and, because time is imaginary, they will give rise to the growing solutions $n \sim e^{\lambda t}$ of Eq.(29) with the eigenvalues $\lambda = -E$.

Suppose that λ_0 is the largest eigenvalue that corresponds to the deepest energy level in the potential well (if there are any). Below a reproduction center will be called *strong* if $\lambda_0\tau_0 \gg 1$ and *weak* if $\lambda_0\tau_0 \ll 1$ (or if there are no bounded states in a potential well that corresponds to this reproduction center). Thus, strong reproduction centers are characterized by exponentially large production of the substance.

Note that the value of the lowest energy level in a potential well can be easily estimated if we know the well depth J, its width r_0 and the medium dimensionality d. This gives us a rough estimate of λ_0 in terms of such three parameters. By using this estimate, we can rewrite the condition $\lambda_0\tau_0 \gg 1$ as $J \gg J^*$ where J^* is a certain function of r_0, τ_0 and d (specific expressions for J^* are given in [3]). Then a center should be classified as strong if $J \gg J^*$ and as weak if $J \ll J^*$.

3. Explosion threshold for strong centers

In a most simple way the explosion threshold for strong centers can be calculated within a *one-center approximation*. Namely, we can neglect the interference of different centers, estimate the total increase of the reproducing substance ΔN_1 on a single center, multiply it by the center concentration m and compare this with the amount of substance that decays per unit time per unit volume. Since

$$\Delta N_1 = n_o\ r_1{}^d\ \exp(\lambda_o \tau_o) \qquad (30)$$

where n_o = <n> and r_1 is the localization radius of the bounded state corresponding to the deepest level $E_o = -\lambda_o$, we have an equation

$$m\ \Delta N_1 = \alpha\ n_o \qquad (31)$$

that gives us a one-center estimate for the explosion threshold

$$m_{cr}{}^{(o)} = \alpha\ r_1{}^{-d}\ \exp(-\lambda_o \tau_o)\ . \qquad (32)$$

To determine the limits of validity of the one-center approximation, we can estimate the magnitude of two-center corrections to the explosion threshold (32).

If we have two reproduction centers 1 and 2 separated not very far in space and in time, then a spreading spot of a substance produced on a first center can reach the region of the second center, so that reproduction for this second center will start not from the average population density at this moment but from some higher density, determined by the influence of the preceding center. Therefore, the total increase of the reproducing substance on these two centers can be expressed as

$$\Delta N = \Delta N_1 + \Delta N_2 + \Delta N_{12}\ . \qquad (33)$$

After averaging over all pairs of reproduction centers we find (see [12,3])

$$\langle \Delta N_{12} \rangle \sim n_o\ r_1{}^{2d}\ m\ \exp(2\lambda_o \tau_0)\ (m/D)^{d/(d+2)}\ . \qquad (34)$$

This correction leads to the *lowering* of the explosion threshold, i.e.

$$m_{cr} = m_{cr}{}^{(o)}\ (1 - \gamma_d\ (\alpha/\alpha^*)^{d/(d+2)}\ )\ , \qquad (35)$$

where

$$\alpha^* = (D/r_1{}^2)\ \exp[-(2/d)\lambda_o \tau_0],\ \gamma_d \sim 1\ . \qquad (36)$$

Eq.(35) holds only if the threshold lowering due to the two-center contributions is relatively small, i.e. if $\alpha \ll \alpha^*$. When $\alpha \sim \alpha^*$ or $\alpha \gg \alpha^*$, we expect that the contributions from various (not only two-center) clasters become significant and they determinine the actual explosion threshold, which would be, then, much lower than the one-center estimate (32).

4. Explosion threshold for weak centers

If the reproduction centers are weak, the one-center approximation is useless. Instead we can start from the mean-field approximation and to determine the explosion threshold from the equation

$$\alpha = \langle f(\mathbf{r},t)\rangle \,, \tag{36}$$

which gives

$$c_{cr}^{(o)} = \alpha/\xi_1 J \tag{37}$$

because $\langle f\rangle = \xi_1 cJ$. Here ξ_1 is some dimensionless constant and we use the dimensionless concentration $c = mr_o{}^d\tau_0$.

The fluctuation corrections to Eq.(37) can be estimated with a help of the diagrammatic perturbation technique for stochastic differential equations [13]. Introducing variations $\delta n = n - \langle n\rangle$ and $\delta f = f - \langle f\rangle$, we find from linearized Eq.(26)

$$0 = -(\alpha - \langle f\rangle)\langle n\rangle + \langle \delta f \delta n\rangle \,, \tag{38a}$$

$$\dot{\delta n} = -(\alpha - \langle f\rangle)\,\delta n + D\,\Delta\delta n + \delta f\,\langle n\rangle + (\delta f\,\delta n - \langle\delta f\delta n\rangle) \,. \tag{38b}$$

After Fourier transformation we obtain

$$\delta n_q = G_q{}^o\,[\,\langle n\rangle \delta f_q + \int (\delta f_{q-q'}\,\delta n_{q'} - \langle\delta f_{q-q'}\,\delta n_{q'}\rangle)\,dq' \tag{39}$$

where the Green function of "free propagation" is

$$G_q{}^o = (-i\omega + \alpha - \langle f\rangle + Dk^2)^{-1} \tag{40}$$

The Dyson equation has the form

$$G_q{}^{-1} = (G_q{}^o)^{-1} - \Sigma_q \tag{41}$$

where Σ_q is given by the following infinite diagrammatic series

$\Sigma_q =$ 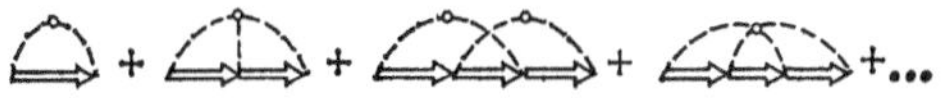

Explosion threshold is reached if the full Green function G_q has a pole at $q = 0$, i.e. if

$$\alpha = \langle f\rangle + \Sigma_o \,. \tag{42}$$

If we take into account only the first diagram in the infinite series for Σ_q , that gives us a corrected value of threshold

$$c_{cr} = c_{cr}^{(o)}[1 - \nu_d\,(J/J^*)] \qquad (43)$$

where $\nu_d \sim 1$. We see that for weak centers ($J \ll J^*$) the correction is small. Estimation of further diagrams shows that all the diagrams with irreducible correlators involving 3 or more points (diagrams 2 and 4 above) remain small for weak centers, while the diagrams with crossing dash lines (such as diagram 3 above) are more dangerous: they are small only provided a more stringent inequality

$$c(J\tau_0)(J/J^*) \ll 1 \qquad (44)$$

holds. Substitution of $c_{cr}^{(o)}$ into (44) leads to the condition

$$\alpha\tau_0(J/J^*) \ll 1\,. \qquad (45)$$

If this condition is not satisfied, fluctuational lowering of the explosion threshold would be significant, despite the fact that all reproduction centers are weak.

Conditions (44) and (45) were derived above in a rather formal way. In the next Section we show that they are related to the effects of large clasters of individual reproduction centers.

5. The role of rare clasters

Suppose we have an isolated claster of weak reproduction centers which is located within a space-time element (ΔV, Δt). The substance production by such claster is described by an equation

$$\dot{n} = f(\mathbf{r},t)\, n + D\, \Delta n \qquad (46)$$

This can be viewed as a Schrödinger equation with imaginary time and a potential $U(\mathbf{r},t) = -f(\mathbf{r},t)$. Since $f(\mathbf{r},t) \geq 0$, any claster defines a certain time-dependent potential well.

The linear differential operator

$$\hat{L} = D\, \Delta + f(\mathbf{r},t) \qquad (47)$$

has a discrete set of positive time-dependent eigenvalues $\{\lambda_j(t)\}$. Let us denote $\lambda(t) \equiv \max \lambda_j(t)$.

In a rather rough way every claster can be specified by two parameters, i.e. by its time-averaged maximal eigenvalue λ and its life-time τ . When $\lambda\tau \ll 1$ such claster remains weak and insignificant. On the other hand, when $\lambda\tau \gg 1$ this claster, formed by an aggregation of weak individual centers, behaves like a single strong reproduction center, leading to exponentially large production of reproducing substance. Evidently, this is possible only if such claster consists of many individual centers.

Note that condition $\lambda\tau \gg 1$ can be written, as well, as $\tau \gg \lambda^{-1}$. It implies that the potential $U(r,t) = -f(r,t)$ will be, generally, a slowly varying function of time, as compared with the characteristic "oscillation period" λ^{-1} in such a potential. Hence, the adiabatic approximation is valid and the total production of the substance by such a claster can be estimated as

$$\Delta N = n_0 \exp\left[\int_0^{\Delta t} \lambda(\tau)\, d\tau \right] \tag{48}$$

By introducing the intensity s of a claster as

$$s = \int \lambda(t)\, dt \tag{49}$$

we obtain from Eq.(48)

$$\Delta N = n_0\, e^s . \tag{50}$$

Suppose that $p(s)$ gives a probability to find a claster of intensity s within a unit space-time element. Then the average increase of population density per unit time due to strong clasters can be estimated as

$$\langle \Delta n \rangle = \int \Delta N(s)\, p(s)\, ds \tag{51}$$

so that the effective reproduction rate due to strong clasters is

$$Q \sim \int e^s\, p(s)\, ds . \tag{52}$$

This contribution can be small as compared with the average reproduction rate $\langle f \rangle$ only provided that $p(s)$ is exponentially small for $s \gg 1$, i.e. if $p(s) \sim \exp[-\Phi(s)]$ with $\Phi(s) \gg 1$.

However, even the exponential rarity of strong clasters cannot guarantee that their contribution would be negligible. Indeed, in this case we have

$$Q \sim \int_1^{\infty} \exp[s - \Phi(s)]\, ds \tag{53}$$

and the magnitude of Q is determined by competition of two exponential factors: exponential rarity of strong clasters might be counterbalanced by exponentially large production of the substance by a single claster.

Note that the condition $\lambda\tau \gg 1$ can be realized in two different ways, i.e. for time-clasters or space-clasters. We define a *time-claster* as a chain of individual centers subsequently created in the same small element of space (Fig.1), for such a claster we have $\lambda \sim \lambda_0$ but $\tau \gg \tau_0$. On the hand, a *space-claster* is a space aggregation of individual centers (Fig.2), such that $\lambda \gg \lambda_0$ but $\tau \sim \tau_0$. Obviously, these are only two limiting situations and the intermediate cases are possible.

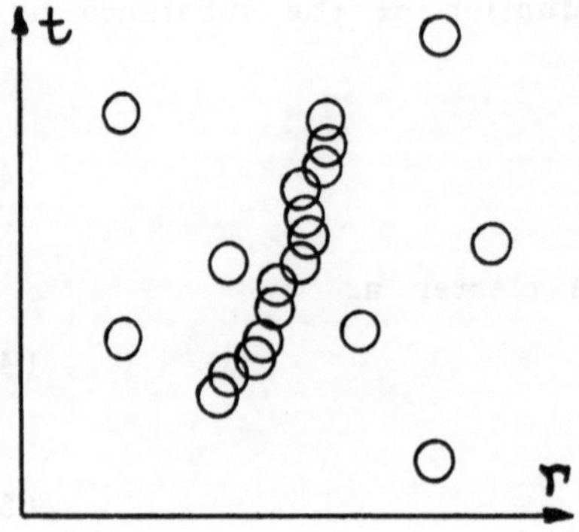

Fig.1

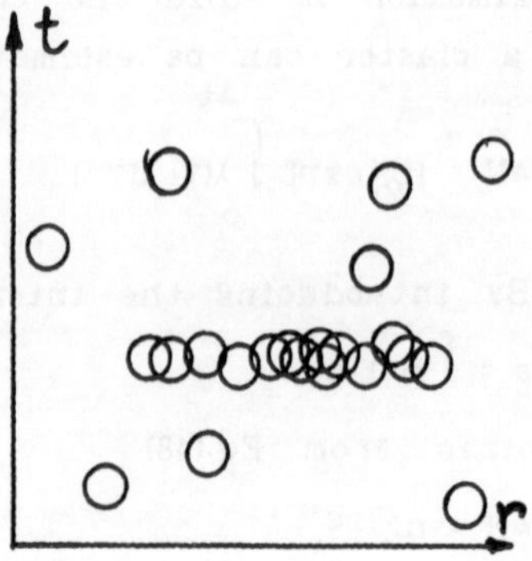

Fig.2

Although all further estimates can be performed in a general case of randomly created reproduction centers, for simplicity we restrict below our discussion to a Gaussian limit $c \gg 1$ in which centers are very weak and largely overlap in space and in time, so that the reproduction rate $f(\mathbf{r},t)$ represents a Gaussian random field. It is convenient to introduce a deviation $\delta f(\mathbf{r},t) = f(\mathbf{r},t) - \langle f \rangle$ that will be a Gaussian noise with a zero average, intensity $S = \xi_2 c J^2$, correlation radius r_0 and correlation time τ_0. Now "strong clasters" represent simply certain very rare anomalously large positive bursts of the random field f.

Let λ_* be a characteristic depth of the lowest level in a"potential well" that corresponds to a *typical* fluctuation $\delta f(\mathbf{r},t)$. Near threshold such typical fluctuations should be weak: $\lambda_* \tau_0 \ll 1$. We have to modify in this case the above classification of rare clasters. Namely, we say now that a time-claster is a long-living typically intensive fluctuation with $\lambda \sim \lambda_*$ and $\tau \sim \tau_0$, while a space-claster is a very intensive fluctuation with a typical life-time, i.e. with $\lambda \gg \lambda_*$ and $\tau \sim \tau_0$.

For a Gaussian random process, the probability of long-living fluctuations with $\tau \gg \tau_0$ and $\lambda \sim \lambda_*$ can be estimated as

$$p(\tau) \sim \exp(-\varphi(\tau)), \ \varphi(\tau) \sim \tau/\tau_0. \tag{54}$$

Hence, their contribution into the effective reproduction rate is

$$Q \sim \int_{1/\lambda_*}^{\infty} \exp(\lambda_* \tau - \varphi(\tau))\, d\tau \tag{55}$$

and it is always exponentially small and negligible, provided that $\lambda_* \tau \ll 1$.

Therefore, the most dangerous rare fluctuations are those with $\lambda \gg \lambda_*$ and $\tau \sim \tau_0$, i.e. the space-clasters. To estimate the probability of such fluctuations, we can use some results of the theory of the Schrodinger equation with random stationary potential. Indeed, within time intervals about τ_0 the fluctuating field $f(\mathbf{r},t)$ remains almost stationary since τ_0 is the correlation time of this random field. Hence, we can calculate $p(\lambda)$ as a probability to find very deep rare levels with energy $E = -\lambda$ in the fluctuational region of spectrum of the Schrödinger operator (or as a density of levels in this region). By using the expressions for the density of levels in the fluctuational region (see [14,15]), we find

$$p(\lambda) = \exp(-\Phi(\lambda)), \qquad (56)$$

where

$$\Phi(\lambda) = a\, \lambda^{2-d/2} D^{d/2} S^{-1} r_0^d\ , \quad \lambda r_0^d/D \ll 1\ , \qquad (57a)$$

$$\Phi(\lambda) = \lambda^2/2S\ , \qquad \lambda r_0^d/D \gg 1\ . \qquad (57b)$$

These expressions hold when $\Phi(\lambda) \gg 1$; $a \sim 1$.

The contribution from such rare fluctuations into the effective reproduction rate is

$$Q \sim \int_{\tau_0^{-1}}^{\infty} \exp(F(\lambda))\, d\lambda\ , \qquad (58)$$

where

$$F(\lambda) = \lambda\, \tau_0 - \Phi(\lambda)\ . \qquad (59)$$

The effective reproduction rate due to rare strong fluctuations can be estimated, using Eqs.(58) and (59), but we should remember that the explicite expressions (57) for $\Phi(\lambda)$ are valid only if $\Phi(\lambda) \gg 1$

For a one-dimensional medium ($d = 1$) and long-living reproduction centers ($\tau_0 \gg r_0^2/D$) the form of the dependence $F(\lambda)$ is shown in Fig.3. When intensity of noise S (and, hence, the concentration of centers c) increases, the maximum of $F(\lambda)$ shifts to higher values of λ and at a certain critical intensity S_{cr} passes the value $\lambda = 1/\tau_0$ which is the lower bound of the integration interval in Eq.(58). When this happens, fluctuations with $\lambda \sim 1/\tau_0$ cease to be exponentially rare (although they are not, still, typical) and this leads to a significant lowering of the explosion threshold. By using Eq.(57a) one can show that this critical intensity Scr *is not* reached when $S \ll D^{1/2}/\tau_0^{3/2} r_o$, but no explicite expression for S_{cr} can be obtained. Since $S \sim cJ^2$, this inequality can be expressed in terms of c , i.e.

$$c\,(J\tau_0)(J/J^*) \ll 1 \qquad (60)$$

When such inequality is satisfied, the contribution from strong reproduction clasters is negligible and the fluctuational lowering of

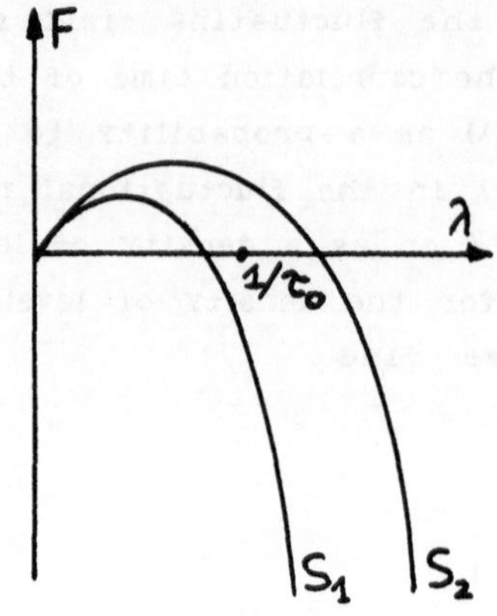

Fig.3

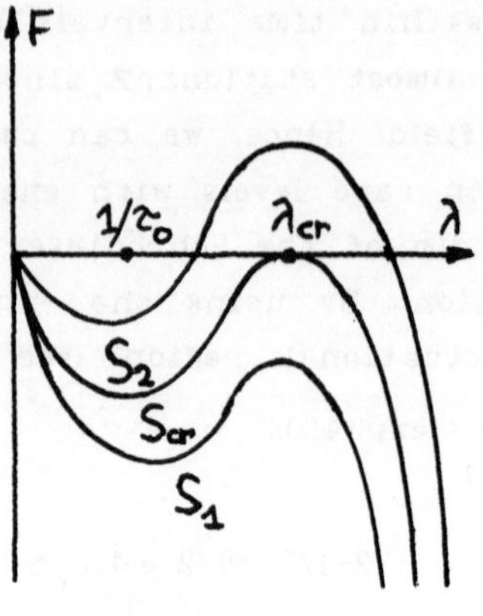

Fig.4

the threshold is small. Note that we have found exactly the same condition (44) that was obtained in Section 4 by using a diagrammatic perturbation technique.

For three-dimensional media (d = 3) and long-living centers ($\tau_0 \gg r_o^2/D$) the dependence of F on λ has a qualitatively different form (Fig.4). When intensity of noise S increases, a maximum of F develops at $\lambda \sim D/r_o^2$ and starts to grow, crossing at some critical intensity S_{cr} the level F = 0 . After that, i.e. for $S > S_{cr}$, exponentially rare strong fluctuations with λ close to D/r_o^2 give the dominant contribution into the effective reproduction rate, which significantly lowers the explosion threshold. The value of S_{cr} can be estimated from Eqs.(57) as $S_{cr} \sim D/\tau_0 r_o^2$. Since $S \sim cJ^2$, we again find that the condition $S \ll S_{cr}$ coinsides with the inequality (44) derived in Section 4.

Further discussion of the role of rare strong fluctuations in two-dimensional media and for short-living reproduction centers can be found in our review [3].

Note that for the infinitely long living reproduction centers ($\tau_0 \to \infty$) the condition (44) is never satisfied. This is quite natural: if the random pattern of reproduction centers is stationary, it is always possible to find somewhere in the medium such a strong claster that its effective reproduction rate exceeds the rate of decay. Hence, for stationary reproduction centers, that are spread randomly and independently one from another in the medium, the explosion threshold in the *infinite* medium should be equal to zero. However, if the medium occupies only a finite volume, it might lack such sufficiently strong claster, so that explosion would start only at some higher critical concentration of centers, depending on a particular sample.

6. A population-settling-down transition

When the nonlinear damping term βn^2 in the generalized Verhulst equation

$$n = - \alpha\, n - \beta\, n^2 + f(\mathbf{r},t)\, n + D\, \Delta n \tag{61}$$

is present, the infinite growth of a population at $c > c_{cr}$ is prevented. Instead of that in the limit $t \to \infty$ a steady state is formed, which is characterized by some positive value of the average population density $\langle n \rangle$. This average population density depends on c and vanishes at $c = c_{cr}$. Hence, we can say that at $c = c_{cr}$ our system undergoes a special sort of a phase transition, namely a *population-settling-down transition*, with a population density playing a role of an order parameter.

It can be remarked that, in some sense, a population-settling-down transition is accompanied by the symmetry breaking. Indeed, let us consider the symmetry properties of the steady states below and above the point of such transition. For $c < c_{cr}$ we have $\langle n \rangle = 0$ and, hence, $n(\mathbf{r},t) \equiv 0$. Obviously, such steady state is invariant under all symmetry operations, including scale transformations. Above the transition point $\langle n \rangle > 0$ and the scale transformation invariancy is lost. Note, however, that the dynamical Eq.(61) itself is never invariant under scale transformations, which makes a difference with the spontaneous symmetry breaking in second order equilibrium phase transitions.

We can expect that the fluctuation phenomena near the point of a population-settling-down transition should be rather unusual. Since the order parameter n of such transition is nonnegative, it follows from $\langle n \rangle = 0$ that $\langle \delta n^2 \rangle = 0$ and, thus, below the point of a transition not only the average value of the order parameter n but, as well, all its fluctuations vanish. Note that for the second order transition in the symmetrical phase we have $\langle \eta \rangle = 0$ but $\langle \delta \eta^2 \rangle > 0$.

Clearly, the behavior of fluctuations might depend on the statistical properties of the external random field $f(\mathbf{r},t)$ in Eq.(61). Below we restrict our discussion to a case when reproduction centers are weak and their strong rare clasters are insignificant.

Within the mean-field approximation, the average population density in the asymptotic steady state is

$$\langle n\rangle = \begin{cases} 0, & c < c_{cr} \\ (\alpha/\beta)(c/c_{cr} - 1), & c > c_{cr}. \end{cases} \tag{62}$$

The mean-field approximation is justified, if

$$(\langle\delta n^2\rangle)^{1/2} \ll \langle n\rangle . \tag{63}$$

By using this approximation, we can calculate $\langle\delta n^2\rangle$ and verify the condition (63).

Particularily, we find that $\langle\delta n^2\rangle = \mu \langle n\rangle^2$, where the factor μ is given by the following expressions:

A. For long-living reproduction centers ($\tau_0 \gg r_o^2/D$ or $l \gg r_o$)

$$\mu = \begin{cases} g\,(r_o/l), & d = 3 \\ g\,\ln(r_c/l)/\ln(l/r_o), & d = 2 \\ g\,(r_c/l), & d = 1 \end{cases} \tag{64}$$

B. For short-living reproduction centers ($\tau_0 \ll r_o^2/D$ or $l \ll r_o$)

$$\mu = \begin{cases} g\,(r_o/l)^2, & d = 3 \\ g\,(r_o/l)^2 \ln(r_c/r_o), & d = 2 \\ g\,(r_o/l)^2\,(r_c/r_o), & d = 1 \end{cases} \tag{65}$$

Here we introduced the notations $l = (D\tau_0)^{1/2}$ and $g = c_{cr}(J\tau_0)(J/J^*)$ Since we assume that strong rare clasters are insignificant this implies (cf. Eq.(44)) $g \ll 1$. The correlation radius

$$r_c = (D/\alpha)^{1/2}\,(c/c_{cr} - 1)^{-1/2} \tag{66}$$

diverges as $c \to c_{cr}$.

We see that in three dimensions ($d = 3$) the mean-field approximation is valid whatever close to the point of transition, provided that the Ginzburg number Gi,

$$Gi = \begin{cases} g & \text{for } l \gg r_o \\ g\,(r_o/l)^2 & \text{for } l \ll r_o, \end{cases} \tag{67}$$

is small ($Gi \ll 1$). For $d = 2$ we find very weak logarithmic divergence and for $d = 1$ we have near the critical point c_{cr} a fluctuational region, where the mean-field approximation is invalid and relative fluctuations are large; this fluctuational region is narrow if $Gi \ll 1$.

Note that for the short-living weak reproduction centers ($l \ll r_o$) the condition $Gi \ll 1$, which ensures the relative smallness of fluctuations above the threshold, is more stringent then the condition $g \ll 1$, which guarantees that the fluctuational lowering of the threshold is small.

In the mean-field approximation the correlation time τ_c, that describes relaxation of population density deviations, is given by the expression

$$\tau_0 = [\alpha (c/c_{cr} - 1)]^{-1} \qquad (68)$$

and diverges as $c \to c_{cr}$

Finally, we can compare the fluctuation properties of the population-settling-down transition with those of the standard second order phase transition. *Both* transitions are characterized by the divergency of the correlation radius and by the phenomenon of critical slowing down. However, the population-settling-down transition has a critical dimension $d_{cr} = 2$, in contrast to $d_{cr} = 4$ for the second order equilibrium phase transitions.

This result seems to be a general feature of all phase transitions with a positively defined order parameter. Indeed, we have found the same behavior for an ecological system with two competing species in Section 1. It is also supported by our studies of systems with annihilation and continuous generation of diffusing particles (see [16]). It should be remarked that Eq.(61) might be considered as a generic equation for the phase transitions with nonnegative real order parameters, i.e. as a counterpart to the standard Ginzburg-Landau equation.

To estimate the fluctuational lowering of the explosion threshold we have used above some theoretical results, found for the one-particle Schrödinger equation with random stationary potential. However, these estimates can be obtained , as well, in a more straightforward way by using a formal solution of stochastic reaction-diffusion equations in terms of path integrals and applying an *optimal fluctuation* technique to determine the probability of exponentially rare events (see [17,18]).

References

1. H.Haken. *Synergetics*, Springer, Berlin, 1978.
2. L.S.Polak, A.S.Mikhailov. *Self-Organization in Nonequilibrium Physicochemical Systems*, Nauka, Moscow, 1983 (in Russian).
3. A.S.Mikhailov, I.V.Uporov. Critical phenomena in systems with breeding, decay and diffusion. Uspekhi Fiz. Nauk *144*, 79 (1984).
4. W.Horstemke, M.Malek-Mansour. The influence of external noise on nonequilibrium phase transitions. Z. Phys. *B24*, 307 (1976).

5. L.Arnold, W.Horsthemke, R.Lefever. White and colored external noise and transition phenomena in nonlinear systems. Z. Phys. *B29*, 867 (1978).
6. E.V.Astashkina, A.S.Mikhailov, A.V.Tolstopyatenko. Noise-induced instability in the Lorenz model. Izv. VUZ. Radiofizika *24*, 1035 (1981).
7. W.Horsthemke, R.Lefever. *Noise-Induced Transitions*, Springer, Berlin, 1984.
8. A.S.Mikhailov. Nonequilibrium phase transition in a biological population. Dokl. Akad. Nauk SSSR *243*, 786 (1978).
9. A.S.Mikhailov. Noise-induced phase transition in a biological system with diffusion. Phys.Lett. *73A*, 143 (1979).
10. A.S.Mikhailov. Effects of diffusion in fluctuating media: a noise-induced phase transition. Z. Phys. *B41*, 277 (1981).
11. A.S.Mikhailov, I.V.Uporov. Noise-induced phase transition and a percolation problem for fluctuating media with diffusion. Zh. Eksp. Teor. Fiz. *84*, 1958 (1980).
12. A.S.Mikhailov, I.V.Uporov. Critical phenomena in media with random reproduction centers. Zh. Eksp. Teor. Fiz. *84*, 1481 (1983).
13. S.Ma. *Modern Theory of Critical Phenomena*, W.A.Benjamin, London, 1976.
14. B.Halperin, M.Lax. Phys. Rev. *148*, 722 (1966).
15. I.M.Lifshits, S.A.Gredeskul, L.A.Pastur. *Introduction into Theory of Disordered Systems*, Nauka, Moscow, 1982 (in Russian).
16. A.M.Gutin, A.S.Mikhailov, V.V.Yashin. Fluctuation phenomena in systems with diffusion-controlled reactions. Zh. Eksp. Teor. Fiz. *92*, 941 (1987).
17. A.Förster, A.S.Mikhailov. Application of path integrals to stochastic reaction-diffusion equations, in: *Selforganization by Nonlinear Irreversible Processes* (eds. W.Ebeling and H.Ulbricht), Springer, Berlin, 1986, p.89.
18. A.Förster, A.S.Mikhailov. Optimal fluctuations leading to transitions in bistable systems. Phys. Lett. *126A*, 459 (1988).

FROM DETERMINISTIC CHAOS TO NOISE IN RETARDED FEEDBACK SYSTEMS

M. Le Berre[+], Y. Pomeau[x], E. Ressayre[+], A. Tallet[+], H.M. Gibbs[*], D.L. Kaplan[*], M.J. Rose[*]

[+]Laboratoire de Photophysique Moléculaire, Bât. 213, Université Paris-Sud, 91405 Orsay, France
[x]Laboratoire de Physique de l'E.N.S., 24, rue Lhomond, 75231 Paris Cedex 05, France
[*]Optical Science Center, University of Arizona, Tucson, Ariz. 85721, USA

I. INTRODUCTION

Deterministic chaos is often considered as antinomic to random noise. Indeed those two concepts have important differences. As is well known, deterministic chaos refers basically to the eventual randomness of the solutions of ordinary differential equations, without any external noise added. On the contrary, random noise in the usual meaning of the word refers to a dynamics that is nonpredictive because it comes from the action of myriad of degrees of freedom, as atoms or molecules in a many body system. However, as was discovered some time ago[1], there is a possibility of going continuously from deterministic chaos to random noise in models of nonlinear dynamics relevant to situations met in optics which can be described by a differential delay equation. A general form of this sort of equation may be taken as

$$dx/dt + dV/dx = g[x(t-r)] \qquad (1)$$

where everything of the l.h.s. is taken at time t, although the r.h.s. is a function of the value of the variable at a previous time, t-r. Note that one could imagine to have more than one delay or even to have both advanced and retarded terms in such a differential-delay equation. Although this could lead to rather interesting things, we shall limit ourselves to the single delay case.

If the delay term on the r.h.s. of (1) is simply taken as zero, then this equation describes a relaxational dynamics and, provided V increases sufficiently fast at large x, as we shall assume it, the long time motion settles down at one of the minima of V.

As the delay increases from zero, the long time motion get chaotic trajectory nearby the bottom of the potential V. It was shown[2,3] for several delay differential equations that the number of degrees of freedom may be tuned continuously and taken as large as wanted : this number is simply proportional to the delay. However the entropy as well as the statistical properties are independant of the delay[1,2,3].

Here we show that it is sometimes possible to increase both the number of degrees of freedom and the entropy in this type of equations, simply by increasing the argument of the delay term. Therefore the delay term on the r.h.s. of Eq.(4) can be seen as a simple noise source. The reason for that is as follows.

Suppose first that f(x) is not monotonic and that any variation of x is amplified by a factor k inside the delay term argument,

$$g(x) = f(kx) \; ; \tag{2}$$

then the r.h.s. [x(t-r)] could have many fluctuations before that the solution x(t) computed on the l.h.s. has time to change and therefore g(t) could be seen as successive independent oscillations. The implicit solution of Eq.(1) is

$$x(t) = x(0)\; H(0,t) + \int_0^t g(x(u-r))\;\; H(u,t)\; du \tag{3}$$

with

$$H(u,t) = \exp - \int_u^t \frac{1}{x(s)} \frac{dV}{dX} ds \qquad (4)$$

One may assume that Eqs.(2,3) has basically a linear dependence with respect to the function f. Accordingly X(t) sums up the contributions from the r.h.s. f(t) ; and as they are independent and more and more numerous when k increases, this sum tends to have a Gaussian distribution. Finally the delay term in the r.h.s. may be considered as a delta correlated noise, at least at the time scale of the variations of X. Below we shall show that this picture is correct at least when the potential V is quadratic (Fig. 1a) which ensures that the l.h.s. of Eq.(1) is simply linear. Then the qualitative behavior of X(t) is a Brownian motion nearby the bottom of the potential.

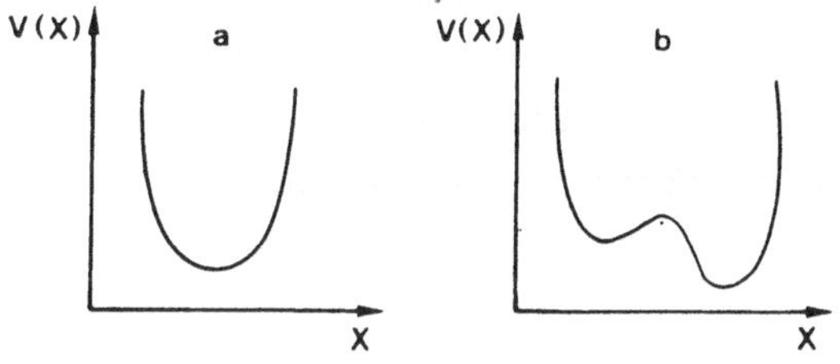

Fig. 1 - a) Quadratic potential $V(X) \propto X^2$;
b) Double well potential. $V(X) \propto X + a X^2 + b X^3 + c X^4$

For that purpose we shall compare the results of a theoretical analysis of the differential delay equation with numerical and experimental results. We intend to come back in a future publication to the case of a more complicated l.h.s. of equation (1), and to examine in particular situations when V is a two well potential (Fig. 1b).

The theoretical analysis shows that it exists an intrinsic time of the system which is the correlation time of f(t). *It characterizes both the number of degrees of freedom and the entropy.*

In Section II, the statistical properties of the solution X(t) are derived, and comparison with experiments is shown.

In Section III, the Lyapunov analysis is described for different retarded systems. A conjecture is stated which relates the decay time of the correlation function to the information dimension and the metric entropy.

II. STATISTICAL PROPERTIES OF SINGLE WELL RETARDED EQUATIONS WITH PERIODIC FEEDBACK

In this section we are dealing with scalar retarded equations like Eq.(3) with quadratic potential and periodic sinusoïdal feedback f(x). This feedback system modelizes the electro-optical device where period doubling route to chaos was first observed by Gibbs et al[4]. The electrical potential obeys the equation

$$\frac{dx}{dt} + x(t) = a\,[1 - R \sin x(t-r)\,] \qquad (5)$$

where time scale is the response time of the whole system, R is a parameter smaller than unity, aR is then the amplitude of the feedback force. The unit of this amplitude is set as imposing a 2π-phase in the periodic feedback.

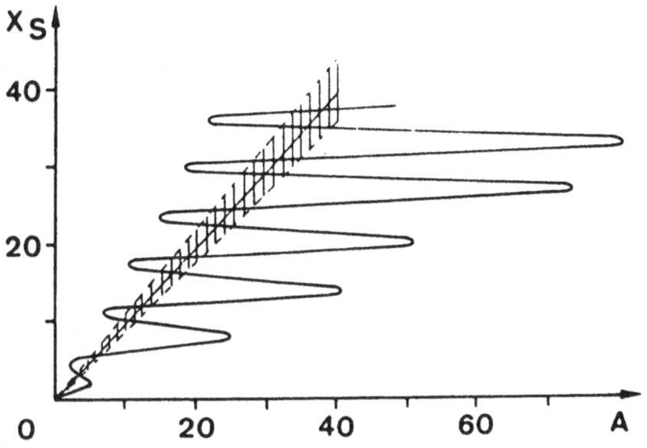

Fig. 2 - Stationary solutions of Eq. (5) in full line for R = 2/3. The dashed area around the bissector corresponds to the numerical standard deviation of the chaotic solution, $a - \sigma_X \leq X \leq a + \sigma_X$

The stationary solutions of Eq.(5) are shown in Fig. 2, but actually they are linearly unstable when both the amplitude of the feedback and the delay r are larger than a few units; Let us now focus on fully developped chaos as observed in the numerics and briefly describe its features. Whatever the delay, the chaotic solutions oscillate irregularly around the average value x = a, with standard deviation reported on Fig. 2 by the dashed area around the bissector. Two examples are reported in Fig. 3a and 3b where the solutions X(t) of Eq.(5) for R = 2/3, oscillate respectively around x = 10 and 100, with standard deviations σ = 1.8 and 5.8. The time dependance of these two solutions look very much the same while the delays are very different (r=3 and 100, respectively).

A qualitative description of the solution is now proposed to help further understanding the mathematics. An enlarged view of the signal in Fig. 3b is shown below in Fig. 3c ; the solution X(t) in full line consists in fast oscillations superimposed on a slow enveloppe schematized by the dashed line. Moreover let us anticipate the statistical analysis and compare the oscillations of the signal X(t) and those of the feedback f(X(t)), both oscillate around x = a (cf Fig. 4), the signal with small amplitude oscillations (cf the dashed area in Fig. 2) while the feedback oscillates widely between 0 and 2a. These wide and fast oscillations of the feedback are the key of the dynamical behavior of such delayed systems, as briefly displayed .

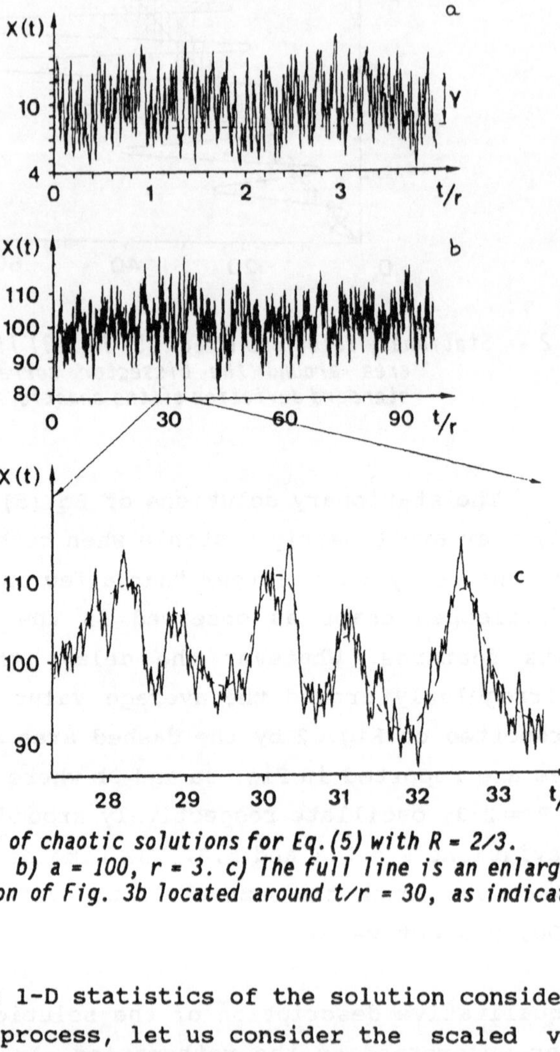

Fig. 3 - Temporal behavior of chaotic solutions for Eq.(5) with R = 2/3. a) a = 10, r = 100. b) a = 100, r = 3. c) The full line is an enlarged view of the portion of Fig. 3b located around t/r = 30, as indicated by the arrows.

To derive the 1-D statistics of the solution considered as a sample of an ergodic process, let us consider the scaled variable $y = x/a$ which obeys an equation like Eqs. (1-2)

$$\dot{y} + y = f[ay(t-r)] \qquad (6)$$

with

$$f[y] = 1 - R \sin y. \qquad (7)$$

In Eq.(5) the parameter a modulates the feedback amplitude while the same parameter modulates the feedback argument in Eq.(6) relative to the scaled variable y.

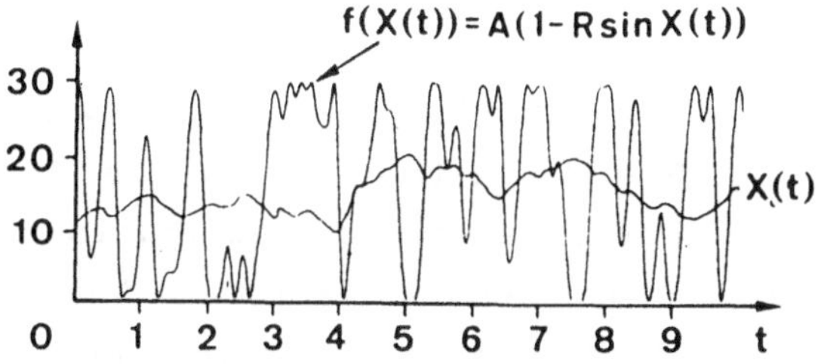

Fig. 4 - Comparison of the signal X(t) and the feedback term f(X(t)) for a = 15, R = 0.98, r = 10

Let us consider a monotonic variation of the signal y(t) schematized on top of Fig. 5. When aR » 1, a very small increment of Y(t-r) equal to π/a corresponds to a full oscillation of f between 1-R and 1+R ; finally the feedback jumps very often between these two extrema, therefore f(t) practically looses its memory after each oscillation.

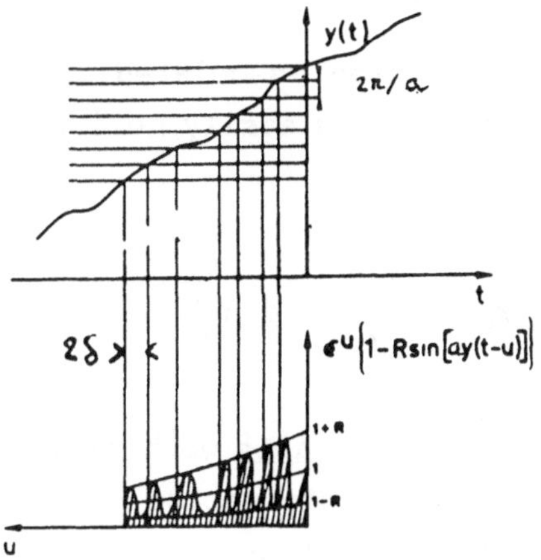

Fig. 5 - Illustration of Eq. (8). On the top y(t) is drawn with increments equal to $2\pi/a$. The integrand in Eq. (8) is the function drawn in the bottom, the dashed area is equal to y(t+r).

Now let us write the solution of Eqs.(6-7) as the very simple implicit expression

$$y(t+r) = \int_{0}^{t+r} e^{-u} \; f\,[ay\,(t-u)\,] + y(0)\, e^{-(t+r)} \tag{8}$$

It appears that the signal at time t+r is exactly equal to the dashed area at bottom of Fig. 5 whatever the delay may be. This area is the sum of an infinite number of independent contributions, with decreasing variance because of the exponential factor in the integrand (Eq. 8). Provided that the mean time interval δ for such an oscillation is much smaller than unity, when $t,r \gg 1$, one can approximate the solution in Eq.(8) by the serie

$$y(t+r) = 1 + \sum_{n=0}^{\infty} e^{-n\delta} \; Y_n \tag{9}$$

with

$$Y_n = - R \int_{\delta_n} \sin\,[a\; y\;(u)\;]\; du \tag{10}$$

The characteristic function associated with the centered 1-D random variable y-1, at any time, is easily derived[1] <u>in the limit of small δ</u>, one obtains

$$\Phi_y(v) = \exp - \left[\frac{v^2}{4\delta} \; < Y_n^{\,2} > \right] \tag{11}$$

which defines <u>a Gaussian variable</u> with variance $\sigma_y^2 = < Y_n^2 > / 2\delta$. At this stage one needs some information on the correlation function $\Gamma_f(u) = R^2 <\sin ay(0) \sin ay(u)>$ in order to derive both the variance σ_y^2 and the memory time δ of the feedback ; this point illustrates the strong connection of all the statistical data in such retarded systems. For the moment let us approximate $\Gamma_f(u)$ by a rectangle of width δ, that gives rise to

$$< Y_n^2 > = R^2 \delta^2 / 2, \qquad (12)$$

from Eq. (10).

The time interval δ for one oscillation of the feedback is derived from the following relation

$$y(t+\delta_t) - y(t) = -\delta_t [y(t) - 1] - \int_{\delta_t} R \sin ay(t-r-u)\, du. \qquad (13)$$

When the increment is equal to π/a, the square average of Eq. 13 leads to

$$\delta \simeq \frac{\pi \sqrt{2}}{a R} \qquad (14)$$

which confirms a posteriori the previous description schematized in Fig.(4) leading to a Gaussian probability distribution in the limit $aR \to \infty$. The variance deduced from Eq.(14) would be $\sigma_y^2 \simeq R/a$.

Let us compare these rough estimations with numerics. The probability distributions of the chaotic solutions $x(t) = ay(t)$ of Eq.(5) are shown in Figs(6) for a feedback amplitude aR increasing from 3.3 up to 67. The Gaussian shape is very well approached as $aR > 15$. This result was already derived by Ikeda and Akimoto[5] from numerical computations of the four first momenta $< X^n(t) >$.

Let us point out that the above derivation leading to a Gaussian statistics in the limit of large feedback is valid for any Eq.(7) with periodic feedback. The case of $f(x) = Dec(x)$ (decimal part of x) was numerically treated by B. Dorizzi and B. Grammaticos (private communication), the chaotic solutions x(t) are shown to behave as Gaussian ones for small feedback amplitude, $a > 5$. This boundary value is related to the period of f(x).

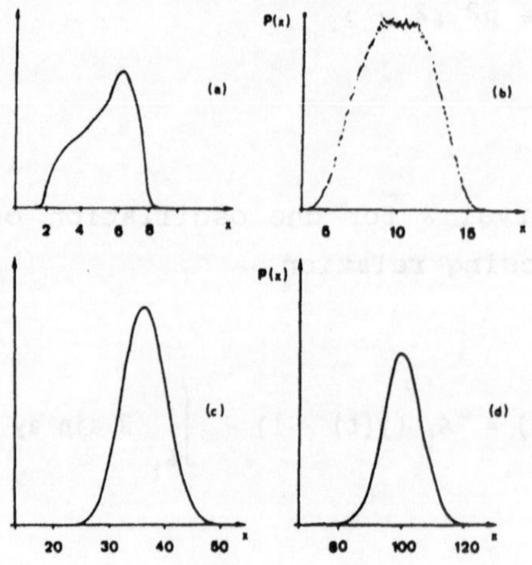

Fig. 6 - Probability distribution for the solution X(t) of (Eq.(5)), R = 2/3.
(a) a = 5, r = 10 ; (b) a = 10, r = 10 ;
(c) a = 36, r = 3 ; (d) a = 100, r = 5.

For the sine feedback, the numerical variance is shown in Fig. 7, it increases linearly with the parameter aR, as predicted above with rough arguments, more precisely one obtains

$$\sigma_X^2 = \frac{a\,R}{2} + 3 \tag{15}$$

in the whole range of Gaussian statistics.

Experimental results also agree with numerics and the above statistical analysis. Actually in experiments the feedback amplitude cannot reach as high amplitudes as the theory does, since the parameter aR is limited to 6 in the Tucson experimental set-up. The two probability distributions shown in Figs. 8 for aR = 4 and 6 agree with our analysis, one can notice that <u>the experimental signal also becomes more and more Gaussian as the feedback strengh increases</u>.

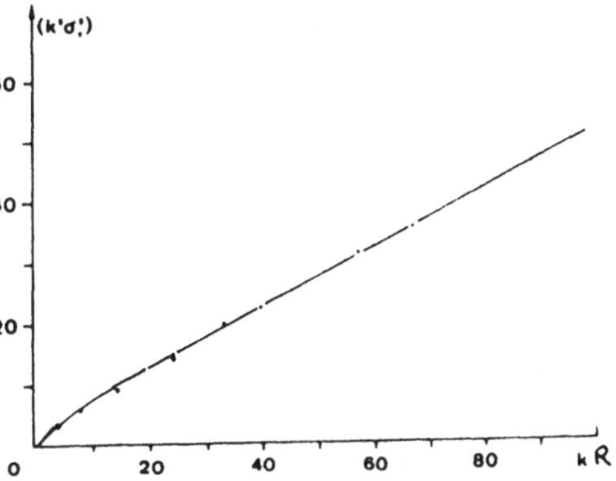

Fig. 7 - Variance of X = kY as a function of kR.

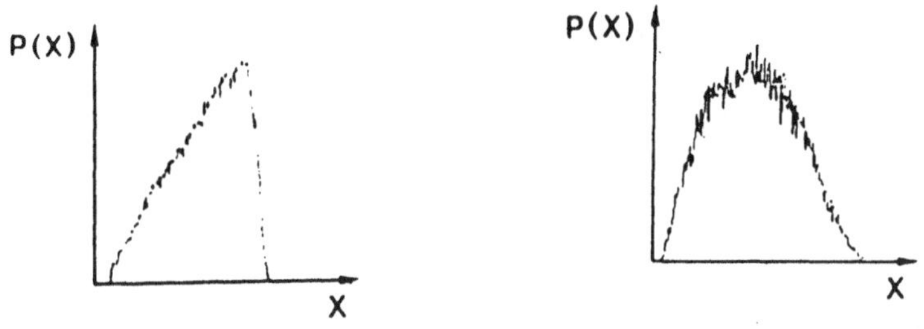

Fig. 8 - Experimental probability distribution for chaotic solutions of $\dot{X} + X = a(1 - R \cos X(t-r))$ for aR = 4 (left), and aR = 6 (right)

This simple system is very attractive from a theoretical point of view since it allows a lot of analytical investigations. Crude derivations for σ^2 and δ are presented above, but paradoxically one can do much better when investigating the two dimensional statistics. From the expansion of Eqs. (9, 10) the increment $Y(t_1) - y(t_2)$ is Gaussian as far as $t_1 - t_2 \gg \delta$, as yet numerically observed[5]. This property complicates a lot the derivation of the correlation function since the usual formula for 2-D Gaussian variables $\{ x(t_1), x(t_2) \}$ are not valid on a 'short' time scale of order δ. However the whole behavior of both correlation functions $\Gamma_x(t)$ for the signal $x(t)$, and $\Gamma_f(t)$ for the feedback f(x(t)) can be analytically predicted, as described extensively in Reference 1. We only summarize here the principal results of this study and show an experimental correlation function.

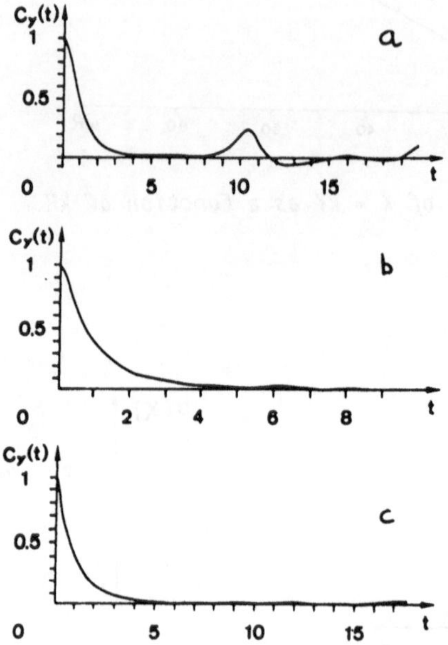

Fig. 9 - Normalized correlation function of y(t)
$\sigma_y^2\, C_y(t) = \langle y(t')\, y(t'+t) \rangle - \langle y \rangle^2$, for R = 2/3.
(a) a = 10, r = 10 ; (b) a = 20, r = 5 ;
(c) a= 36, r = 3

Because of the rather complex temporal behavior of the solution x(t), let us suppose that both the delay r and the amplitude of the feedback aR are larger than unity. The function C_x exhibits three regimes, (cf Fig. 9) : It is quadratic near the origin on a time scale $t_c = (2/aR)^{1/2}$ further it decays exponentially as $e^{-|t|}$, actually with a decay time equal to the response time of the system and finally it has secondary bumps at t = t, 2r,... This exponential decay of C_x for $t > t_c$ is the signature of a Markovian process, which corresponds in Fig. 3c to the "enveloppe" of x(t) schematized by the dashed curve. The small oscillations which appear in the full line of Fig. 3c have a very short time scale (of order δ, given by Eq. 14) but they are too weak to bring about a loss of memory in the signal $x(t)$; they only reflect the strongly oscillatory behavior of the feedback.

For non Gaussian signal ($aR < 15$) it appears a third memory time which seems natural in such retarded system : $y(t)$ is correlated with $y(t+r)$, that gives rise to successive positive or negative bumps with decreasing height located at $t = r, 2r, \ldots$ as predicted analytically. It is shown that these bumps disappear as aR increases (cf Fig. 9c). While the fast fluctuations in $x(t)$ have no effect on C_x, they have a fundamental role in the dynamical definition of the deterministic chaos, as it will be seen in section III. The decay time of the feedback correlation $\Gamma_f(\theta)$ can be derived directly from a MacLaurin expansion of $\Gamma_f(\theta)$. It involves successive derivatives $C_x^{(2k)}(0)$ which are calculated by derivations and autocorrelations of each side of Eq. 6. it gives

$$\Gamma_f(\theta) \propto 1 - \left(\frac{aR\,\theta}{2}\right)^2 - \frac{1}{3}\left(\frac{aR\,\theta}{2}\right)^4 + \ldots \quad (16)$$

which displays the memory time of the feedback

$$\delta = 2 / aR \quad (17)$$

introduced above to show the Gaussian character of $y(t)$. The expression in Eq. (17) agrees very well with the half widths (defined at $\Gamma_f(t) = 1/e$) of the numerical correlations shown in Fig. 10 ; it also confirms our rough estimate in Eq. 14, up to a factor 2. An experimental curve Γ_f is shown in Fig. 11 corresponding to $aR = 5.3$, $f(x) = 1-R\cos x$. The width of the central peak as well as the secondary negative bump at $\theta = r$ agree both with the theory. The experimental widths of Γ_f are shown in Fig. 12.

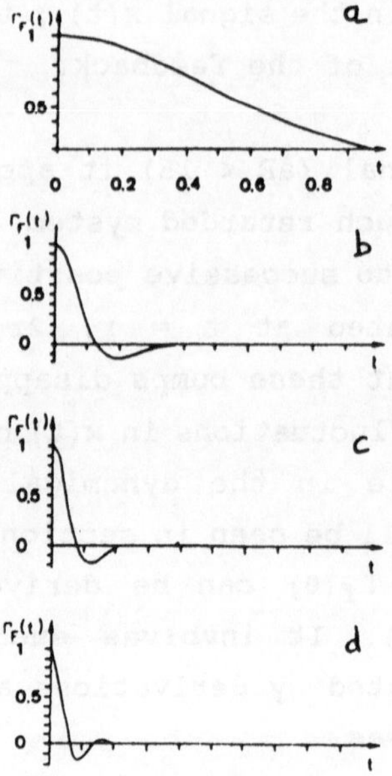

Fig. 10 - Normalized correlation function $\Gamma_f(t) \propto R^2 < \sin ky(0) \sin ky(t) >$ for $R = 2/3$. (a) $a = 6$, $r = 8.7$; (b) $a = 36$, $r = 5$; (c) $a = 60$, $r = 5$; (d) $a = 100$, $r = 5$.

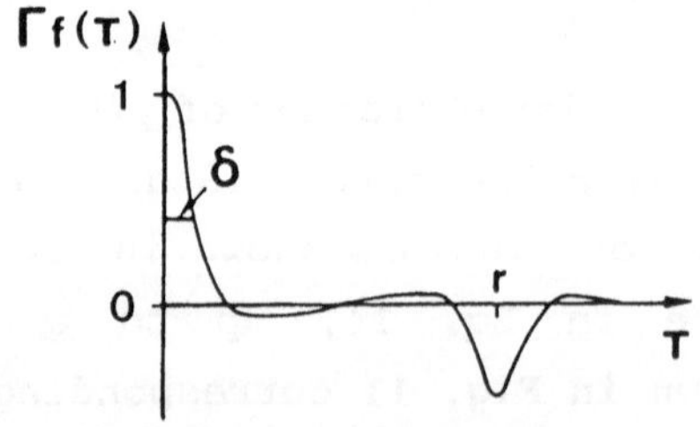

Fig. 11 - Experimental correlation function $\Gamma_f(T)$ of the feedback $f(X) = 1 - R \cos X$ for $aR = 5.3$

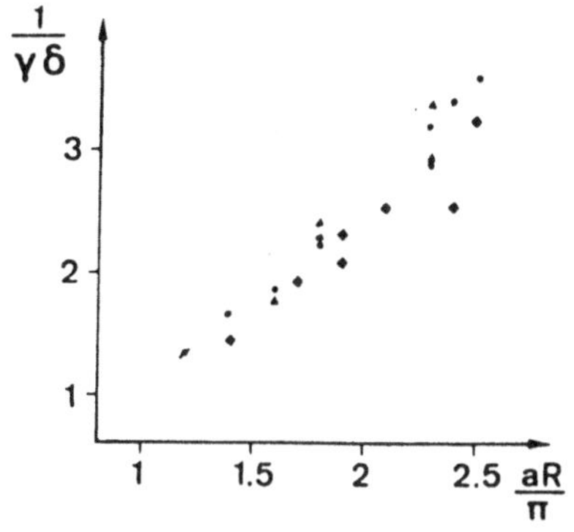

Fig. 12 - Widths of Γ_f (defined at 1/e). Circles correspond to theory ; triangles correspond to measurement from time signals f(x(t)) ; diamond corresponds to power spectrum transform measurements. Uncertainly in the abscissa is about 5 %.

In conclusion, the 2D statistical study has displayed <u>four increasing time scales</u> when aR » 1 and r » γ^{-1}, (γ^{-1} is the dissipative time)

$$\gamma^{-1}\delta = \frac{2}{aR}, \quad \gamma^{-1}t_c = \sqrt{\frac{2}{aR}}, \quad \gamma^{-1}, \quad r\,\gamma^{-1} \tag{18}$$

<u>in agreement with numerics and experiments</u>.

III. CORRELATION DECAY AND LYAPUNOV EXPONENTS

In Section II the decay times in the correlation function Γ_x and Γ_f have been shown to be independent of the delay ; on the other hand the usual deterministic analysis leading to the Lyapunov spectrum has already given some invariant quantities in other retarded systems[2,3]. While it is an involved task[6,7] in any system, we will show that the decay time of Γ_f has a direct connection with the Lyapunov exponents, moreover this property will be generalized to other retarded systems, and set as a conjecture.

From the set of Lyapunov exponents $\{\lambda_i\}$, one defines two quantities, the Lyapunov dimension $d_L = j + |\lambda_{j+1}|^{-1} / \Sigma\ \lambda_i$ ($i \leq j$ where j is the largest integer such that $\Sigma\ \lambda_j \geq 0$), which is the dimension of an hypervolume which neither contracts nor expends with time ; and the entropy $h = \Sigma\ \lambda_i$ defined as the sum of <u>positive</u> Lyapunov exponents, which is the exponential expansion rate of an hypervolume constructed only on the expanding eigendirections (or the decay rate of the information contained in a measurement, in the language of information theory).

Farmer[2] has observed in the scalar retarded Mackey-Glass equation that <u>d_L increases linearly with the delay r</u>, while <u>h is nearly independent of r</u>, when the delay is large enough with respect to the dissipative time γ^{-1}. Later on the same was observed for the all-optical plane-wave ring cavity[3]. Finally for the electro-optical device, our numerical derivations of the Lyapunov spectrum gives[1]

$$d_L = 0.45\ aR\ r, \tag{19}$$

$$h = aR / 10 - 0.4, \tag{20}$$

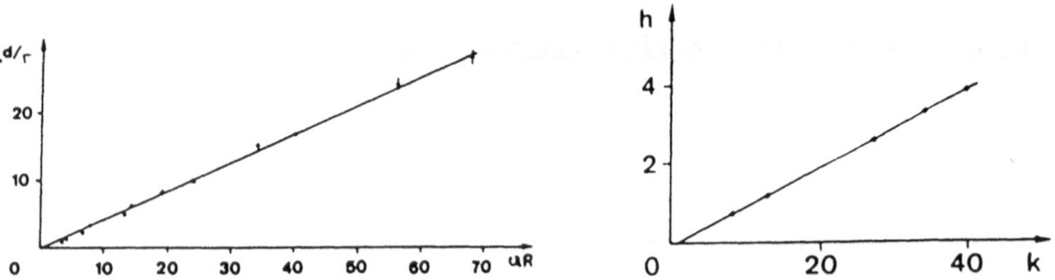

Fig. 13 - Left : Lyapunov dimension d of the chaotic solution. The ratio d/r is plotted as function of the parameter aR = k. Right : Entropy h.

as shown in Fig. 13. Both simple expressions in Eqs. (19, 20) can easily be expressed in terms of the decay time of Γ_f (Eq. 17)

$$d_L \simeq r / \delta \qquad (21)$$

$$h^{-1} \simeq 5\,\delta \qquad (22)$$

which gives some new insight in the dynamics of the hybrid : the interaction between the system and its feedback can be seen as a set of 'kicks' of mean duration δ, after each time interval δ the system is ready to undergo a new kick of the feedback and so on during the time interval r. Loosely speaking, the r/δ kicks are independent events, it follows that Eq. (21) interprets the number of independant kicks during a delay r as the effective number of degrees of freedom (d_L) in a retarded system. On the other hand, the expansion time h^{-1} of a small hypervolume constructed upon the dilating directions is seen to be equal to a few kicks duration, as seen in Eq. (22).

These conclusions on the hybrid were investigated in the two other systems mentionned above, the ring cavity and the Mackey-Glass equation. Equation (21) was found to be still valid in both systems ; surprisingly this later relation is even valid for delay of order unity (where the dimension d_L is no longer proprotional to the delay, other parameters being kept constant). For example the attractor in Fig. 14 (ring cavity device), has a dimension $d_L = 2.58$, and a correlation time $\delta = 0.35$ for a delay $r = 0.9$, while $d_L = 3.85$, $\delta = 0.26$ for $r = 1$, then both correspond to $d_L = r/\delta$.

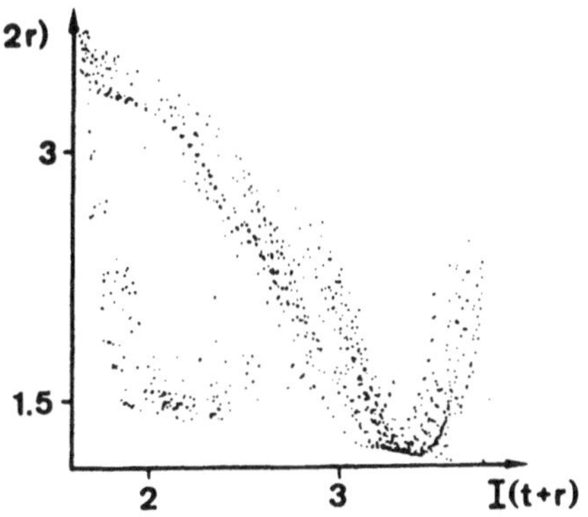

Fig. 14 - Poincarré section I(t+2r) as a function of I(t+r) for I(t) = 1.5. With notation of Reference 3, Eqs. (2-4), parameters are $\alpha l = 4$, $\alpha l\,\Delta/2 = 6\pi$, $E_0 = 1.7$, $R = 0.95$, delay $r = 0.9$.

The relation between the entropy and the correlation time is of the same type as Eq. (22) in the two other systems, one obtains

$$h^{-1} = 10\ \delta \qquad (23)$$

Finally the statistical and deterministic studies meet themselves on the following double conjecture for retarded equations : The dimension d_L is nearly equal to the delay divided by the decay time of the feedback δ, the information loss time h^{-1} is few times this memory time δ (the factor depending on the system).

REFERENCES

1 B. Dorizzi, B. Grammaticos, M. Le Berre, Y. Pomeau, E. Ressayre, A. Tallet, Phys. Rev. A 35, 328 (1987).

2 J.D. Farmer, Physica (Amsterdam) 4D, 368 (1982).

3 M. Le Berre, E. Ressayre, A. Tallet and H.M. Gibbs, Phys. Rev. Lett. 56, 274 (1986).

4 H.M. Gibbs, F.A. Hopf, D.L. Kaplan and R.L. Shoemaker, Phys. Rev. Lett. 46, 474 (1981) ; Phys. Rev. A25, 2172 (1982). For an extensive bibliography in optics, see the book of H.M. Gibbs, 'Optical Bistability : Controlling Light with Light', Academic Press, New York, 1985.

5 K. Ikeda and O. Akimoto, in Coherence and Quantum Optics V, Rochester, 1983, ed. by L. Mandel and E. Wolf, (Plenum, N.Y. 1984).

6 D. Ruelle, Phys. Rev. Lett. 56, 405 (1986) ; J. Stat. Phys. 44, 281 (1986)

7 R. Badii, K. Heinzelmann, P.F. Meier and A. Politi, Phys. Rev. 37, 1323 (1988).

Non-local and non-linear problems in the physics of disordered media

Etienne Guyon, Stéphane Roux, Alex Hansen
Laboratoire d'Hydrodynamique et Mécanique Physique, UA CNRS 857
E.S.P.C.I., 10 rue Vauquelin 75231 Paris Cédex 05, France

Abstract

Many concepts issued from the field of statistical physics have enriched our understanding of the physical properties of random materials. This text outlines some of these recent developments. We will recall some basic characteristics of percolation and will extend them to problems displaying non-local ordering and to problems where the existence of threshold behaviors at the local scale lead to strong heterogeneous behaviors in systems with an initial small disorder.

1 Introduction

The present book is centered around the study of dynamical systems, such as lasers and convective instabilities, far from equilibrium. Due to the non-linear couplings involved, one observes spatio-temporal patterns often unsteady and chaotic which are out of scale with the basic microscopic structures present in equilibrium or near-equilibrium conditions. The physics of heterogeneous materials in its classical form (1) as exemplified by the Clausius-Mossotti treatment of the dielectric constant of a composite, made of spherical inclusions with a dielectric constant different from the matrix they are imbedded into, is as far as can be from the project of the present book. In this particular example, one can define a macroscopic dielectric constant of an homogeneous equivalent material by simple averaging of the local dielectric behavior, in the spirit of mean field treatments and no structural property is included. A lot of work has been devoted to the extension of self-consistent approximations to include spatial correlations and, therefore, to introduce geometrical and structural information. However, very quickly, the procedure becomes untractable.

We refer to "large disorder" to characterize materials in which the existence of a multiplicity of length scales (like would be the case if the dielectric composite was composed of an assembly of inclusions presenting a very large distribution of sizes) precludes the possibility to perform local averaging. The tools of percolation (2) and of fractals(3) have been used with success in such cases to various problems (electrical, hydrodynamic, mechanical) (ref.4; courses 1,2,3). We will apply and extend these approaches here by considering several classes of large disorder effects appearing even in systems with small heterogeneities.

- The first one is characterized by the fact that the ordered state results from non-local interactions, this being due to the fact that only partial transmission of order takes place along a single continuous channel unlike ordinary percolation.
- The second one deals with systems where the local laws are non linear. In a given range of geometrical parameters, the macroscopic laws do not reflect directly these non linearities but rather the progressive change of the global geometry due to the effect of the local non linearities. This change can be reversible, as in the mechanics of packed grains where the number of inter-granular contacts increases with applied stresses, or irreversible as in the fracture of a material in which there is an amplification of the local disorder.

Let us recall briefly some characteristic features of percolation and fractals using a simple experiment due to Lenormand (5): A model porous medium is made by etching a transparent flat plate according to a periodic pattern of channels and then cover it by an upper flat plate. The widths of the channels have been distributed at random. The resulting geometry is that of a periodic continuous network of pores. The medium is initially filled with a fluid wetting the pores. It is"invaded" next by a second, immiscible, fluid which does not wet the pores. In quasistatic conditions, (more precisely if the flow rate v is such that the capillary number defined as $Ca = v\eta / \gamma$, where η is a fluid viscosity, γ a surface tension, is much smaller than 1), the non-wetting fluid can be injected in the medium under a large enough pressure to overcome the capillary pressure difference. More precisely, for a given excess pressure Δp_o, the only pores to be invaded will be those in contact with the injection surface and which have a radius $r > r_o$, where

$$\Delta p_o = \frac{2\gamma}{r_o \cos\theta}$$

θ is the contact angle of the meniscus separating the two phases and the pore wall. For large enough Δp_o , only finite clusters of pores in contact with the injection wall will be invaded. They grow as Δp_o increases (Fig. 1a). There is a critical value Δp_c called breakthrough pressure such that, for the first time, a continuous path of non wetting fluid is established across the medium. This problem is a simple and direct application of invasion percolation of bonds : The bonds are labelled by the random variable r. By increasing Δp_o (or decreasing r_o), one increases progressively the number of active (to the non-wetting fluid) bonds. Δp_c (and r_c) corresponds to the percolation threshold where there is a transition from a short range regime where only finite clusters (those attached to the injection face) exist to one where long-range connectivity order is created due to the formation of the breakthrough path. In an infinite network, for $\Delta p_o > \Delta p_c$ (or $r_o < r_c$) there is a finite probability $P(\Delta p_o)$ that a given site far away from the wall has been invaded by the non-wetting fluid. $P(\Delta p_o)$ is the order parameter associated with the control variable p and is such that

$$P(\Delta p_o) \propto (\Delta p_o - \Delta p_c)^\beta$$

β (= 5/36 in 2 D ; ≈ 0,42 in 3 D) is the critical exponent for percolation and is universal (independen of the lattice or disorder) . There also exists a singular correlation length

$$\xi(\Delta p) \propto [\Delta p_o - \Delta p_c]^{-\nu}$$

which measures the largest distance of the invasion from the inlet face for $\Delta p < \Delta p_c$ ($\nu = 4/3$ in 2D; ≈ 0.89 in 3 D). At threshold, ξ is infinite and the percolation infinite cluster is fractal at all scales (because there is no longer any finite reference length). For $\Delta p_o > \Delta p_c$, the infinite cluster is fractal only up to a length scale $\xi(\Delta p_o)$ and is seen as a homogeneous structure above this length. The value of the fractal dimension d_F can be obtained by a simple cross-over argument by evaluating the number of bonds of the infinite clusters over a reference volume ξ^d in two different ways :

- as the lowest limit of a homogeneous problem : $\xi^d . P(\Delta p_o)$
- as the highest limit of a fractal one ξ^{d_F}, by direct definition of the fractal dimension. It follows from a finite size scaling argument that $d_F = d - \beta / \nu$ ($d_F = 1.89$ in 2 D ; ≈ 2.51 in 3 D) as has been checked in real and numerical experiments(2).

The geometrical and transport properties of such percolation problems have been investigated in great detail. They are simple because the connectivity property is local. It has no dynamics in it and is, in fact, an equilibrium statistical problem.

Let us consider how the opposite problem where the invading phase is injected under high flow rate such that the capillary number Ca is no longer much smaller than unity. Capillary effects are no longer dominant and the penetration is controlled by the ratio of the viscosities of the invader (η_1) and invaded (η_2) phase.

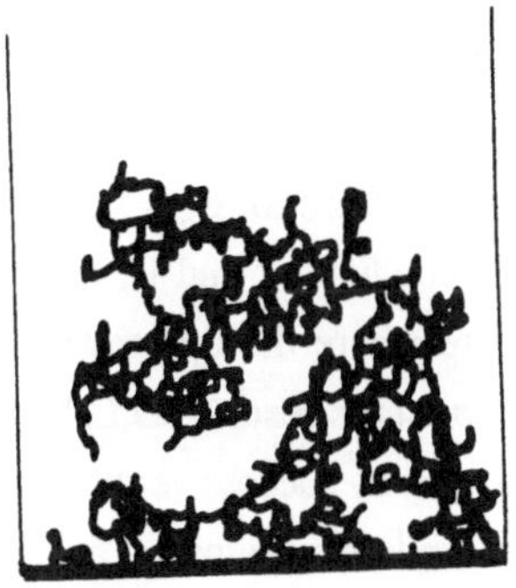

Figure 1 a — Invasion percolation (Ca <<1) in a model porous medium — Figure 1 b D. L. A.

The figure 1b has been obtained for a large value of the ratio η_2/η_1 and shows a quite different pattern from figure 1a (in particular its fractal dimension $d_F \approx 1.7$ is smaller than for invasion percolation)(6). This problem corresponds to the so-called diffusion limited agregation (D.L.A.) model (7). The common structure of both problems can be established in the following way: If η_1 is small, we can neglect the pressure gradients due to viscous flow in phase 1 . At the boundary with phase 2, the pressure of the invading fluid p_1 is constant and is equal to the injection pressure. On the other hand, in the invaded phase 2, the Darcy law, which expresses the proportionality between the average flow rate and the pressure gradient, applies:

$$v = -\frac{k}{\eta_2}\,\mathbf{grad}\ p_2$$

where k is the geometry-dependent permeability of the porous medium. From the equation continuity of the flow div $v = 0$, it follows that $\Delta p_2 = 0$.

The progression of the invader can thus be framed into an electrostatic problem (figure 2 a).

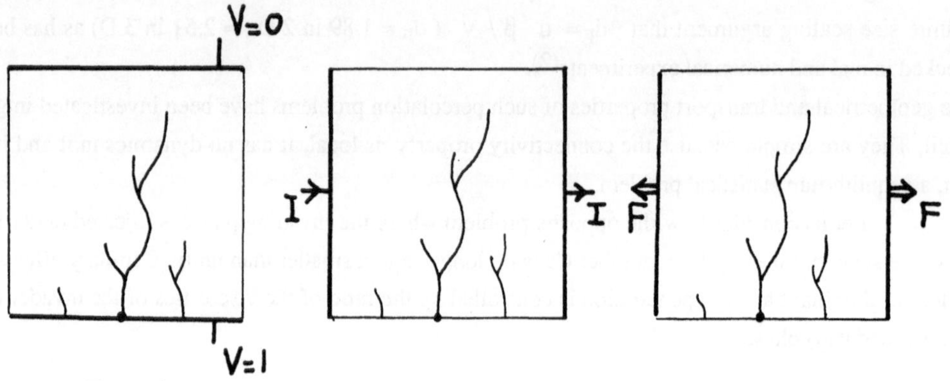

Figure 2 a   Figure 2 b   Figure 2 c

A voltage difference V is applied between two electrodes, a growing one (or cathode) V and a fixed one at a potential V=0. In conditions where one can neglect convection currents (if enough ionic changes are present in the solution to screen the field), the growth rate of the cathode is equivalent to the progression of the front in diphasic flow and is proportional locally at the surface to $\mathbf{E_s} = -\,\mathbf{grad}\ V$. (Note that, in the model, the growing tree is assumed to be perfectly conducting and is, thus, an equipotential). This leads to the fractal structure observed on the figure where the tip growth and associated shielding properties of the inner lying part are classical electrostatic results. D.L.A was first observed and explained in the different problem of growth of aggregates by successive sticking of individual Brownian particles. It has been observed in many situations of random growths in Laplacian fields(7).

2 Central force percolation

2.1 Electrical and mechanical percolation

The basic characteristics of percolation as introduced through the example of immiscible two-phase flow in a porous medium at low flow rate have direct electrical analogs. The pores filled with non-wetting phase are replaced by conductors, those filled with the wetting one by insulators. At percolation threshold, there is a transition from an insulating to conducting phase such that just above threshold the conductance of the lattice

$$\sigma(p) \propto (p - p_c)^t \qquad (t \approx 1.3 \text{ in } 2\,D;\ \approx 2.0 \text{ in } 3\,D)$$

where p is the proportion of conducting bonds. The same critical exponent t would be obtained in principle if the porous medium had been invaded by conducting Hg injected in an insulating porous phase. The hydraulic conductivity, or permeability of the injected phase should also show a similar critical behavior with the same exponent t.

A mechanical analog of the problem is an incomplete lattice of springs (instead of resistors) (8). If one also introduces angular rigidity at each node between two active bonds, the electrical percolation threshold and the mechanical rigidity threshold are identical and correspond to the loss of long-range connectivity. It also follows that all other geometrical characteristics of both problems are identical. One can also argue that the critical behavior of the elastic modulus E , such that

$$E \propto (p - p_c)^{\tau}$$

is directly related to the electrical one (more precisely , $\tau = t + 2\nu$, the presence of the correlation length exponent ν being due to the fact that the elasticity is dominated by propagation of moments which takes place over the characteristic length ξ).

2.2 Central force percolation

We will now consider the more original case of central-force percolation which is characterized by the fact that connectivity is not enough to insure long-range order. We consider a two-dimensional structure made of nodes joined by elastic bars (or springs) freely rotating around their endpoints. The problem of the rigidity of a such a mechanical system as shown on figure has been adressed quite early by Maxwell and Cremona for deterministic structures and is still a subject of interest in mathematics(9).

Following the initial work of Feng and Sen on central-force percolation (10), we consider a regular triangular lattice of springs rotating freely at their nodes. In the random dilution model, a proportion (1-p) of randomly distributed bonds are missing while another proportion p is present. In the dual problem of random reinforcement, a proportion p of bars is infinitely rigid while the other ones have a finite elastic constant.

In the first case, the rigidity goes to zero at a well defined p* in the limit of an infinite lattice with a power law characterized by a critical index f:

$$E \propto (p - p^*)^f$$

Similarly, in the random reinforcement problem, the elastic modulus diverges as p tends to p* from below as

$$E \propto (p^* - p)^{-g}$$

p* is larger than the usual connectivity threshold p_c and cannot be obtained from simple geometrical arguments. For this reason, and because of the numerical difficulties encountered in this case, there has been some controversies concerning the value of p* and, thereby, of the critical exponents(10,11,12). The determination of the rigid or non-rigid character of a randomly diluted lattice is a non-local problem. A simple way to rephrase this property is to note that a bond transmits only a partial information : if the displacement at one site is imposed, then, one bond away from the site, the displacement is not completely determined; only its component along the axis of the bond is known. This is in contrast with previously introduced systems with angular elasticity. In this case where rigidity and connectivity are equivalent, the complete information (displacement and orientation) could be transmitted through a single bond. The basic requirement of invariance of the energy of the system under rigid displacements imposes the introduction of additional degrees of freedom to the sites : when angular elasticity is present, one needs both the location and orientation of each bond: a total of 3 degrees of freedom in 2D instead of 2 for central force problems. The displacement alone at the level of one site is required for central-force systems.

Figure 3

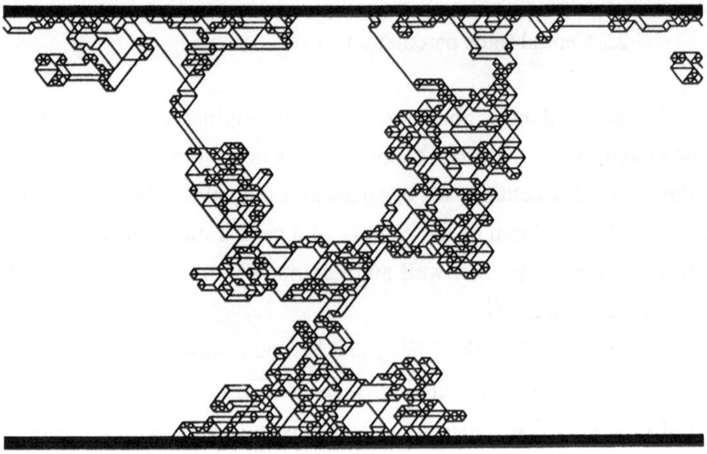

What is the influence of the non-local character of the rigidity on the critical behavior? The only informations available today come from numerical simulations of the problem. The situation is not completely clear. However the most recent results(11, 12) obtained for differents cases and geometries suggest the following picture: most of the critical exponents measured for central-force percolation have their analog counterpart in standard percolation (e.g, f is the counterpart of τ). The numerical values of the exponents are found to be indistinguishable for both types of percolation problems. This holds for the exponents governing the elastic behavior (such as f and g) but also for the structure of the subset of bonds of the infinite cluster which carry a non-zero force (backbone) whose fractal dimension d_B is again identical to that of usual percolation. An example of such a backbone structure is shown on Figure 3 next to the corresponding one for usual percolation. In fact, the complete distribution of forces in the backbone is expected to be universal. It is characterized by a continuous function and not a single number (a property called multifractality)(13). This continuous function seems to be related to the analog spectrum of scalar transport problems, as also conjectured for elasticity of systems with angular elasticity in the usual percolation framework. Some additional critical exponents were also obtained which match perfectly their value for usual percolation.

Therefore, we presently conclude that non-locality in central force percolation has no influence at all on the critical behavior. A reason to believe this result is to imagine that there exists an intermediate length scale at which angular elasticity is restored. Thus, under a real space renormalisation group treatment, one would find that the central force percolation problem falls into the universality class of usual percolation.

However, going a little closer into the details, the situation is richer than it appears at first sight. For central-force percolation, new specific features arise in between the connectivity and rigidity thresholds: Wang and Brooks-Harris[14] have conjectured the existence of a new phase just below the rigidity threshold, p^*, where the system is rigid for applied torques (splay-rigidity) but not for forces (all bonds are free to be translated provided they are far away from a border)- think for instance at an industrial design ruler which may be moved but not rotated. This splay-rigid phase exists only in a very narrow region below p^*, and disappears at a second threshold p^{**}. It is at present not ruled out that p^* and p^{**} might be identical (limit of infinite systems) and numerical simulations are still controversial.

III 3) Extensions

Between the connectivity and the rigidity threshold, the central force percolation lattice can undergo large deformations. This non-linear regime has not been much studied[15,16]. For a given value of p between p_c and p^*, there exists a given displacement that the lattice can undergo without becoming rigid. Let us call $\lambda(p)$ the maximal extension d that the lattice can support. In the (p,d) plane, the curve $d = \lambda(p)$ separates two phases: one floppy, and one rigid.

For a given value p such that $p_c < p < p^*$, the central-force correlation length $\xi'(p)$ gives roughly speaking, the size of the largest rigid clusters in the lattice. $\xi'(p)$ diverges as p tends to p^* as

$$\xi'(p) \propto (p^*-p)^{-\nu'}$$

where ν' is the central-force percolation exponent. Above the size $\xi'(p)$, the clusters are connected together ($p>p_c$) but they are free to move with respect to each other. We suggest, in addition, that the structure of the lattice for length scales larger than ξ' can be modelled by a usual percolation problem. This second percolation problem is in general far from its critical threshold, unless p approaches p_c. Moreover the variation of $\xi'(p)$ is smooth in the vicinity of p_c ; thus, we expect that the second correlation length $\xi''(p)$, defined for the second percolation problem diverges, as p tends to p_c, according to

$$\xi''(p) \propto (p-p_c)^{-\nu}$$

where ν is the usual percolation correlation length exponent. Therefore, for a given value of p, we expect three different regimes according to the size L of the system considered.

A: $L < \xi'(p)$: Central-force percolation regime

B: $\xi'(p) < L < \xi''(p)$: Usual percolation regime

C: $\xi''(p) < L$: Disordered but homogeneous system

The two last regimes are particularily interesting. Starting with a displacement $d=\lambda(p)$, only a very tenuous subpart of the system will be stretched. Increasing the applied force, this subsystem will get denser and denser. It will give an additional non-linear contribution to the non linearity resulting from large deformations.

The picture presented above is just a rough qualitative speculative view, and a lot of work is to be done in this field in order to explore the complete phase diagram. Such singular behaviors are presently studied experimentally as a direct numerical study would be very delicate, since it would combine the difficulties of non-linear problems with the extreme sensitivity of critical phenomena. Some experimental results on simple models assemblies of bars have been obtained very recently by Ben Ayad and Gilabert (15). They concern mainly the behavior of $\lambda(p)$ in the range $p_c < p < p^*$. In particular the observed scaling of $\lambda(p)$:

$$\lambda(p) \propto (p^*-p)^a \qquad \text{close to } p^*$$

$$\lambda(p) \propto (p-p_c)^{-b} \qquad \text{close to } p_c$$

is in reasonable agreement with the above-mentioned picture.

A molecular dynamic study of the rupture problem of atoms interacting via a central-force Lennard-Jones potential first suggested the study of this problem (16). We also note the work of Thorpe (17) using a central-force lattice with bonds in tension (their equilibrium length being less than the lattice spacing). The system is maintained rigidly by the edges like a tennis racket. In this problem again, the mechanical threshold is that of continuity. The stretching deliberately locates the system in a non-critical region and allows to construct a self-consistent model to account for the tangent elastic properties observed.

The problems of stretched lattices are of importance in practical problems such as unwoven textiles. The individual bonds acquire rigidity only when they are stretched . An opposite case is that of packed granular arrays(18,19). The grain are kept in contact under pressure and are free to rotate locally as required for central force percolation to apply. The situation is quite the opposite from the previous example because the system gains rigidity under pressure (like in vacuum-packed coffee) and not in extension. We will consider in the next paragraph the "diode"-like non-linear effects we have just described.

Another open non linear problem of interest is that of buckling. At rigidity threshold, there exist long straight lines of bonds which contribute to the rigidity. It is well known that under an infinitesimal compressive force the structure will buckle. Consequently, these bonds will no longer insure rigidity. Such an effect would be absent in the generic problem of a triangular lattice with disordered bond angles and lengths.

3 Non linear effects

3.1 Reversible effects

We consider now the "diode"-like character of the individual contacts , i.e. the absence of force in one direction of the applied displacement..

a) Experiments

The mechanical properties of random packings of parallel cylinders having nearly the same diameter submitted to a uniaxial pressure have been studied together with their photoelastic properties and illustrate the content of the present chapter.

-The observation of the array between crossed polaroids shows that the strained cylinders form a subset of branched and continuous lines extending between the planes where the deformation is imposed(20). A large fraction of the cylinders is unstrained. When the stress increases, the fraction of strained cylinders also increases. The structure is strongly reminiscent of that described by Kardar, Parisi and Zhang(21) which applies to optimal lines for polymers in a random potential . The correspondence is not fortuitous although rather involved. The photoelastic lines correspond to minimal continuous lines drawn across cylinders with fluctuating diameters or positions.

-The macroscopic mechanical response is strongly non linear (22) and the relation between the force F and the displacement Δh is observed to be

$$F/F_0 = (\Delta h / h_0)^m$$

The value of the exponent $m \cong 3.5$ is much larger than the value for an individual contact between two cylinders (in the so-called Hertz problem between two spheres, the local value is 3/2). The strong non-linearity originates from the increase of the number of contacts with increasing force which causes the progressive hardening of the structure.The value of m is not affected by the removal of a fraction of unstressed cylinders (defined by photoelastic observations) but, on the other hand, decreases very strongly if the cylinders removed were highly stressed ones(22). Such an experiment supports the determinant role of an active mechanical sublattice.

The same non-linear macroscopic response has been observed when studying the conductivity of such a system (with conducting cylinders) under pressure: because stressed contacts are also good electrical contacts (according to Hertz law, their area of contact increases with force), the "active" mechanical sublattice carries the largest part of the current.

A numerical model of this problem has been studied recently. It consists in an array of parallel cylinders of slightly fluctuating radii(19). When two cylinders are in contact, they interact according to Hertz law. At the beginning of the simulation, a large compression is imposed so that all cylinders touch their neighbors. Then, progressively, the external force is released, and the force-displacement characteristic is recorded. The average behavior shows a power-law regime similar to the one observed experimentally. This regime was checked to correspond to a progresive decrease of the number of stressed cylinders that participated to the rigidity of the array. In addition, this transition regime did not appear to be very sensitive on the local behavior : if instead of Hertz law, a linear relation was considered for local contacts, then the transition regime was qualitatively unchanged.

These results, together with the numerical difficulties inherent to non-linear problems, suggest to study in full detail a simpler case, which already display similar features: the study of an array of electrical diodes, which share the asymmetry in the response characteristic with the above problem.

b) Diode networks(23

We use this term with a different meaning as initially introduced by Redner[24] (This author was dealing with bonds where the current could pass only in one direction. Although directedness can be included in the present model, it is not *a priori* related to the properties discussed below.

We start with a regular lattice of non linear resistors having piece-wise non linear i(v) characteristics such as in the example of figure 4a. In this example, the voltage thresholds v_g, such that the conductance is zero below v_g and constant above it, are distributed at random. The overall characteristics on figure 4b shows 3 regimes:

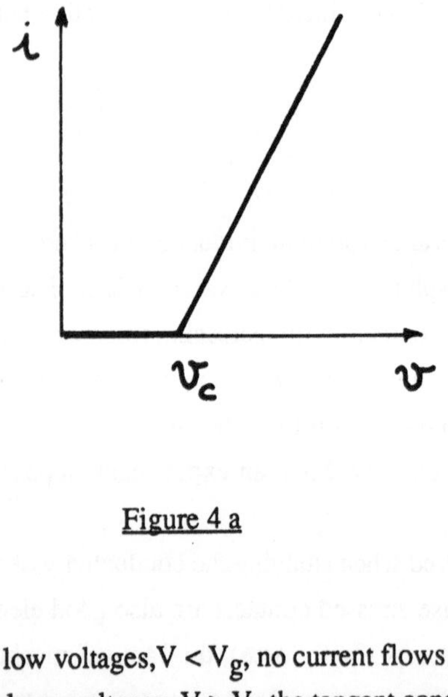

Figure 4 a

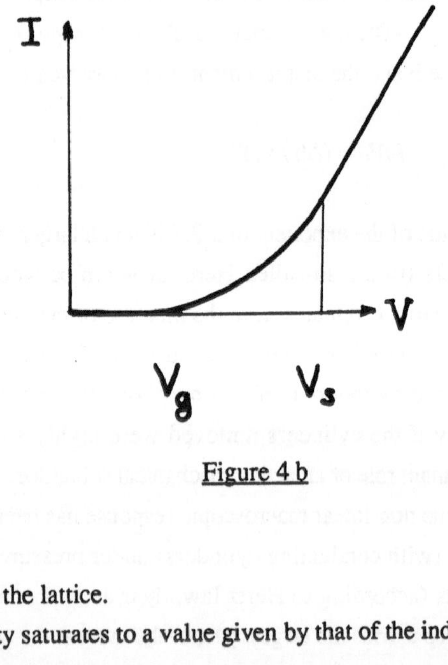

Figure 4 b

-At low voltages, $V < V_g$, no current flows across the lattice.

-For large voltages, $V > V_s$ the tangent conductivity saturates to a value given by that of the individual element in their passing state.

-The intermediate regime is the one we are interested in where the non-linear behavior is due to the progressive enrichment of the lattice of active bonds as V increases. In this regime, a power-law behavior has been observed in numerical simulations of 2D systems [24] for a uniform distibution of thresholds between 0 and 1.

$$I \propto (V-V_g)^{\alpha}$$

with $\alpha = 2.0$. (The bounds of the interval on which the thresholds are sampled in are not restrictive and a similar behavior holds for any interval.) The power law extends over the all range $V_g<V<V_s$ and not just in a narrow critical regime. This suggests the use of mean-field type arguments: A small increase of V by dV in this range is felt uniformly across all bonds. If the thresholds v_g of the bonds are distributed uniformly, the number of bonds which become actived is proportional to dV . In a mean field treatment, the conductance Σ also increases as dn (and dV) and the current as ΣdV. Thus $I \propto (V-V_g)^2$.
The same type of behavior has been also found in an analog experiment using a square lattice of Zener diodes(25) (the directed character turned out to be unimportant for the overall I(V) characteristics all the diodes were assemblied in the same half plane direction). It is also expected in the problem of a porous medium filled with bentonite (a threshold fluid such that no flow takes place in a channel if the pressure gradient is below a certain threshold).
V_g can be exactly computed beforehand, by noting, that for a path P to be conducting, the potential drop between its two ends V must be larger that the sum of the potential gaps v_{gi} of all its bonds i. Therefore V_g is thus the minimum over all paths of this sum (26):

$$V_g = \min_P \left(\Sigma_{i\in P}\, v_{gi}\right)$$

The criterion can be compared with that for ordinary percolation as presented in part 1. In this case, we also consider all possible paths P and look for the maximum value of the local parameter r_i on this path. The threshold is given by the value r_c such that

$$r_c = \min_P \left(\max_{i\in P}\, r_i \right)$$

which corresponds to a different (extreme) weighting on the bonds of the continuous paths.

c) Generalization to other non linear laws

The above problem can be put in correspondence with a different type of local law: the individual bonds have now a uniform and constant resistance when the voltage drop they are subjected to is less than a random threshold uniformly distributed between zero and one; they have zero tangent resistance above it (fig.5). The correspondence is similar to that between the classical insulator-conductor problem and superconductor-conductor one in percolation and can be made more precise using an argument expressed by Straley (27) and using properties of dual lattices. As one increases the applied voltage V, some bonds acquire a zero resistance, thus reducing the over all resistance until a continuous path of zero resistance bonds spans the lattice at V_g. A mean-field argument also predicts a power law behavior below V_g given by

$$(I_g - I) \propto (V_g - V)^{\beta}$$

where I_g and V_g are the intensity and voltage drop at the point where the system has zero tangent resistance with $\beta = 1/2$.

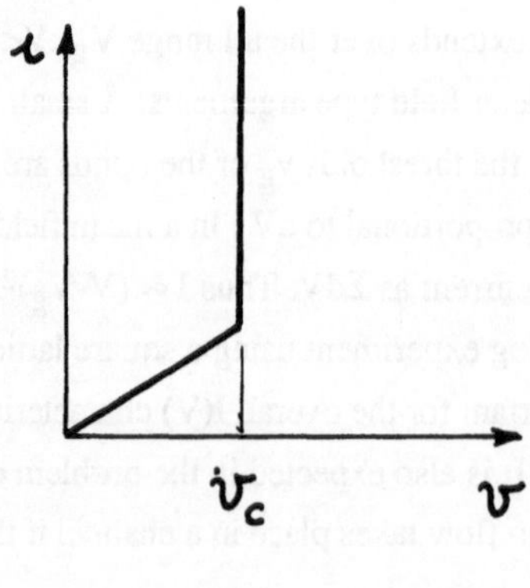

Figure 5 a

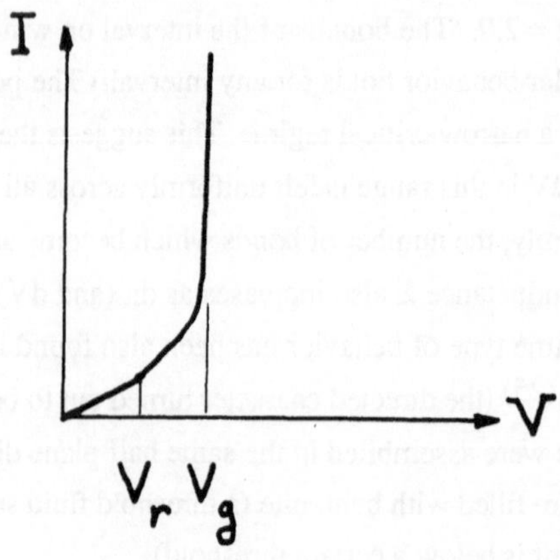

Figure 5 b

Two other problems can be constructed by simply exchanging the roles of current and voltage in the two above problems. These situations are representative of real cases of material science; for example by exchanging the role of current with force and that of voltage with displacement in figure 4 one can represent the transition from purely elastic deformation to plastic flow in an heterogeneous material as the stress increases. Of particular present interest is the application to the flux flow characteristics of superconductors in a magnetic field where the smooth transition has been attributed to a distribution of local pinning forces in the same spirit as the present work[28].

3.2. Irreversible behavior [29,30,31]

a)breaking of model lattices

We consider now problems in which the non linearities lead to an irreversible modification of the elements and we will review briefly some recent results in this new subject. Dielectric breakdown and fractures are two examples of such problems. The rupture of heterogeneous materials induced by a distribution of local breaking strengths provides a nice example for mechanical studies. One can also find electrical analog examples (failure of a network) or porous medium ones (certain problems of plugging by material carried by flow). The dual problem of creation of shorts in a lattice has also been studied and has also many applications. A network of fuses is a simple electrical model which displays a number of characteristic features met in real materials. It makes use of a regular lattice of fuses characterised by their resistances r and their critical current i_c above which the resistance is irreversibly open. Three classes of disorder can be introduced:

-suppression of a fraction of bonds at random (random depletion)

-random distribution of the resistances in a full lattice of fuses (random conductances)

-random variation of the critical currents i_c (random fuses)

In the first two situations, there is a distribution of currents which determines which bond will break first as the applied voltage V increases even if all the i_c are equal.

-In the limit of a large disorder of the critical currents, the bonds will break sequentially according to the list of increasing values of i_c (an infinite disorder would correspond to a probability distribution function varying as 1/x around x=0 where x is the value of the breaking current distributed between 0 and 1). This is very similar to a percolation rule apart from the fact that the bonds cut at random must belong to the backbone which carries the current. The modification of the current distribution induced by a first broken bond will lead to the increase of the current distribution of currents in the neighboring bonds(29).

-If the disorder is infinitely small, once an initial crack has been initiated for a large enough current there will be an immediate propagation of it in a direction transverse to the applied field . If we consider a randomly diluted lattice, the largest crack present in the distribution as given by the tail of the statistical distribution of clusters will lead to breaking. This leads to non-trivial size dependence of the breaking strengths(31). However screening and amplification effects for the crack propagation from the tip of a preeexisting crack due to nearby clusters are important as soon as the dilution of initial cracks is not small (32).

-Coming back to the first case of large initial disorder in the bond breaking strengths, when there will be enough bonds broken the distribution of currents (or forces) will be strongly non uniform and the bonds to be broken will no longer follow the natural order of bond strengths. This leads to a rapid and irreversible amplification of the disorder via the propagation of a large crack without modification of the other ones. This is indeed observed in simulation experiments but a full and controlled study on real systems remains to be done.

An interesting new result has been obtained in a series of numerical simulation of the breaking of 2 dimensional lattices of size L x L both for the conducting fuse problem and for the mechanical ones (with and without angular elasticity)(33). Different distributions of breaking strengths were also used. It was found that, in all cases, the current-voltage (or force-current) characteristics could be expressed with a single scaling expression

$$I/L^{\alpha} = f\,(V/L^{\beta})$$

with $\alpha = \beta \approx 0.75$. The values of the size scaling exponents α and β indicates that the breaking strengths decrease for largest samples is due to the effect of the larger fluctuations for largest sizes which is beyond all classical treatments of damage. It is also a well-known experimental fact by mechanicians of rupture. We intend to check this type of law on three dimensional materials to which the numerical approach will also be extended.

b)<u>D.L.A. type fracture</u>

This problem has been introduced in part 1. It also can be applied to the study of the propagation of a single random crack in a random medium. This process is best obtained in stress-corrosion processes in which the progression of a crack is assisted by the presence of a chemical agent progressing to the tip of the crack and lowering the surface energy needed to break additional bonds.The similarity between the figures obtained in such problems and classical fractal patterns as the D.L.A. ones introduced in part one suggests a deeper correspondence. A numerical effort has been initiated in this direction(33). In some simple geometries of fracture (anti-plane mode) it is also known that the Laplacian solution for the deformation field induced by a fracture in a stress field is equivalent to an electrical one where the current replaces the applied stress and a similar criterion can be expressed for breaking bonds in both cases. The figures 2 show the duality correspondences which must be made to go from a D.L.A. pattern to the crack one: Between figures 2a and 2b one replaces the zero current vertical lines by equipotential ones between which one passes a current I. The crack corresponds to an insulating line in the continuous conducting medium. The local breaking probability is proportional to the current at the tip. Between figure 2b and 2c , is made the correspondance between the electrical and mechanical variables.

Similar patterns have been found in hydrodynamic flows in 2 different instances:

-Daccord et al (34) have studied the Hele-Shaw flow of water displacing more viscous solution of water and polymer. Unlike ordinary results with immiscible Newtonian fluids which show characteristic finger structures, here the patterns are made of narrow branched lines very similar to D.L.A. ones. We believe that rather than the usual Saffman-Taylor mechanism, one should invoke some kind of hydrodynamic "rupture" resulting to the highly non-Newtonian behavior of the displaced liquid. Similar results have been obtained with water chasing a clay (bentonite) solution in a similar geometry(35)

-The same group in St Etienne(36) has followed the dissolution of a block of calcite in which water was circulated. The initial heterogeneities of the pores (actually, this process is also essential in the diagenesis of soluble rocks) are amplified as a function of time. Figure 6 represents a molding of a dissolution pattern obtained by radial injection of pure water in a cylindrical plaster cylinder.By using the same type of argument as for the figure 1. b, one can understand why the resulting dissolution pattern resembles D. L. A. patterns . This results from having a constant pressure

Figure 6 (courtesy of G. Daccord ; Dowell Schlumberger)

in the dissolved region and a Darcy flow solution for the fluid in the undissolved porous medium. We had carried independently some experiments of the dissolution of sintered glass beads in an hydrofluoric solution. The time evolution of the global permeability to the dissolving fluid shows a strongly non-linear variation associated to the process.

4 Conclusions

This presentation has outlined a few recent effects in the physics of random media, some of which being still object of controversy and all of them requiring more work in the confrontation between modeling and experimental studies on real materials. They are all open to experimental modelisation where the non-linear characteristics are often more amenable to experimental tests than numerical simulations. These examples outline the new impacts brought about by statistical physics to these problems. Whereas classical approaches treat in a detailed fashion the properties at a local level and solve the global one assuming periodicity or at least regularity in the distribution of heterogeneities, the emphasis is put here on the effect of the global geometries and the existence of a multiplicity of length scales which control the macroscopic properties in the case of large disorder systems.

Propagation of cracks provided a first case of dynamical behavior in such systems. Organised spatial patterns can also been found . An example is the structure of cracks formed at the surface of dry clay beds. The study by Skjeltorp [37] of drying monolayers of silica particle suspensions is a beautiful illustration of the problem. It is also studied in our laboratory on the real example of crack patterns formed on cement walls in which the 3-dimensional development of the cracks should define the length scale in the plane[38]. But other dynamical structures resulting from the interaction between the organisation of random matter and the applied external field would deserve study in particular in suspended beds of particles (suspensions, fluidised beds) and granular arrays[39].

Acknowledgments

This review has benefited from the contributions and the discussions with Daniel Bideau, Henry Crapo, Alain Gilabert, Hans Herrmann, Jean Paul Troadec

References

(1) R. Landauer, ETOPIM A.I.P. conf. proc. Vol.40 eds. J.C. Garland & D.B. Tanner (New York, 1978) p.1

(2) D. Stauffer, Introduction to percolation theory (Taylor & Francis; London 1985)

(3) B.B. Mandelbrot, The fractal geometry of nature (Freeman; San Francisco (1982)

(4) Chance and matter eds J. Souletie, J. Vannimenus & R. Stora (North Holland; Amsterdam 1987).

(5) R. Lenormand, C. Zarcone & A. Sarr, Jour. Fl. Mech. 135 337 (1983)

(6) R. Lenormand, C. Zarcone & E. Teboul, Jour. Fl. Mech. 189 165 (1988)

(7) e. g. T. A. Witten in (4)

(8) e. g. E. Guyon in (4)

(9) H. Crapo & W. Whiteley, The geometry of rigid structures ; Enc. of Math. Cambridge Univ. Press To appear

(10) S. Feng & P.N. Sen, Phys. Rev. Lett. 52 216 (1984)
M.A. Lemieux, P. Breton & A.M.S. Tremblay, Jour. Phys. Lett.. Lett. 46 L1 (1985)

(11) S. Roux & A. Hansen, Eur. Lett. 6 301 (1988)

(12) A. Hansen & S. Roux, preprint

(13) L. de Arcangelis, in Mixing and Disorder eds E. Guyon, J.P. Nadal & Y. Pomeau (Kluwer 1988)

(14) J. Wang & A. Brook Harris, Phys. Rev. Lett. 552 459 (1985)

(15) A. Gilabert, M. Ben Ayad, S. Roux & E. Guyon, preprint

(16) B.K. Chakraverty, D. Chowhury & D. Stauffer, Z. Phys. B 32 343 (1986)(1986)

(17) N. Tang & M. Thorpe, Phys. Rev. B 36 3798 (1987)

(18) H. Crapo, D. Bideau & J. P. Troadec, in collaboration with the authors of this review, consider the extension of this study to the transport properties of granular materials (to appear)

(19) D. Stauffer, H.J. Herrmann & S. Roux, J. de Phys. 48 347 (1987)

(20) L. Oger, J.C. Charmet, D. Bideau, J.P. Troadec C.R.A.S 302 II 277 (198

(21) M. Kardar, G. Parisi & Y. C. Zhang Phys. Rev. Lett. 56 889 (1986)
Y. C. Zhang Phys. Rev. Lett. 59 2125 (1987)

(22) T. Travers, D. Bideau, A. Gervois, J.P. Troadec & J. C. Messager, J. Phys. A 19 L 1033 (1988)
T. Travers, Thèse Rennes (1988)

(23) S. Redner, Phys. Rev. B 25 3242 (1982)

(24) S. Roux & H.J. Herrmann, Eur. Lett. 4 1227 (1987)

(25) A. Gilabert, S. Roux & E. Guyon, J. de Phys. 48 1609(1987)

(26) S. Roux, A. Hansen & E. Guyon, J. de Phys. 48 2125 (1987

(27) J. P. Straley, Phys. Rev. B 15 5733 (1977)

(28) P. de la Cruz, J. Luzuriaga, E. N. Martinez & E. J. Osquiguil, Phys. Rev.B 36 6850 (1987)

(29) M. Sahimi,& J. D. Goddard, Phys. Rev. B. 33 7848 (1986)
S. Roux, A. Hansen, H.J. Herrmann & E. Guyon, J. Stat. Phys. in press
B. K. Chakrabarti, in Rev. in Solid State Science (World Sci. Singapoore)

(30) L. de Arcangelis, A.Hansen, H. J. Herrmann & S. Roux, preprint

(31) P. M. Duxbury, P. D. Beale & P. L. Leath, Phys. Rev. Lett. 57 1052 (1986)

(32) A. Gilabert, C. Vanneste, D. Sornette & E. Guyon, J. de Phys. 48 763 (1987)

(33) E. L. Hinrichsen, A. Hansen & S. Roux, preprint

(34) J. Nittman, G. Daccord & H.E. Stanley, Nature 314 141 (1985)

(35) H. Vandamme, C. Laroche, & L. Gatineau, Rev. Phys. APP. 22241 (1987)

(36) G. Daccord, Phys. Rev. Lett. 58 479 (1987)

(37) A. T. Skjeltorp, in Time dependant effect in disordered materials R. Pynn and T. Riste edit. Plenum B 167 page 1
P. Meakin Thin Sol. Films 151 165 (1987)

(38) J.C. Charmet, L.Oger & P. Acker priv. comm.

(39) J. Rachenbach & P. Evesque C.R.A.S. to appear

CONVECTION IN BINARY MIXTURES : PROPAGATING AND STANDING PATTERNS

M. Lücke
Institut für Theoretische Physik, Universität des Saarlandes, D–6600 Saarbrücken, West Germany

1. INTRODUCTION

Convection in binary fluid layers[1–3] heated from below shows rich bifurcation properties out of the quiescent conductive state and a variety of states with different structural and temporal properties[3–11]. There are stationary bifurcations — forwards, backwards, and tricritical — into stationary overturning convection (SOC) rolls. But also stationary squares are stable under certain circumstances and, furthermore, a periodic switching between these two competing patterns can occur. There are Hopf bifurcations — again forwards, backwards, and tricritical — into oscillatory roll patterns of travelling waves (TW) and standing waves (SW) and also transient growth of TW, SW and modulated TW (MTW) roll convection. The convective TW state can be spatially localized or extended. There is codimension–two (CT) point bifurcation with competition between standing and propagating roll patterns and wave number jumps at onset. The travelling roll pattern generates via phase differences between the TW's of velocity, temperature and concentration global lateral currents of heat and concentration. Furthermore, the TW velocity field presumably generates Reynolds stresses that in turn cause mean lateral fluid flow. The Lagrangian particle dynamics in the TW velocity field is far from being dull.

Here we review a recent Galerkin approach[12–15] that explains, elucidates, and reproduces many of these exciting convective properties and, furthermore, yields insight that has not been available before.

2. THE SYSTEM

Consider a horizontal layer of a binary fluid heated from below in the homogeneous gravitational field, $\mathbf{g}=-g\,\mathbf{e}_z$. The governing hydrodynamic field equations in Oberbeck–Boussinesq approximation for the *deviations* from the horizontally homogeneous conductive state read

$$\nabla\cdot\mathbf{u}=0 \tag{2.1a}$$

$$(\partial_t+\mathbf{u}\cdot\nabla)\mathbf{u} = -\nabla p+\sigma[(1+\psi)\theta+\zeta]\mathbf{e}_z+\sigma\nabla^2\mathbf{u} \tag{2.1b}$$

$$(\partial_t+\mathbf{u}\cdot\nabla)\theta = Rw+\nabla^2\theta \tag{2.1c}$$

$$(\partial_t+\mathbf{u}\cdot\nabla)\zeta = \nabla^2[L\zeta-\psi\theta]. \tag{2.1d}$$

Here $\mathbf{u}=(u,v,w)$ is the velocity field with vertical component w, p is the pressure field, θ the temperature, and

$$\zeta = c - \psi\theta \tag{2.2}$$

is used instead of the concentration field such that the diffusive concentration current density[1-3,14] is

$$\mathbf{j}_c = -L\nabla\zeta. \tag{2.3}$$

The fact that the concentration current is driven also by the temperature gradient is called the Soret effect.

Lengths are scaled by the layer thickness d, times by the vertical diffusion time d^2/κ with κ being the thermal diffusivity. Temperatures are scaled by $\nu\kappa/(\alpha g d^3)$ and concentration by $\nu\kappa/(\beta g d^3)$ where ν is the kinematic viscosity while $\alpha=-\rho^{-1}\partial\rho(T,p,c)/\partial T$ and $\beta=-\rho^{-1}\partial\rho(T,p,c)/\partial c$ are expansion coefficients of the fluid. The Prandtl number, $\sigma=\nu/\kappa$, and the Lewis number, $L=D/\kappa$, being the ratio of mass diffusivity D and thermal diffusivity κ are material parameters of the fluid. Finally there are two control parameters: The Rayleigh number $R=(\alpha g d^3/\nu\kappa)\Delta T$ measures the thermal stress due to the imposed temperature difference ΔT between the bottom and the top of the layer. The separation ratio ψ[16] can be viewed to measure how much the contribution, $\sigma(1+\psi)\theta\,\mathbf{e}_z$, from the temperature deviation θ to the buoyancy force in (2.1b) is enhanced ($\psi>0$) or depressed ($\psi<0$) in the mixture relative to a hypothetical one-component fluid with $\psi=0$, $\zeta=0$ and the same θ. A somewhat related interpretation of the role of ψ is based on the vertical gradients of the conductive temperature, $\partial_z T_{cond}=-R$, and of the conductive concentration profile $\partial_z c_{cond}=-\psi R$. Thus $\psi>0$ ($\psi<0$) increases (decreases) the conductive mass density gradient, $\partial_z\rho_{cond}=(\sigma\kappa^2/gd^3)(1+\psi)R$, and thereby reduces (increases) the stability of the layer against generation of convective flow.

We shall consider here horizontal boundaries at z=0,1 that are perfectly heat conducting, $\theta=0$, and impervious to concentration currents, $\partial_z\zeta=0$, as realized experimentally by good conducting plates, e.g., of copper or sapphire. Note that the vanishing vertical concentration current implies an additional coupling of contration gradients and temperature gradients at the horizontal boundaries: $\partial_z c=-\psi\partial_z\theta$. The idealized boundary condition, $\zeta=0$ or equivalently c=0, that is used[17-23] in the double-diffusive or thermohaline problems implies in general a finite vertical concentration current through the horizontal boundaries and most importantly ignores the boundary-induced coupling between c and θ. The latter was shown[12-14] to change linear as well as nonlinear convective properties dramatically in comparison to the permeable boundary situation.

In the following we shall use for mathematical convenience free-slip boundary conditions on the velocity field instead of the realistic no-slip ones. That causes basically a global upwards shift of the stability curves in the ψ–R plane and also a shift of critical wave numbers. Both effects can be compensated by considering properly reduced quantities. In addition the convective amplitudes and Nusselt numbers are slightly larger.

3. STANDING AND PROPAGATING ROLL PATTERNS

3.1 Mode truncation

To describe convective rolls with axes along y with a minimal set of modes we truncate the expansion of the laterally periodic hydrodynamic fields as follows[12]

$$w(x,z,t) = \left[\hat{w}_{101}(t)e^{-ikx}+c.c.\right]f(z) \tag{3.1a}$$

$$u(x,z,t) = -\frac{1}{k}\left[i\hat{w}_{101}(t)e^{-ikx}+c.c.\right]\partial_z f(z) \tag{3.1b}$$

$$\theta(x,z,t) = \left[\hat{\theta}_{101}(t)e^{-ikx}+c.c.\right]\sqrt{2}\sin(\pi z) + \hat{\theta}_{002}(t)\sqrt{2}\sin(2\pi z) \tag{3.1c}$$

$$\zeta(x,z,t) = \left[\hat{\zeta}_{100}(t)e^{-ikx}+c.c.\right] + \zeta_{001}(t)\sqrt{2}\cos(\pi z). \tag{3.1d}$$

Here $f(z)=\sqrt{2}\sin(\pi z)$ for free–slip boundaries. The truncation is motivated by taking first the basic critical modes (thick lines in Fig. 1), i.e., the lowest modes growing in the linearized problem above threshold. They are (101) for θ and $\mathbf{u}$ and (100) for ζ. Then we include modes $\hat{\theta}_{002}$ and $\hat{\zeta}_{001}$ (thin lines in Fig. 1) that are excited via the convective nonlinearities on the left hand side of (2.1b–2.1d) by the basic modes. We exclude modes that represent higher lateral harmonics.

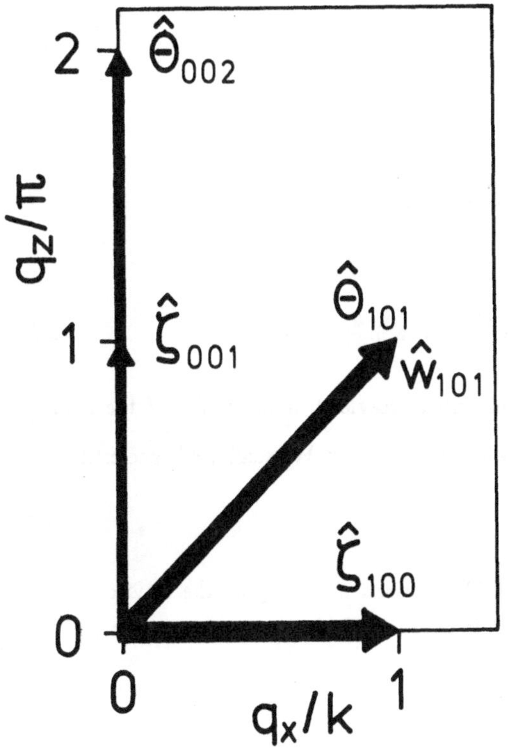

FIG. 1 Wave vectors $\mathbf{q}=(n_x k,0,n_z\pi)$ retained in the expansion of the fields $\mathbf{u}$, θ, and ζ in the first quadrant of the q_x–q_z plane. Thick lines denote basic critical modes $\hat{w}_{101}$, $\hat{\theta}_{101}$, and $\hat{\zeta}_{100}$. Thinner lines show modes $\hat{\theta}_{002}$ and $\hat{\zeta}_{001}$ that are coupled via the convective nonlinearities to the former.

3.2 Model equations

Projecting the field equations (2.1) onto the 8 modes of the Galerkin expansion (3.1) one obtains[12] the following system of coupled nonlinear ordinary differential equations

$$\tau\dot{\mathbf{X}}(t) = -\sigma\hat{q}^2\mathbf{X}(t)+\sigma(\hat{k}^2/\hat{q}^2)\left[(1+\psi)\mathbf{Y}(t)+(8/\pi^2)\mathbf{U}(t)\right] \tag{3.2a}$$

$$\tau\dot{\mathbf{Y}}(t) = -\hat{q}^2\mathbf{Y}(t)+[r-Z(t)]\mathbf{X}(t) \tag{3.2b}$$

$$\tau\dot{Z}(t) = -b\left[Z(t)-\mathbf{X}(t)\cdot\mathbf{Y}(t)\right] \tag{3.2c}$$

$$\tau\dot{\mathbf{U}}(t) = -(L/3)\hat{k}^2\mathbf{U}(t)+\hat{q}^2\psi\mathbf{Y}(t)+V(t)\mathbf{X}(t) \tag{3.2d}$$

$$\tau\dot{V}(t) = -(b/4)\left[LV(t)+(8/3)\psi Z(t)+2\mathbf{X}(t)\cdot\mathbf{U}(t)\right]. \tag{3.2e}$$

Real and imaginary parts of the basic critical modes $\hat{w}_{101}$, $\hat{\theta}_{101}$, and $\hat{\zeta}_{100}$ are combined into two–component vectors

$$\mathbf{X}=(X_1,X_2)=(1/q_c^0)(\mathrm{Re}\,\hat{w}_{101}, \mathrm{Im}\,\hat{w}_{101}) \tag{3.3a}$$

$$\mathbf{Y}=(Y_1,Y_2)=(q_c^0/R_c^0)(\mathrm{Re}\,\hat{\theta}_{101}, \mathrm{Im}\,\hat{\theta}_{101}) \tag{3.3b}$$

$$\mathbf{U}=(U_1,U_2)=(\pi q_c^0/2\sqrt{2}\,R_c^0)(\mathrm{Re}\,\hat{\zeta}_{100}, \mathrm{Im}\,\hat{\zeta}_{100}). \tag{3.3c}$$

The secondary modes $\hat{\theta}_{002}$ and $\hat{\zeta}_{001}$ are real

$$Z= -(\pi\sqrt{2}/R_c^0)\hat{\theta}_{002}; \qquad V=(\pi^2/2\sqrt{2}\,R_c^0)\hat{\zeta}_{001}. \tag{3.3d}$$

The constants

$$(k_c^0)^2=\pi^2; \quad (q_c^0)^2=(k_c^0)^2+\pi^2; \quad R_c^0=(k_c^0)^{-2}(q_c^0)^6; \quad \tau=(q_c^0)^{-2}; \quad b=4\pi^2\tau \tag{3.3e}$$

are defined by the critical wave number, k_c^0, and critical Rayleigh number, R_c^0, of the model at ψ=0, i.e., of a one component fluid with the same Prandtl number. We shall use henceforth

$$r=R/R_c^0; \qquad \hat{k}/k_c^0; \qquad \hat{q}^2=(k^2+\pi^2)/(q_c^0)^2. \tag{3.3f}$$

as reduced Rayleigh number and wave numbers. The Nusselt number, i.e., the ratio of laterally averaged vertical convective and total heat currents is given within the approximation (3.1–3.3) by

$$N(t)=1+(2/r)Z(t). \tag{3.4}$$

3.3 Convective threshold

In the ψ–r plane of control parameters there are two (families of) bifurcation lines[12]

$$r_{stat}(\psi;\hat{k})=(\hat{q}^6/\hat{k}^2)\left[1+\psi+(8/\pi^2)(\psi/\hat{L})\right]^{-1} \tag{3.5}$$

$$r_{osc}(\psi;\hat{k})=(\hat{q}^6/\hat{k}^2)(1+\sigma)(1+\hat{L})(1+\hat{L}/\sigma)\left[1+\sigma+\psi(1+\sigma-8/\pi^2)\right]^{-1} \tag{3.6}$$

with $\hat{L}=(\hat{k}^2/3\hat{q}^2)L$. These stability curves agree quite well with recent numerical results obtained for no–slip, impermeable boundary conditions. At $r_{stat}(r_{osc})$ the conductive state $\zeta=\theta=0=\mathbf{u}$ looses stability against growth of a stationary (oscillatory) convective pattern of straight rolls with wave number $\hat{k}$. For an externally fixed $\hat{k}$ these stability boundaries of the conductive state (shown *schematically* in Fig. 2) meet at a codimension–two point $\psi_{CT}(\hat{k})$[12] such that there the Hopf frequency, $\omega_H(\psi;\hat{k})$[12], vanishes. However, the critical wave numbers, k^c_{stat} and k^c_{osc}, i.e., the wave numbers for which $r_{stat}(\psi;\hat{k})$ and $r_{osc}(\psi;\hat{k})$ are minimal are not the same when r^c_{stat} and r^c_{osc}. This property, first discussed[12,13] within the Galerkin model (3.2) and found later also in numerical stability analyses[24], is caused by the impermeability of the boundaries — for the unphysical permeable boundaries one has $k^c_{stat}=k^c_{osc}$. Thus, when in experiments the wave number is not fixed externally there is a jump[13]

$$\hat{k}^c_{stat}-\hat{k}^c_{osc}=(1+1/\sigma)(L/3)+\mathcal{O}(L^2) \tag{3.7a}$$

in wave number and a jump to a finite, minimal oscillation frequency[13]

$$\omega_{min}=(\pi^2/3)(1+1/\sigma)^{1/2}L^{3/2}\left[1+\mathcal{O}(L^{1/2})\right] \tag{3.7b}$$

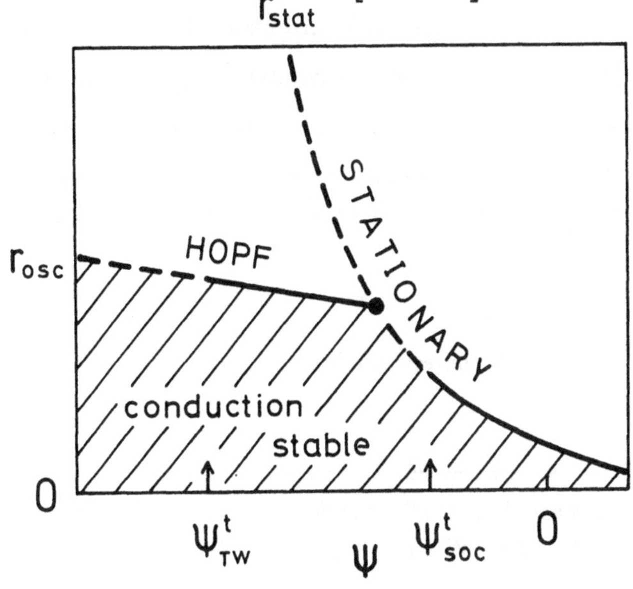

FIG. 2 Schematic bifurcations lines r_{osc} and r_{stat} of oscillatory and stationary convection with fixed wave number out of the conductive state. For fixed $\hat{k}$ the Hopf frequency vanishes at the CT point (full dot). In a N vs r diagram TW (SOC) convection bifurcates forwards out of the conductive state across the full parts of the line labelled r_{osc} (r_{stat}) and backwards across the dashed parts. At the tricritical values ψ^t the initial slope, $\partial N/\partial r$, diverges.

when ψ is varied through the value[13]

$$\tilde{\psi} = -(\pi^2/72)(1+1/\sigma)L^2[1-(1/3)(1+1/\sigma)L+\mathcal{O}(L^2)] \tag{3.7c}$$

at which r^c_{stat} and r^c_{osc} intersect in the ψ–r plane. While the L–dependence of (3.7) is the same as for the numerically obtained results with no–slip conditions the jumps (3.7a,b) under such conditions are larger, typically by a factor of about two, than in our free–slip model[25].

Note that the experimental Lewis numbers are very small, $L \simeq 10^{-2}$. Therefore the value $\tilde{\psi}$ where $r^c_{stat}=r^c_{osc}$ is so small that barodiffusion[16], on the one hand, and non–Oberbeck–Boussinesq effects for the thermodiffusion ratio[16] entering into ψ, on the other hand, have to be taken into account to obtain realistic stability boundaries close to the codimension–two point.

We finally mention that the critical wave number, $\hat{k}^c_{stat}$, for onset of stationary convection is strongly dependent on ψ — it drops to zero at positive $\psi_0=L(16/\pi^2)/(1-L\,16/\pi^2)$ — while $\hat{k}^c_{osc}$ is independent of ψ.

3.4 Stationary overturning convection (SOC) rolls

Besides the trivial conductive solution with vanishing mode amplitudes the model (3.2) has at most two further fixed point solutions not counting symmetry degeneracy. They are simply obtained analytically[12] for arbitrary r, ψ, $\hat{k}$, L, σ and they represent standing patterns of stationary overturning convective rolls. In a N vs r bifurcation diagram there is for r>0 a tricritical value, ψ^t_{SOC} [12] which is negative and quite small — about $-2{\times}10^{-6}$ for $^3He-^4He$ mixtures and about $-2{\times}10^{-7}$ for ethanol–water mixtures. For $\psi>\psi^t_{SOC}$ a SOC state grows with increasing r in a forwards bifurcation at r_{stat} out of the conductive state (c.f. Sec. 4 for details). At $\psi=0$ the SOC state is identical to that of the standard three–mode Lorenz model for a one component fluid. In particular $N_{SOC}(\psi=0)=1+2(r-r_{stat})/r$ and, furthermore, the concentration field becomes homogeneous for $\psi=0$. For $\psi<\psi^t_{SOC}$ the SOC state bifurcates at r_{stat} backwards into a lower unstable branch, forms a saddle node at r_s, and bends over into a stable upper branch that eventually looses stability at larger r[12,13] (in Sec. 3.5.3 we show a schematic bifurcation diagram with the SOC state close to the saddle node). For $\psi<-\hat{L}/(\hat{L}+8/\pi^2)$ where $r_{stat}\rightarrow\infty$ the SOC state is disconnected from the conductive state.

3.5 TW and SW convection

At r_{osc} nonlinear steady state TW and SW convective roll solutions bifurcate out of the conductive state.

3.5.1 Travelling waves The complete TW solutions of (3.2) have been obtained in closed analytical form as functions of r, ψ, $\hat{k}$, L, σ and their stability was determined from the eigenvalue spectrum of an 8×8 matrix[14]. The steady state TW modes $\hat{w}_{101}(t)$, $\theta_{101}(t)$, and $\zeta_{100}(t)$ are complex of the form $A(t)=|A|e^{i(\omega t+\varphi_A)}$ with constant amplitudes $|A|$, common phase velocity

ω, and constant but different (initial) phases φ_A. The real modes $\hat{\zeta}_{001}$ and $\hat{\theta}_{002}$ and with the latter also the Nusselt number are constant in the TW solution. The convective field patterns

$$w(x,z,t)=2|\hat{w}_{101}|\cos(kx-\omega t-\varphi_w)f(z) \tag{3.8a}$$

$$\theta(x,z,t)=2|\hat{\theta}_{101}|\cos(kx-\omega t-\varphi_\theta)\sqrt{2}\sin(\pi z) + \hat{\theta}_{002}\sqrt{2}\sin(2\pi z) \tag{3.8b}$$

$$\zeta(x,z,t)=2|\hat{\zeta}_{100}|\cos(kx-\omega t-\varphi_\zeta) + \hat{\zeta}_{001}\sqrt{2}\cos(\pi z) \tag{3.8c}$$

of $\mathbf{u}$, θ, ζ — and similarly of the pressure — travel with constant velocity ω/k to the right, say, but have constant nonzero relative phases.

While the TW's of w, θ, and ζ are plane ones the surfaces of constant phase in the concentration wave are bent as a result of $c=\psi\theta+\zeta$ being a superposition of fields with different z–profiles. A recent numerical linear stability analysis[24] suggests that higher modes might give rise to a small bending of the surfaces of constant phase also in θ and w. In Fig. 3 we show spatial TW intensity profiles in the x–z plane. The intensity of the deviations

$$\Delta T(x,z)=R(1-z)+\theta(x,z) \tag{3.9a}$$

$$\Delta C(x,z)=\psi\Delta T(x,z)+\zeta(x,z) \tag{3.9b}$$

$$\Delta\rho(x,z)=-(\sigma\kappa^2/gd^3)[\Delta T(x,z)+\Delta C(x,z)] \tag{3.9c}$$

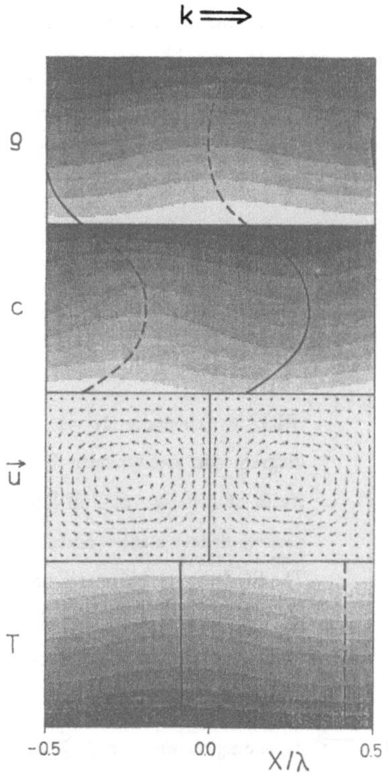

FIG. 3 Intensity plots of TW convection in the x–z plane. The magnitudes of mass density ρ, concentration c, and temperature T are shown with ten gray levels (black $\hat{=}$ large, white $\hat{=}$ small). The velocity field $\mathbf{u}$ is represented by arrows with lengths proportional to $|\mathbf{u}|$. Positions of crests (valleys) in the TW's are shown by full (dashed) lines to indicate form and positions of surfaces of constant phase in the different waves. Parameters L=0.015, σ=18.4 are for ethanol–water mixtures with ψ=–0.5, r=1.53, N=1.016. From ref. 14.

of temperature, concentration, mass density from their reference values T_0, C_0, ρ_0 at the upper plate are shown with ten different gray levels. To identify the lines of constant phase in the TWs the positions of lateral minima and maxima are marked by dashed and full lines, respectively. The temperature wave lags behind the wave in the vertical velocity field, w, by a phase angle, $\varphi_w-\varphi_\theta=\arctan(\omega\tau)$, that is largest at the convective threshold, r_{osc}. The concentration wave runs ahead of the w wave.

3.5.2 Standing waves In the SW solution of (3.2) the modes $\hat{w}_{101}(t)$, $\hat{\theta}_{101}(t)$, and $\hat{\zeta}_{100}(t)$ oscillate with zero mean in the complex amplitude plane with a common frequency along a straight line through the origin. Their amplitudes periodically revert their sign. The modes $\zeta_{001}(t)$ and $\theta_{002}(t)$ and with it the Nusselt number oscillate with twice the frequency around a finite mean. All field patterns are fixed in space. The amplitude of the velocity field goes through zero such that all rolls simultaneously revert their turning directions.

Since the SW amplitudes $\hat{w}_{101}(t)$, $\hat{\theta}_{101}(t)$, and $\hat{\zeta}_{100}(t)$ move along a straight line the SW solution lives in a five-dimensional subspace of (3.2) and thus coincides with the oscillatory solution of a five-mode Galerkin model for standing patterns where **X**, **Y**, **U** are replaced by scalars.

3.5.3 Bifurcation properties In Fig. 4 we show schematically the N vs r bifurcation diagram of SOC and TW solutions at negative separation ratios. There is a tricritical value ψ^t_{TW} [14] close to the CT point such that for $\psi^t_{TW}<\psi<\psi_{CT}$ ($\psi<\psi^t_{TW}$) a stable (unstable) TW solution bifurcates forwards (backwards) out of the conductive state. As an interesting aside we mention that the analogous eight-mode Galerkin model for *permeable* boundary conditions[23] does not have finite-amplitude TW solutions. This again demonstrates the significance of the boundary condition of the concentration.

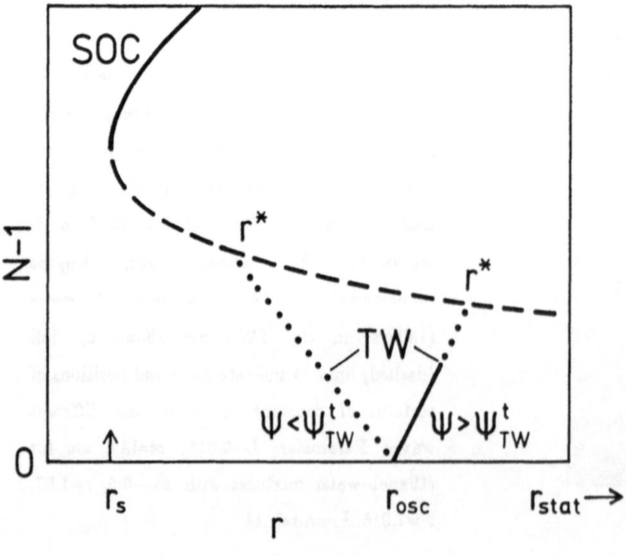

FIG. 4 Schematic N vs r bifurcation diagram for steady state TW and SOC states at negative ψ. Full lines (dots and dashes) denote stable (unstable) solutions. The SOC branch bifurcates backwards out of the conductive state at r_{stat}. The bifurcation of the TW state at r_{osc} is subcritical, tricritical, or supercritical depending on whether $\psi \lesseqqgtr \psi^t_{TW}$.

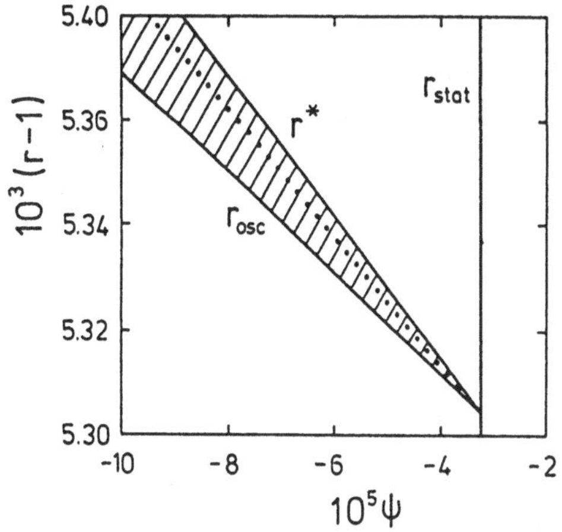

FIG. 5 Domain of existence (shaded area) of TW solutions in the $r-\psi$ plane close to the intersection of r_{osc} and r_{stat} for fixed wave number $\hat{k}=1$, $L=0.015$, $\sigma=18.4$. The TW state branches in a forwards bifurcation off the conductive state at r_{osc}, becomes unstable against growth of a modulated TW at the dotted line, and ends with zero frequency in the unstable SOC state at r^*. From ref. 14.

The square of the TW's oscillation frequency decreases linearly with growing distance $|r-r_{osc}|$ from its Hopf value at threshold, $r=r_{osc}$. The TW solution ends with zero frequency at r^* on a SOC branch. The squared velocity field amplitude grows linearly and the Nusselt almost linearly with $|r-r_{osc}|$ which is a result of the severe truncation (3.1) of the modes coupled to the TW.

The forwards bifurcating TW loses stability against amplitude modulations by a second frequency roughly halfway between r_{osc} and r^*. The range of existence of the forwards bifurcated TW narrows down to zero when approaching the CT point. In Fig. 5 we show for a fixed wave number $\hat{k}=1$ the vicinity of the CT point in ethanol–water mixtures on an expanded scale. With r_{osc} and r_{stat} approaching each other the shaded interval (r_{osc}, r^*) in which the TW solution exists shrinks to zero at the CT point. This presumably will suppress interesting CT bifurcation dynamics.

The SW solution is unstable close to the oscillatory threshold in agreement with general results[26] based on symmetry related arguments. The latter show that the SW (TW) solution is stable only if (i) SW *and* TW bifurcate supercritically and (ii) the SW (TW) branch lies above the TW (SW) branch. A detailed discussion of the SW solution of (3.2) will be given in ref. 25.

3.5.4 Transient growth Linearizing (3.2) around the conductive fixed point one finds that shortly above the Hopf bifurcation threshold, $r>r_{osc}$, convective amplitudes starting from infinitesimal initial values grow in an oscillatory way with a frequency ω and a growth rate γ determined by the relevant Hopf eigenvalue. The different types of growth behaviour — TW, SW, or MTW — are best visualized in the plane of the complex mode amplitudes[12–14, 23]. The initial growth dynamics being linear the transient type is fixed by the initial values. Randomly chosen ones typically lead to MTW transients in which the complex numbers $\hat{w}_{101}(t)$, $\hat{\theta}_{101}(t)$, and $\hat{\zeta}_{100}(t)$ rotate outwards on elliptic spirals such that the amplitudes are modulated with 2ω. TW transients correspond to circular spirals and SW transients are found in the limit where the ellipse becomes flat. It is straightforward to derive (relations between) the initial conditions for the different transients from the linearized equations[12–14]. In experiments reported so far the initial convective field amplitudes seem to have been related in such a way as to produce TW transients. It is only in the TW transient that the Nusselt number does not oscillate while growing.

Finally we should like to stress that transient SW, TW, MTW convection has to be clearly distinguished from the nonlinear steady state limit cycle states. The former can grow from initial infinitesimally small deviations from the conductive fixed point whenever $r>r_{osc}$ and the nature of such a transient remains unchanged as long as nonlinearities are unimportant. The nonlinear steady state SW or TW solutions of (3.2) bifurcate either super– or subcritically out of the conductive state depending on ψ while a steady MTW solution does not branch off the latter but rather exists only with finite amplitudes[12,13].

3.6 Lateral currents of heat and concentration generated by a TW

In the steady state as well as in the transiently growing TW solution there are finite, constant phase differences between the wave fields of temperature and velocity and also between concentration and velocity. These phase differences drive mean lateral currents of heat, $q_x=(1/R)\langle\theta u\rangle$, and concentration, $j_x=(1/R)\langle cu\rangle$, that are proportional to the sines of the respective phase differences. It is convenient to express the above reduced current densities[14] in terms of the Nusselt number and the phase velocity ω/k

$$q_x(z)=(\tau\omega/k)(N-1)\sqrt{2}\sin(\pi z)\partial_z f(z) \qquad (3.10a)$$

$$j_x(z)=-(\tau\omega/k)(N-1)[\chi-\psi\sqrt{2}\sin(\pi z)]\partial_z f(z) \qquad (3.10b)$$

with

$$\chi=(\pi\sqrt{2}/4)(\sigma+L/3)^{-1}\left[(1+\sigma)(1+\psi)-(8/\pi^2)\psi\right]. \qquad (3.10c)$$

As a result of the θ and c dependence of the mass density ρ there is also a lateral mass current $\langle\rho u\rangle$. Its mass transport in proper units is small[14] compared to that one associated with j_x.

Note that the TW generated lateral transport of heat and concentration is proportional to the product of the TW's propagation velocity and its vertical convective heat transport N−1. The z–profiles of the currents (3.10) have inversion symmetry at the midplane, z=1/2, so that the total currents through the entire cross–section of the fluid layer vanish. But in the upper half of the layer there is a global time independent lateral current of heat (concentration) opposite (parallel) to the propagation direction of the TW and vice versa in the lower half of the layer. It would be interesting to know, for example, whether in experimental realizations of ***spatially localized*** TW states concentration accumulates in the lower half of the fluid at one end of the localization region and gets depleted on the other end and vice versa in the upper half and whether the associated changes in the vertical gradients are detectable. An alternative might be that the concentration currents are deformed near the ends of the localization region such as to form a circular concentration current. Similar questions have to be answered also for the lateral heat current.

3.7 Do TW's generate mean fluid flow?

An interesting question is whether the oscillatory velocity field $\mathbf{u}=(u,0,w)$ of a travelling roll pattern generates a finite mean Reynolds stress, $\langle wu \rangle$. That in turn could drive according to the laterally averaged Navier–Stokes equation (2.1b)

$$(\partial_t-\sigma\partial_z^2)U=-\langle\partial_x p\rangle-\partial_z\langle wu\rangle \tag{3.11}$$

a steady mean lateral fluid flow, U(z). Here the brackets $\langle ... \rangle$ indicate a lateral spatial average or equivalently a time average[14].

In the Galerkin truncation (3.1, 3.2) all terms in (3.11) vanish by construction. However, one can (i) derive a generally valid exact formula[14] for the mean Reynolds stress of a TW and (ii) obtain from it an estimate[14] for U(z). To that end we decompose the TW velocity field

$$w(x,z,t)=w(x-\frac{\omega}{k}t,z)=\sum_{n=-\infty}^{\infty} w_n(z)e^{-i(kx-\omega t)n} \tag{3.12a}$$

where

$$w_n(z)=|w_n(z)|e^{i\varphi_n(z)}=w_{-n}^*(z) \tag{3.12b}$$

is the n'th lateral Fourier coefficient and $\varphi_n(z)=-\varphi_{-n}(z)$ its phase. Because of incompressibility the coefficients of u are given by

$$u_n(z)=\partial_z w_n(z)/(ikn)\ . \tag{3.12c}$$

Here u and w are defined as the fluctuating parts of the TW field so that $w_0=u_0=0$. From (3.12) one immediately finds that the mean Reynolds stress

$$\langle wu\rangle=\frac{2}{k}\sum_{n=1}^{\infty}\frac{1}{n}|w_n(z)|^2\partial_z\varphi_n(z) \tag{3.13}$$

vanishes only when all phases, $\varphi_n(z)$, are independent of z. Recent numerical calculations[24] of the exact eigenfunctions of the linear problem indicate that, at least for linear, i.e., transiently growing TW solutions there is a small z–dependence of $\varphi_1(z)$. With this result we estimated[14] the mean flow velocity in typical experimentally observed stationary TW states to be about 1 % of the pattern propagation velocity. Recent numerical simulations[27] of TW's showed similar mean flow velocities.The contribution of U(z) to the lateral transport of heat, concentration,and mass in comparison with the currents of Sec. 3.6 is presently investigated.[27]

3.8 Lagrangian particle dynamics in a TW velocity field

An interesting problem[14] that has also been addressed experimentally recently[28] is the Lagrangian particle dynamics in the time–dependent velocity field of a TW and the associated mixing behaviour of the fluid. Consider for the sake of simplicity the velocity field

$$w(x,z,t)=\cos[\pi(x-ct)]\ f(z) \tag{3.14a}$$

$$u(x,z,t)=-(1/\pi)\sin[\pi(x-ct)]\ \partial_z f(z) \tag{3.14b}$$

of a TW of wavelength $\lambda=2$ moving with velocity c to the right. Here $f(z)=\mathscr{C}_1(z-1/2)$ is, e.g., the first Chandrasekhar function which looks like $\sin(\pi z)$ except that it goes to zero quadratically at z=0,1 in accordance with no–slip boundary condition requirements. The Lagrangian equations of motion, $\dot{x}=u$ and $\dot{z}=w$, of a particle in the above time–dependent velocity field (3.14) are non autonomous and complicated so the particle motion can be expected to be rather complex. However, in the frame of reference Σ' that is comoving with the TW according to

$$x'=x-ct \tag{3.15a}$$

the velocity field is stationary

$$w'(x',z)=\cos(\pi x')\, f(z) \tag{3.15b}$$

$$u'(x',z)=-(1/\pi)\sin(\pi x')\, \partial_z f(z) - c. \tag{3.15c}$$

Since in this frame the particles move along streamlines, i.e., isolines of the streamfunction

$$\phi(x',z)=-(1/\pi)\sin(\pi x')\, f(z)-cz \tag{3.16}$$

the particle dynamics and the mixing is much easier visualized[14].

In Fig. 6 we show at successive equidistant times the positions of particles in the comoving frame, Σ', that were at t=0 distributed homogeneously over a left turning roll ($-1 \lesssim x' \lesssim 0$) and a right turning roll ($2 \lesssim x' \lesssim 3$). In the upper part we show some representative streamlines for c=0.4. Particles that start on closed streamlines in Σ' move in the laboratory frame together with the TW to the right while circulating around the comoving elliptical fixed points of Σ'. All others starting on open streamlines move in Σ' to the left along the open streamlines which meander around the regions of closed streamlines. Those particles moving in Σ' to the left with a mean lateral velocity $<-c$ move also in the laboratory to the left, i.e., into the direction opposite to the TW. On the other hand, particles moving in Σ' along the open streamlines to the left with a mean lateral velocity $>-c$ move in the laboratory frame in the same direction as the TW, albeit slowlier.

With decreasing propagation velocity c the region occupied by closed streamlines grows. However for every finite c there are open streamlines and also backwards motion in the laboratory frame. With increasing c the closed streamlines shrink and the number of particles that can comove with the TW decreases. Above the threshold velocity $c=(1/\pi)\max\partial_z f(z)$ which is about 1.6 for a vertical profile given by the first Chandrasekhar function all streamlines are open in Σ'. In the laboratory frame there is still particle motion to the right in such a situation, however, only with velocity $<c$.

One should keep in mind that the Lagrangian motion gives immediate information about the mixing behaviour of the velocity field but not about its mass flow and mass transport properties. In fact with respect to the latter the Lagrangian motion is utterly misleading. For example, it is false to infer from the motion of particles to the left, opposite to the TW, that there is a mean backflow[28] or a mean backwards mass transport generated by the TW field (3.14). By construction vanishes the mean fluid flow! Also the mass transported through *any* area element ds during one oscillation period T of the TW is zero,

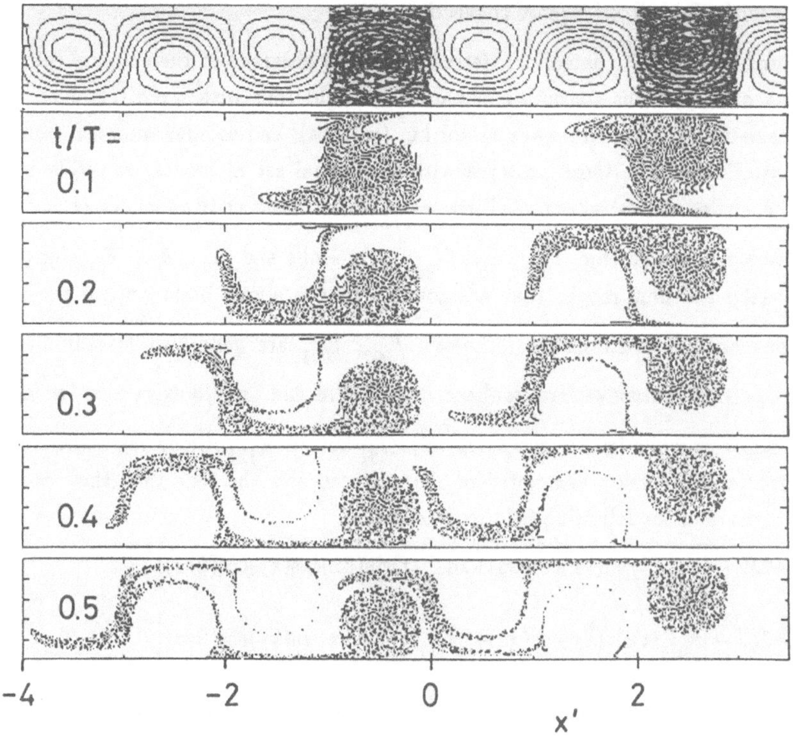

Fig. 6 Lagrangian particle motion in the TW velocity field (3.14). Top row shows streamlines in the frame Σ' comoving with the TW (c=0.4, λ=2) to the right. Below we show positions of 5000 particles in Σ' that were initially distributed homogeneously over the shaded regions of the top row occupied by a left ($-1 \lesssim x' \lesssim 0$) and a right ($2 \lesssim x' \lesssim 3$) turning roll.

$$\rho \int_0^T dt\, \mathbf{u}(x-ct,z)\cdot d\mathbf{s}=0. \tag{3.17}$$

The Lagrangian dynamics is such that during one oscillation cycle of the TW as many particles move through any fixed area element $d\mathbf{s}$ in the laboratory frame with velocity parallel to $d\mathbf{s}$ as in opposite direction. This property is not obvious in figures that display particle positions at successive times in the laboratory frame[28] or in the comoving frame.

4. COMPETITION BETWEEN ROLL AND SQUARE PATTERNS

Recent experiments[9,10] done in circular and square shaped containers at positive separation ratios showed that immediately above onset of convection square patterns were stable. Then, with increasing r there were pattern oscillations with rolls and squares appearing and disappearing alternatingly and for still higher driving r there were stationary rolls stable. We shall review in this section the results of a Galerkin model[15] that reproduces and explains the main experimental features. A detailed discussion can be found in ref. 15.

4.1 Galerkin model for roll and square convection

Moses and Steinberg[28] (hereafter referred to as MS) observed in their square cell containers standing convective patterns consisting of parallel straight rolls with wave vector $k\mathbf{e}_x$ (hereafter referred to as x–rolls), rolls with wave vector $k\mathbf{e}_y$ (y–rolls), and squares oriented parallel to the x–y sidewalls. To describe these patterns with a minimal set of modes we retain in the field expansion the modes shown in Fig. 7. First we take the basic critical modes (thick arrows in Fig. 7) for each set of rolls, e.g., $\hat{u}_{101}$, $\hat{\theta}_{101}$, $\hat{\zeta}_{100}$ for x–rolls and $\hat{u}_{011}$, $\hat{\theta}_{011}$, $\hat{\zeta}_{010}$ for y–rolls. In addition we keep the first modes that are coupled to the above basic critical modes via the nonlinear convective interactions in (2.1b–d): $\hat{\theta}_{002}$, $\hat{\zeta}_{001}$ are generated already by pure roll patterns; in addition the convective nonlinearities $(\mathbf{u}\cdot\nabla)\mathbf{u}$ and $(\mathbf{u}\cdot\nabla)\theta$ drive also the modes $\hat{u}_{112}$ and $\hat{\theta}_{112}$ respectively while $(\mathbf{u}\cdot\nabla)\zeta$ generates only $\hat{\zeta}_{001}$ but not $\hat{\zeta}_{111}$. Using the reality of the fields and the mirror and inversion symmetry of the patterns and the fact that they do not move laterally one arrives at the following field representation

$$u_1(\mathbf{x},t)=-2\sqrt{2}\,\frac{\pi}{k}\left[\hat{u}_{101}(t)\sin(kx)\cos(\pi z)+2\hat{u}_{112}(t)\sin(kx)\cos(ky)\cos(2\pi z)\right] \tag{4.1a}$$

$$u_2(\mathbf{x},t)=-2\sqrt{2}\,\frac{\pi}{k}\left[\hat{u}_{011}(t)\sin(ky)\cos(\pi z)+2\hat{u}_{112}(t)\cos(kx)\sin(ky)\cos(2\pi z)\right] \tag{4.1b}$$

$$u_3(\mathbf{x},t)=2\sqrt{2}\left[\hat{u}_{101}(t)\cos(kx)+\hat{u}_{011}(t)\cos(ky)\right]\sin(\pi z)+4\sqrt{2}\hat{u}_{112}(t)\cos(kx)\cos(ky)\sin(2\pi z) \tag{4.1c}$$

$$\theta(\mathbf{x},t)=2\sqrt{2}\left[\hat{\theta}_{101}(t)\cos(kx)+\hat{\theta}_{011}(t)\cos(ky)\right]\sin(\pi z)+\sqrt{2}\left[4\hat{\theta}_{112}(t)\cos(kx)\cos(ky)+\hat{\theta}_{002}(t)\right]\sin(2\pi z) \tag{4.1d}$$

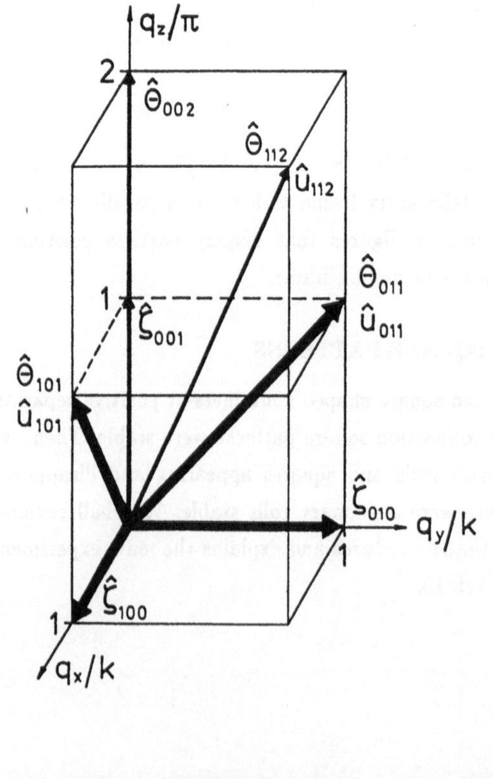

Fig. 7 Wave vectors $\mathbf{q}=(n_x k, n_y k, n_z \pi)$ for roll and square convection in the first octant of $\mathbf{q}$–space. Thick lines denote basic critical modes for x–rolls and y–rolls. Thinner lines show modes that are coupled via convective nonlinearities to the former.

$$\zeta(\mathbf{x},t)=2\hat{\zeta}_{100}(t)\cos(kx)+2\hat{\zeta}_{010}(t)\cos(ky)+\sqrt{2}\hat{\zeta}_{001}(t)\cos(\pi z) \tag{4.1e}$$

for free–slip, impermeable horizontal boundary conditions. Note that all mode amplitudes in (4.1) are real and that the lateral positions of the nodes in (4.1) are fixed.

The equations for the 10 mode amplitudes read

$$\tau\dot{\mathbf{X}}= -\sigma\hat{q}^2\mathbf{X}+\sigma\frac{\hat{k}^2}{\hat{q}^2}\left[(1+\psi)\mathbf{Y}+\frac{8}{\pi^2}\mathbf{U}\right]+\begin{bmatrix}X_2\\X_1\end{bmatrix}S \tag{4.2a}$$

$$\tau\dot{\mathbf{Y}}= -\hat{q}^2\mathbf{Y}+(r-Z)\mathbf{X}+\begin{bmatrix}X_2\\X_1\end{bmatrix}T \tag{4.2b}$$

$$\tau\dot{\mathbf{U}}= -\frac{L}{3}\hat{k}^2\mathbf{U}+\hat{q}^2\psi\mathbf{Y}+V\mathbf{X} \tag{4.2c}$$

$$\tau\dot{Z}= -b[Z-\mathbf{X}\cdot\mathbf{Y}] \tag{4.2d}$$

$$\tau\dot{V}= -b\left[\frac{L}{4}V+\frac{2}{3}\psi Z+\frac{1}{2}\mathbf{X}\cdot\mathbf{U}\right] \tag{4.2e}$$

$$\tau\dot{S}= -2\sigma\hat{d}^2S+\sigma(1+\psi)\frac{\hat{k}^2}{\hat{d}^2}T-\frac{4}{3}\frac{\hat{q}^2}{\hat{d}^2}X_1X_2 \tag{4.2f}$$

$$\tau\dot{T}=rS-2\hat{d}^2T-\frac{b}{4}(X_1Y_2+X_2Y_1). \tag{4.2g}$$

Here the critical modes have been combined into two–component vectors

$$\mathbf{X}=\begin{bmatrix}X_1\\X_2\end{bmatrix}=\frac{1}{q_c^o}\begin{bmatrix}\hat{u}_{101}\\\hat{u}_{011}\end{bmatrix};\quad \mathbf{Y}=\begin{bmatrix}Y_1\\Y_2\end{bmatrix}=\frac{q_c^o}{R_c^o}\begin{bmatrix}\hat{\theta}_{101}\\\hat{\theta}_{011}\end{bmatrix};\quad \mathbf{U}=\begin{bmatrix}U_1\\U_2\end{bmatrix}=\frac{\pi q_c^o}{2\sqrt{2}\,R_c^o}\begin{bmatrix}\hat{\zeta}_{100}\\\hat{\zeta}_{010}\end{bmatrix}. \tag{4.3a}$$

and

$$Z=-\frac{\pi\sqrt{2}}{R_c^o}\hat{\theta}_{002};\quad V=\frac{\pi^2}{2\sqrt{2}\,R_c^o}\hat{\zeta}_{001};\quad S=\frac{\pi\sqrt{2}}{q_c^{o2}}\hat{u}_{112};\quad T=\frac{\pi\sqrt{2}}{R_c^o}\hat{\theta}_{112}. \tag{4.3c}$$

The critical quantitities k_c^o, q_c^o, R_c^o, τ, b of (4.2) for $\psi=0$ are identical to those given in (3.3e) for the model (3.2). The reduced quantities r, $\hat{k}$, $\hat{q}$, are defined in the same way as in (3.3f) and $\hat{d}^2=(k^2+2\pi^2)/q_c^{o2}$. The Nusselt number, N(t), is given by 1+(2/r)Z(t) as in Section 3.

For S=0, T=0 the equations (4.2a–e) are identical to (3.2), however, the physical meaning of the mode amplitudes is different. In Section 3 we described a set of x–rolls that could propagate laterally — X_1 and X_2 were real and imaginary parts of a complex mode amplitude that allowed pattern propagation via a time–dependent phase. Here, however, X_1 and X_2 are real amplitudes of standing x– and y–rolls, respectively.

The convection patterns reported by MS had a wavelength $\lambda\simeq 2$ corresponding to the critical wave number, $k_c^o=\pi$, in a reference no–slip one–component system at $\psi=0$. We therefore fix also in our free–slip model $\hat{k}=k/k_c^0=1$. Then the threshold for onset of convection of (4.2) is

$$r_{stat}(\hat{k}=1) = \left[1+\psi+(24/\pi^2)\psi/L\right]^{-1}. \qquad (4.4)$$

The depression (4.4) of the threshold in the mixture relative to the one–component fluid, $r_{stat}(\hat{k}=1, \psi=0)=1$, agrees quite well with the experiments.

4.2 Relation to shadowgraph pictures

Most of the experimental information on the spatial structure of convective patterns stems from shadowgraph pictures. Within our model (4.1–4.3) one can easily identify and interpret the lateral intensity distributions of shadowgraph pictures and relate them to intensity patterns of the hydrodynamic fields. Since shadowgraph patterns are besides Nusselt numbers an additional and convenient means to display and compare our results with experiments we establish here the relation to our model variables.

Assuming small variations of the refractive index to be proportional to variations, $\Delta\rho$, of the mass density around its mean, ρ_0, one finds that the lateral intensity modulation ΔI of a shadowgraph picture is given by a vertical average

$$\Delta I(x,y,t) = -A\int_0^1 dz\, \Delta\rho_{conv}(\mathbf{r},t) \qquad (4.5a)$$

over the density variation

$$\Delta\rho_{conv}/\rho_0 = -(\kappa\nu/gd^3)[(1+\psi)\theta+\zeta] \qquad (4.5b)$$

caused by convection. The laterally homogeneous conductive part yields only a constant. The constant A is positive since increasing the mass density increases the refractive index and the light intensity. Within our mode truncation

$$\Delta I(x,y,t) = -A'[p_1(t)\cos(kx)+p_2(t)\cos(ky)] \qquad (4.6a)$$

can be expressed in terms of a positive constant A' related to A[15] and in terms of two shadow–graph amplitudes

$$p_i(t) = (1+\psi)Y_i(t)+U_i(t) \quad i=1,2. \qquad (4.6b)$$

Thus p_1 and p_2 measure particular q_x– and q_y–Fourier amplitudes of the convective temperature and ζ field. On the other hand, the shadowgraph amplitudes p_1 and p_2 measure the light intensity of stripes parallel to the y– and x–axis, respectively, in a shadowgraph pattern. Thus, $p_1=p_2$ corresponds to a square convective pattern while $p_2=0$ ($p_1=0$) represents convective patterns of x–rolls (y–rolls). For finite $p_1 \neq p_2$ the shadowgraph intensity pattern (4.6) is a superposition of two crossed line patterns the intensities of which reflect the size of the q_x– and q_y–Fourier amplitudes of the convective fields. We also can infer the structure and spatial phase of the convective velocity field from the shadowgraph picture: For example a dark stripe in y–direction at x=0 implies $p_1>0$ ↔ reduced refractive index ↔ reduced density ↔ warm ↔ upflow. The opposite holds for a bright stripe.

4.3 Convection in the absence of sidewall forcing

In the Galerkin model (4.1–4.3) there are no symmetry breaking imperfections that distort amplitudes or favour particular phases of the convective fields as lateral sidewalls do in experiments. The rather drastic effects of sidewall forcing on the convective behaviour in square shaped containers are reviewed in Section 4.4. Here we first discuss the convective solutions of the unperturbed system (4.1–4.3).

There are eight convective fixed points which have been determined analytically[15]. Their projection onto the plane of shadowgraph amplitudes is shown in Fig. 8 with the associated patterns in the insets. The four fixed points for y–rolls (1 and 1') and x–rolls (3 and 3') are equivalent to each other. Also the four solutions 2,2',4,4' for square convection are equivalent among each other. The symmetry operations in real space by which the field patterns of each set are transformed into each other are rotation by $\pi/2$ and translations by $\lambda/2$ along x and/or y. Primed and unprimed convective states are related to each other by reverting everywhere the flow direction.

In Fig. 9 we show the bifurcation diagram of the Nusselt number. The two stationary convective states of rolls and squares, respectively, bifurcate at r_{stat} out of the conductive state, N=1. In the Soret regime immediately above onset convection is stationary and very feeble for both patterns, the slope of N is very small, and the buoyancy force in (2.1b) is dominated by the concentration field[15]. Convection with sizeable amplitudes starts only around the threshold value, r=1, of the one–component reference system. In the Rayleigh regime, r>1, the buoyancy force is dominated by the temperature contribution to the density variation and rolls convect heat more

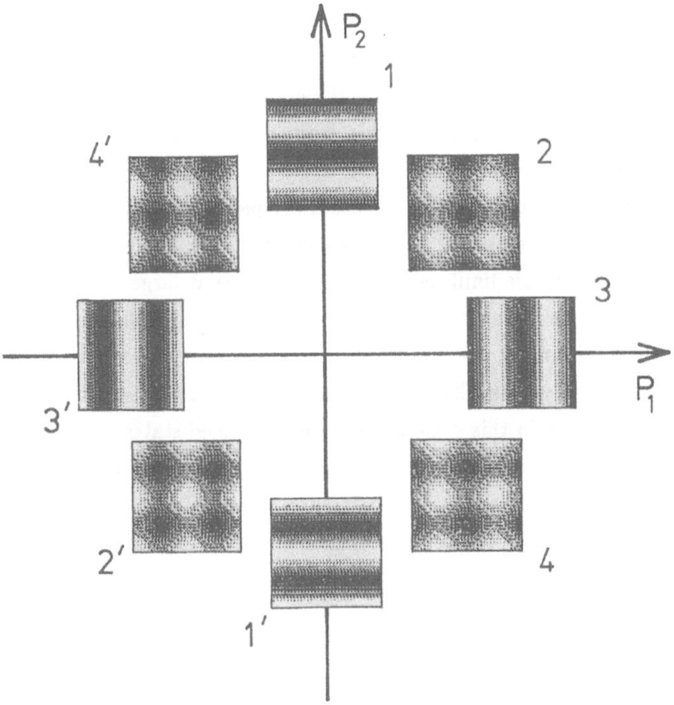

Fig. 8 Convective fixed points projected onto the plane of shadowgraph pattern amplitudes. The light intensity distributions for the corresponding flow patterns are shown in the insets with 10 gray levels. Dark areas imply upflow. From ref. 15.

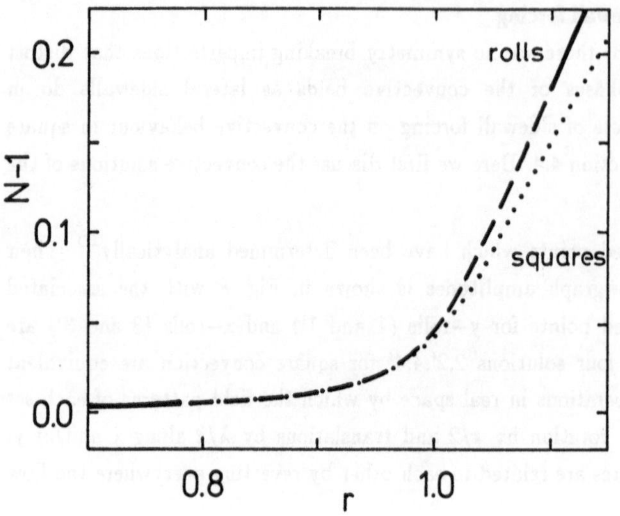

Fig. 9 Nusselt number vs Rayleigh number for stationary convection in the absence of imperfections. Squares (dots) are always unstable, however, only very slightly in the Soret regime r<1. Rolls are very slightly (strongly) stable in the Soret regime (Rayleigh regime r>1). In the intermediate interval around r=1 rolls are unstable (dashed line). There pattern oscillations between rolls and squares occur. Parameters are ψ=0.1, σ=23, L=0.018 with a convective threshold $r_{stat}(\hat{k}=1)$=0.068.

efficiently than squares. In the far Rayleigh regime stationary roll (square) convection is strongly stable (unstable).

In the Soret regime stationary roll (square) convection is very weakly stable (unstable) with one real eigenvalue being very close to zero for each pattern. This property of the Soret regime may be visualized by a mexican hat potential with a very slightly indented groove: The four roll fixed points (1,1',3,3') would be situated in four downwards indentures on the p_1, p_2 axes while the four square fixed points would correspond to the four local maxima in between on the diagonals. We also mention that without the S,T modes the mexican hat would be truly rotational symmetric with a circle of fixed points in the groove corresponding to a sequence of patterns varying continuously between those of Fig. 8 each being marginally stable with a zero eigenvalue. Note, however, that the Galerkin equations are not derivable from a potential — the above picture serves only as a convenient means for discussion.

In the Soret regime the roll and square fixed points have also a complex conjugate eigenvalue with a negative real part that grows with increasing r. At a threshold value (r=0.78 in Fig. 9) it causes a forwards Hopf bifurcation into a stable limit cycle which undergoes at large r in the far Rayleigh regime a hysteretic transition to the stable roll state. Around r=1 the limit cycle's projection onto the plane of shadowgraph amplitudes has almost circular shape. Therein each of the eight states of Fig. 8 appear successively with alternating roll and square patterns along the orbit, say, 1–2–3–4–1'–2'–3'–4'–1. Since in this cycle primed and unprimed states, e.g. 1 and 1', appear alternatingly after half a period the flow direction (e.g., the turning direction of the rolls) is reverted globally every half period. However, such pattern oscillations with periodically reverting flow directions have not been observed in the experiments of MS for reasons explained below.

4.4 Effect of sidewalls

4.4.1 Sidewall forcing in experimental square cells Inspecting the shadowgraph pictures of MS one finds[15] that only patterns with one particular phase relation marked by upflow and downflow positions relative to the sidewalls occur in the stationary and oscillatory regimes. The flow direction of the roll and square states is always the same — only the amplitudes of the x–rolls and y–rolls vary. Thus in the experimental cell there are sidewall forces that favour, say, the roll states 1 and 3 with upflow near the lateral boundaries. Furthermore, and consequently, the square state 2 that is basically a superposition of x–rolls (3) and y–rolls (1) with upflow near the sidewalls is favoured over square pattern 2' with downflow near the walls. As an aside we remark that any sidewall forcing with a bias for a particular flow direction tends to suppress square patterns 4 and 4' in a square container since the flow configurations of these patterns are not the same near the x– and y–sidewalls.

In conclusion one thus finds from examinations of the shadowgraph pictures of MS that in their square cell containers there are sidewall forces that favour patterns 1,2,3 in the first quadrant of Fig. 8.

4.4.2 Incorporating symmetry breaking sidewall forces into the model Since the above described sidewall forces are absent in our Galerkin model we tried to incorporate their phase selection effect phenomenologically by adding a small constant inhomogeneous term to one of the equations of motion for the pattern carrying modes. We found good agreement with experiments for a force $\boldsymbol{\xi}$ that drives the modes $\mathbf{U}$ of the ζ–field. Thus we replaced eq. (4.2c) (with $\hat{k}=\hat{q}=1$) by

$$\tau\, \dot{\mathbf{U}}(t) = -(L/3)\, \mathbf{U}(t) + \psi\, \mathbf{Y}(t) + V(t)\, \mathbf{X}(t) + \boldsymbol{\xi}. \tag{4.7}$$

We used $\boldsymbol{\xi}=\xi(1,1)$ since the forcing of x– and y–sidewalls in a square cell are the same. The size of ξ is a fit parameter (c.f. further below).

In the picture of the mexican hat potential the forcing with positive ξ tilts the hat along the 135° diagonal thereby lowering the groove in the first quadrant of Fig. 8 so that square pattern 2 and roll patterns 1 and 3 with upflow near the sidewalls become favoured. In fact, stationary square patterns become stable only by the sidewall forcing. This is consistent with the fact that MS found well developed regular square patterns only in square shaped cells, i.e., in systems for which in our model the forces $\xi_1=\xi_2$ provide the largest bias for squares — for rectangular cells with side lengths $L_y \ll L_x$ one would have $\xi_1 \ll \xi_2$ which favours roll patterns 3 or 3'.

4.4.3 Changes induced by the sidewall forcing In the far Rayleigh regime well above r=1 the mode amplitudes are large. There nonlinear convective mode–coupling forces dominate over the sidewall forcing — the groove of the mexican hat has deep minima for the roll states — and consequently the stationary roll solution remains unchanged except for the fact that fixed points 1,3 are slightly favoured over 1',3'. Also the threshold Rayleigh number r_2 (c.f. Fig. 10) down to which rolls are stable depends only slightly on ξ.

In the Soret regime, however, where the mode amplitudes are very small the sidewall forcing stabilizes square state 2 and destabilizes all other patterns — the unforced system's mexican hat potential has such a miniscule maximum at position 2 in the groove that already a small tilt produces a minimum there.

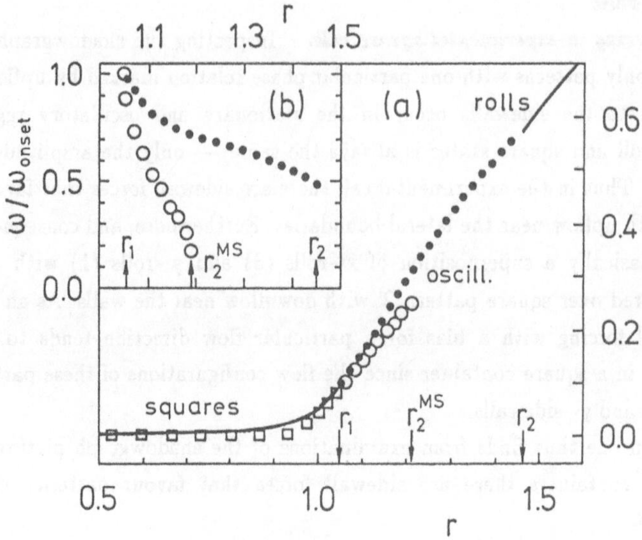

Fig. 10 Mean Nusselt number and reduced oscillation frequency (inset) versus r in the presence of sidewall forcing. Full lines (dots) refer to stationary (oscillatory) convection in the model for ψ=0.1, L=0.018, σ=23. Open squares (circles) are experimental results of MS for stationary squares (oscillations). Oscillations (stationary rolls) first appear at $r_1(r_2)$. The forcing ξ=0.005 was chosen in the model such that its Hopf bifurcation, r_1, coincides with the experimental onset of oscillations in the large square cell of MS. From ref. 15.

Increasing r beyond a Hopf threshold value r_1 (c.f. Fig. 10) stationary squares become unstable in a Hopf bifurcation to a limit cycle. Instead of running through all eight patterns in the groove of the hat the system now oscillates as in the experiments between roll patterns 1 and 3 with square state 2 appearing in between as shown in Fig. 11. With increasing r the oscillation frequency decreases from its value at the Hopf bifurcation threshold towards a final finite value at r_2 where a transition with small hysteresis into a stationary roll state occurs.

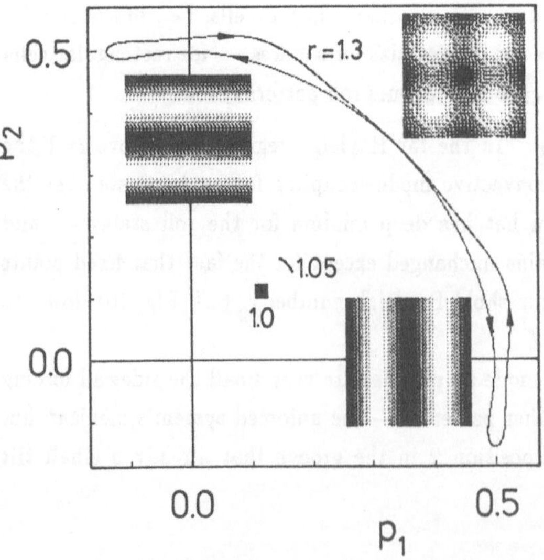

Fig.11 Limit cycle trajectories projected onto the p_1–p_2 plane of shadowgraph pattern amplitudes. The full square represents the stationary square convective state at r=1. Parameters as in Fig. 10. From ref. 15.

The existence range, $r_2^{MS}-r_1$, of experimental oscillations as far as we could read it off from the figures of MS is smaller than the range, r_2-r_1, in our model. Furthermore, the experimental oscillation period is about a factor of ten longer than that of our model. Both differences might be due to our horizontal free–slip boundary conditions that facilitate pattern oscillations with periodic reorientation of the rolls in comparison with the experimental no–slip boundaries. The Nusselt number oscillates with twice the pattern oscillation frequency provided that the x–roll state and the y–roll state are identical. Imperfections breaking this symmetry of the square cell under rotation of $\pi/2$ would cause (small) deviations from the precise double frequency oscillation of N(t). As in the experiments the initial sinusoidal oscillation of N(t) becomes more and more deformed[15] when r increases beyond the Hopf threshold.

Acknowledgements The results presented here are based on work done together with S. J. Linz, H. W. Müller, and J. Niederländer. Stimulating discussions with them and support by the Deutsche Forschungsgemeinschaft are gratefully acknowledged.

REFERENCES

1. L. D. Landau and E. M. Lifschitz, *Fluid Mechanics* (Pergamon Press, 1959).
2. G. Z. Gershuni and E. M. Zhukhovitskii, *Convective Stability of Incompressible Fluids* (Keter Press, Jerusalem, 1976).
3. J. K. Platten and J. C. Legros, *Convection in Liquids* (Springer, Berlin, 1984).
4. For a longer list of references, see, e.g., ref. 14.
5. G. Ahlers and I. Rehberg, Phys. Rev. Lett. **56**, 1373 (1986); G. Ahlers, D. S. Cannell, and R. Heinrichs, Nucl. Phys. B (Proc. Suppl.) **2**, 77 (1987); T.S. Sullivan and G. Ahlers, Phys. Rev. Lett. **61**, 78 (1988).
6. H. Gao and R. P. Behringer, Phys. Rev. A **34**, 697 (1986), A **35**, 3993 (1987).
7. C. M. Surko, P. Kolodner, A. Passner, and R. W. Walden, Physica D **23**, 220 (1986); P. Kolodner, C. M. Surko, A. Passner, and H. L. Williams, Phys. Rev. A **36**, 2499 (1987); P. Kolodner, D. Bensimon, and C. M. Surko, Phys. Rev. Lett. **60**, 1723 (1988).
8. E. Moses and V. Steinberg, Phys. Rev. A **34**, 693 (1986); (E) A **35**, 1444 (1987); V. Steinberg, E. Moses, and J. Fineberg, Nucl. Phys. B (Proc. Suppl.) **2**, (1987).
9. P. Le Gal, A. Pocheau, and V. Croquette, Phys. Rev. Lett. **54**, 2501 (1985).
10. E. Moses and V. Steinberg, Phys. Rev. Lett. **57**, 2018 (1986).
11. O. Lhost and J. K. Platten, to be published in Phys. Rev. A (1988).
12. S. J. Linz and M. Lücke, Phys. Rev. A **35**, 3997 (1987); (E) A **36**, 2486 (1987).
13. S. J. Linz and M. Lücke, in *Propagation in Nonequilibrium Systems*, ed. J. E. Wesfreid, H. R. Brand, P. Manneville, G. Albinet, and N. Boccara (Springer, Berlin, 1988), p. 292.
14. S. J. Linz, M. Lücke, H. W. Müller, and J. Niederländer, Phys. Rev. A **38**, Dec. (1988).
15. H. W. Müller and M. Lücke, Phys. Rev. A **38**, Sept. (1988).
16. S. J. Linz and M. Lücke, Phys. Rev. A **36**, 3505 (1987).
17. G. Veronis, J. Mar. Res. **23**, 1 (1965); J. Fluid Mech. **34**, 315 (1968).
18. R. S. Schechter, I. Prigogine, and J. R. Hamm, Phys. Fluids **15**, 379 (1972).
19. H. R. Brand, P. C. Hohenberg, and V. Steinberg, Phys. Rev. A **30**, 2548 (1984).
20. J. K. Platten and G. Chavepeyer, Int. J. Heat Mass Transfer **18**, 1071 (1975).
21. E. Knobloch, D. R. Moore, J. Toomre, and N. O. Weiss, J. Fluid Mech. **166**, 409 (1986); A. E. Deane, E. Knobloch, and J. Toomre, Phys. Rev. A **36**, 2862 (1987); A **37**, 1817 (1988).
22. M. G. Velarde, in *Evolution of Order and Chaos*, ed. H. Haken (Springer, Berlin, 1982).
23. M. C. Cross, Phys. Lett. A **119**, 21 (1986); G. Ahlers and M. Lücke, Phys. Rev. A **35**, 470 (1987).
24. E. Knobloch and D. R. Moore, Phys. Rev. A **37**, 860 (1988); M. C. Cross and K. Kim, Phys. Rev. A **37**, 3909 (1988).
25. S. J. Linz, Ph. D. thesis, Universität Saarbrücken (1988), unpublished.
26. E. Knobloch, Phys. Rev. A **34**, 1538 (1986); and earlier work cited therein.
27. W. Barten, M. Lücke, W. Hort, and M. Kamps, unpublished.
28. E. Moses and V. Steinberg, Phys. Rev. Lett. **60**, 2030 (1988).

TIME-DEPENDENT PHASE TRANSITIONS

Paul Mandel and H. Zeghlache
Optique nonlinéaire théorique, Université Libre de Bruxelles,
Campus de la Plaine, CP 231, Bruxelles 1050, Belgium

T. Erneux
Department of Engineering Sciences and Applied Mathematics,
Northwestern University, Evanston IL 60208, USA

CONTENT

1. INTRODUCTION.

Since the work of Graham & Haken [1968,1970a & b], Kazantsev, Rautian & Surdutovich [1968] and Degiorgio & Scully [1970] it is known that the laser first threshold can be described as a second order phase transition which can be derived from a Landau-Ginzburg theory. Therefore it also falls into the class of problems which can be studied in the terms of bifurcation theory. In these notes it is the point of view of the bifurcation theory which we shall adopt because it will enable us to make a number of statements which are rather independent of the precise system which we study.

In the next two sections we shall give an extremely simple example of a time-dependent phase transition and discuss the problematics of such a phenomenon. This will show analytically how the intuitive idea of stability can be drastically modified when a control parameter varies in time, even very slowly. As an example let us consider the simple problem

$$\frac{dI}{dt} = -I^2 + (A-1)I \tag{1.1}$$

which has two steady state solutions: $I=0$ and $I=A-1$ like the intensity in the semiclassical laser theory. These steady solutions are shown on Fig.1(a). It is easy to check that $I=0$ is stable for $A\leq 1$ whereas $I=A-1$ is stable for $A\geq 1$. This result can be derived by direct integration of (1.1). However if we sweep the control parameter A linearly in time according to the law $A\equiv A(t)=A(0)+vt$ we obtain the result shown on Fig.1(b). Comparing the two graphs shows how dramatic the effect of the sweep can be. The purpose of these lecture notes is to propose an analytical study of the modified stability properties when a control parameter is varied slowly in time. All the following sections will deal with the same problem but each time with the addition of one aspect which has been idealized (that is, usually neglected) in the second and third sections.

2. STATIC BIFURCATION.

A rich though still oversimplified description of a tuned single mode unidirectional ring laser in the homogeneous limit is contained in the set of equations for the reduced variables [Haken,1985]:

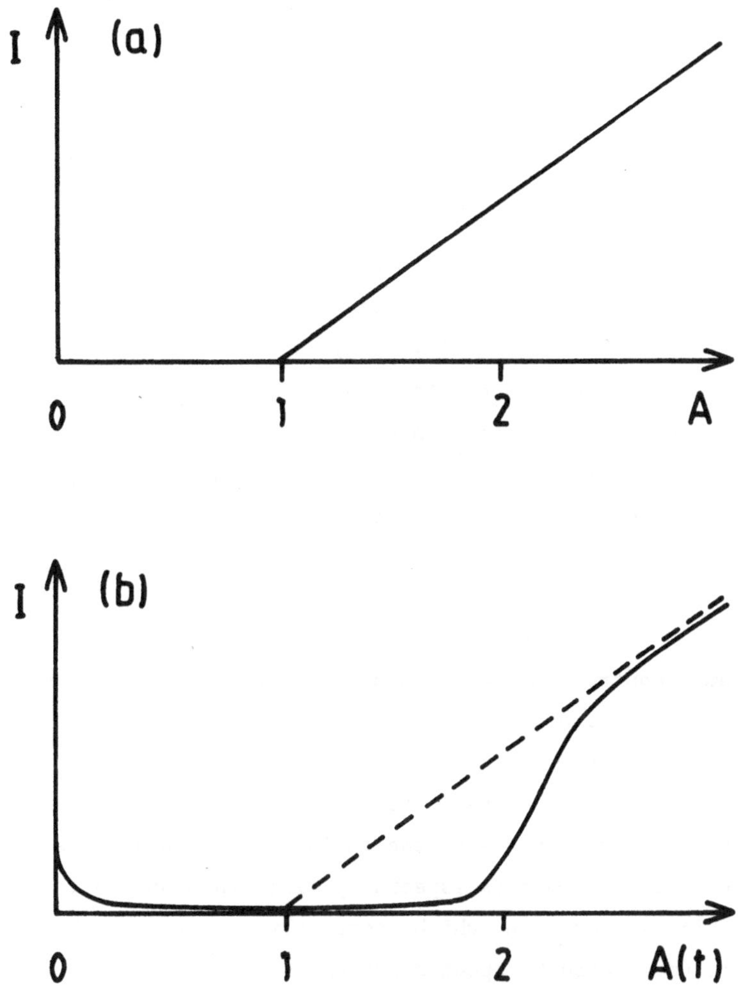

Figure 1. (a):steady solutions of $dI/dt=-I^2+(A-1)I$
(b):solution of $dI/dt=-I^2+(A-1)I$ when $A=A(0)+vt$ with $v=0.1$, $A(0)=0$ and $I(0)=0.5$.

$$E_t \equiv dE/dt = -E + AP$$

$$P_t = \gamma_p(-P + EF) \qquad (2.1)$$

$$F_t = \gamma_f(-F + 1 - EP)$$

In these equations E is the electric field in the cavity, P and F are related to the atomic polarization and population difference, respectively. The time is scaled to the cavity decay rate; the same scaling has been introduced to define the reduced atomic decay rates γ_p and γ_f. Finally the parameter A is of central importance. It is known as the optical pump parameter and can be written as the ratio of the low intensity gain due to stimulated emission (α) to the cavity linear loss factor (κ):

$$A = \alpha/\kappa \qquad (2.2)$$

It will be our control parameter in a bifurcation theory approach of the laser problem since it combines all remaining relevant parameters. The laser first threshold is given by the condition $A=1$. Lasing action thus begins when the gain overcomes the losses. Since α is proportional to the square of the atom-field coupling constant, the lasing condition $A>1$ also defines a strong coupling domain. This is especially clear when one notices that the linear Fokker-Planck equation which correctly describes the black-body radiation field (dominated by spontaneous emission) has solutions which are defined for $A<1$ and diverge exactly at $A=1$.

The steady solutions of Eq.(2.1) for the intensity $I=E^2$ are $I=0$ and $I=A-1$. A linear stability analysis of the trivial solution $I=0$ gives the characteristic equation

$$(\lambda+\gamma_f)[\lambda^2+\lambda(\gamma_p+1)+\gamma_p(1-A)]=0 \qquad (2.3)$$

whereas the linear stability of the lasing solution $I=A-1>0$ leads to the characteristic equation:

$$\lambda^3+\lambda^2(1+\gamma_p+\gamma_f)+\lambda\gamma_f(A\gamma_p+1)+2\gamma_p\gamma_f(A-1)=0 \qquad (2.4)$$

Eq.(2.3) indicates that the trivial solution is stable below threshold ($A<1$) and unstable above threhold ($A>1$). When $A=1+\epsilon$, $0<\epsilon<<1$ the finite intensity solution is stable but it may become unstable for higher values of the pump parameter A. When this second laser threshold exists, it leads through a Hopf bifurcation to periodic solutions which may be either stable or unstable. In the latter case the laser has a chaotic intensity. As shown by Haken [1975] a simple change of variables leads from the Eqs.(2.1) to the Lorenz equations. The main difference between the laser equations and the Lorenz equations is the domain of parameter which can be accessed experimentaly.

In these notes we shall be mainly concerned with the laser first threshold which takes place at $A=1$ and corresponds to the exchange of stability between the trivial and the finite steady intensity solutions. As is clear from general considerations on critical points and obvious from a glance at the two characteristic equations (2.3) and (2.4), the laser first threhold is characterized by a vanishing root, i.e., an infinite relaxation time. This fact is the source of critical fluctuations at the threshold. By this we mean that the fluctuations are anomalously weakly damped. This fact has sometime been called critical slowing down in other contexts and has the same physical content.

3. TIME-DEPENDENT PERFECT BIFURCATION.

In order to study in a finite amount of time the laser first threshold, or more generally a first order phase transition, it is temptating to sweep the control parameter across the bifurcation point. In the laser case a classic example is furnished by the experimental result of Ruschin and Bauer [1979] who displayed the critical intensity fluctuations at threshold by reducing with a constant rate the gas pressure in a laser with saturable absorber. This amounts to replace A by $A(t)=A(0)+vt$ where $A(0)$ is the initial value of the control parameter and v the sweep rate. For a simple laser the natural choice is $A(0)<1$ with $v>0$ and the equations of motion become:

$$E_t=-E+A(t)P$$

$$P_t=\gamma_p(-P+EF) \tag{3.1}$$

$$F_t=\gamma_f(-F+1-EP)$$

The main difference is that now we are dealing with a set of non-autonomous differential equations which does not have in general a steady state since the control parameter is not steady. However the trivial solution remains an exact solution of the equations Eqs.(3.1) even when $v\neq 0$. In the good cavity limit (γ_p and $\gamma_f \gg 1$) we may adiabatically eliminate the atomic variables P and F which leads to

$$E_t=E(-1+\frac{A(t)}{1+E^2}) \tag{3.2}$$

Since $E=0$ is an exact solution, we may linearize (3.2) around $E=0$ and perform a standard stability analysis of this trivial solution. Let

$$E=\eta x+O(\eta^2) \quad 0<\eta\ll 1$$

for all times including the initial time. The linearized equation becomes

$$x_t=x[-1+A(t)]\equiv x\mu(t) \tag{3.3}$$

The exact solution of (3.3) is

$$x(t)=x(0)\exp[\int_0^t \mu(t')dt'] \tag{3.4}$$

As in the static case we define the stability domain by the condition that the argument of the exponential be negative. When it cancels the boundary of stability is reached and when it is positive the solution is unstable. In practical terms, the instability of the solution x(t) signal the fact that x(t) is no longer O(1) and that E(t) is therefore no longer $O(\eta)$. Thus the dynamical solution leaves the vicinity of $E=0$ and, in fact, begins a jump transition towards a finite intensity solution. The dynamical stability condition is therefore

$$\int_0^{t^*} \mu(t')dt'=0 \tag{3.5}$$

which is an implicit equation for the critical time t^* at which the dynamical solution becomes unstable. If we define an intermediate time $\bar{t}$ as the time at which the static bifurcation point is reached: $A(\bar{t})=1$ we may express the dynamical instability condition (3.5) as a balance condition:

$$\int_0^{\bar{t}} \mu(t')dt' = -\int_{\bar{t}}^{t^*} \mu(t')dt' \qquad (3.6)$$

The lhs integral describes an accumulation of stability for $t<\bar{t}$ since in this domain μ is negative whereas the rhs integral describes an accumulation of instability since μ is positive for $t>\bar{t}$. Thus (3.6) expresses the balance between an accumulated stability and an accumulated instability.

It is important to realize that the simplification which lead from the full set (3.1) to the single equation (3.3) is by no means responsible for the result (3.5). As shown previously [Mandel & Erneux, 1984; Mandel, 1986] the same analysis can be performed on the set of equations (3.1). The result of this analysis is that the critical time t^* defined by (3.5) is a <u>lower</u> bound for the critical time t^* derived from the full set (3.1).

In the case of a linear sweep we have $\mu(t)=A(0)-1+vt$ and the dynamical bifurcation point is reached when (3.6) is verified, i.e., when

$$t^*=2\bar{t}, \qquad A(t^*)-A(\bar{t})=A(\bar{t})-A(0) \qquad (3.7)$$

Except for the fact that t^* is a function of the sweep rate:

$$t^*=2[1-A(0)]/v \qquad (3.8)$$

the relation between t^* and $\bar{t}$ as well as the relations between $A(t^*)$, $A(\bar{t})$ and $A(0)$ are all independent of the sweep rate: no matter how fast or how slowly the sweep takes place, the relations (3.7) remain true.

<u>4. TIME-DEPENDENT BIFURCATION: CONSTANT IMPERFECTION</u>.

In many cases, the existence of a zero solution to the bifurcation equation results from an idealization of the problem. For instance, the laser equations (2.1) are semi-classical equations which neglect the contribution of spontaneous emmission. One of the consequences of spontaneous emmission is that the zero solution is replaced by an extremely small intensity. The success of the semiclassical theory rests, in a large part, on the fact that this intensity of spontaneous origin is orders of magnitude smaller than the intensity of stimulated origin above the first laser threshold. In this section we shall analyze how even an exponentially small deviation to the zero solution may drastically alter the dynamic bifurcation diagram.

Let us consider the generalization of the bifurcation equation (3.3):

$$x_t = x[-1+A(t)]+\delta = x\mu(t)+\delta, \quad \mu(t)=\mu_i+\epsilon t, \quad 0<\epsilon<<1, \quad \mu_i=A(0)-1<0 \tag{4.1}$$

The parameter δ is usually called the imperfection parameter. If we introduce a change of variable $\tau=\mu(t)$, Eq.(4.1) becomes

$$\epsilon x_\tau = \tau x+\delta \tag{4.2}$$

whose solution is

$$x(\tau)=x(\mu_i)\exp[(\tau^2-\mu_i^2)/2\epsilon]+(\delta/\epsilon)\exp(\tau^2/2\epsilon)\int_{\mu_i}^{\tau}\exp(-s^2/2\epsilon)ds \tag{4.3}$$

with $\mu_i=\mu(0)<0$ so that $\tau=0$ corresponds to the time at which the steady bifurcation is crossed when $\delta=0$. In the limit $\epsilon \to 0$ the dominant contribution to the integral appearing in (4.3) is $\sqrt{2\pi\epsilon}$. Defining an exponentially small imperfection through

$$\delta=\delta_0(\epsilon)\exp(-k^2/2\epsilon), \quad k=O(1), \quad \delta_0(\epsilon)=\epsilon^p, \quad p>0 \tag{4.4}$$

we obtain from (4.3) the result

$$x(\tau)\simeq x(\mu_i)\exp[(\tau^2-\mu_i^2)/2\epsilon]+\epsilon^{p-1/2}\sqrt{2\pi}\,\exp[(\tau^2-k^2)/2\epsilon] \tag{4.5}$$

In a nonlinear theory additional terms in powers of δ are expected [Mandel & Erneux, 1987].

On the basis of the result (4.5) we can already make some comments on the influence of an imperfection on the delayed bifurcation. Two cases have to be distinguished:

(i)either $\mu_i^2<k^2$ and it is the first of the two exponentials in (4.5) which diverges first, thus defining the position of the dynamical bifurcation. In this case the delay is maximum, depends on the initial value of the control parameter since it is given by $\tau=-\mu_i$ which is precisely the result obtained in (3.8).

(ii)or $\mu_i^2>k^2$ and it is the second exponential in (4.4) which diverges first. In this case the delay is reduced since it occurs at $\tau=k<-\mu_i$. Furthermore the delay no longer depends on the initial value of the control parameter.

Thus there exists a critical value of the imperfection defined as

$$\delta_c=\epsilon^p\exp(-\mu_i^2/2\epsilon), \quad p>1/2, \quad \mu_i=O(1) \tag{4.6}$$

which separates two domains: (i)when $\delta<\delta_c$ the delay is maximum, independent of the imperfection but dependent of the initial value of the control parameter; (ii) when $\delta>\delta_c$ the delay occurs at $\tau=k$, depends on the imperfection but no longer on the initial condition of the control parameter. Still in both cases the delay can be O(1) on the ϵ scale.

5. TIME-DEPENDENT BIFURCATION: PERIODIC IMPERFECTION.

In a number of situations, the imperfection is time-dependent; we shall analyze an example of this class of problems in this section by considering a periodic imperfection with zero mean:

$$x_t=x[-1+A(t)]+\delta\cos(\sigma t)\equiv x\mu(t)+\delta\cos(\sigma t), \quad \mu(t)=\mu_i+\epsilon t, \; 0<\epsilon<<1, \; \mu_i<0 \tag{5.1}$$

As in the previous section this equation has to be considered as an approximate equation which derives from a set of nonlinear equations during a linear stability analysis.

Introducing again the time $\tau=\mu(t)$ leads to

$$\epsilon x_\tau=\tau x+\delta\cos[\sigma(\tau-\mu_i)/\epsilon] \tag{5.2}$$

whose solution is

$$x(\tau)=x(\mu_i)\exp[(\tau^2-\mu_i^2)/2\epsilon]+(\delta/\epsilon)\exp(\tau^2/2\epsilon)\int_{\mu_i}^{\tau}\exp(-s^2/2\epsilon)\cos[\sigma(s-\mu_i)/\epsilon]ds \tag{5.3}$$

We shall use the definition (4.4) for the exponentially small δ. Depending on the relative magnitude of σ and ϵ, various cases have to be considered.

5a. If $\sigma \ll \epsilon$ the cosine function in (5.3) can be replaced by unity and the imperfection is practically constant. Hence (5.3) reduces to (4.3) and the results of section 4 apply here.

5b. If $\sigma=O(\epsilon)$, we introduce a new frequency

$$\omega=\sigma/\epsilon=O(1)$$

so that the cosine function in (5.3) can be approximated by $\cos(\omega\mu_i)$ in the limit $\epsilon \to 0$. This therefore leads to the solution

$$x(\tau)\simeq x(\mu_i)\exp[(\tau^2-\mu_i^2)/2\epsilon]+\epsilon^{p-1/2}\sqrt{2\pi}\,\exp[(\tau^2-k^2)/2\epsilon]\cos(\sigma\mu_i/\epsilon) \tag{5.4}$$

As in section 4 the magnitude of μ_i relative to k will determine the main properties of the dynamical bifurcation.

(i) either $\mu_i^2<k^2$ and it is the first of the two exponentials in (5.4) which diverges first, thus defining the position of the dynamical bifurcation. In this case the delay is maximum, depends on the initial value of the control parameter and is given by $\tau=-\mu_i$ which is precisely the result obtained in (3.8).

(ii) or $\mu_i^2>k^2$ in which case the dynamical bifurcation will take place at $\tau=k$. Furthermore $x(\tau)$ will diverge to $+\infty$ if $\cos(\sigma\mu_i/\epsilon)>0$ whereas it will diverge towards $-\infty$ if $\cos(\sigma\mu_i/\epsilon)<0$.

5c. If $\sigma=O(\sqrt{\epsilon})$ we introduce another frequency via

$$\Omega=\sigma/\sqrt{\epsilon}=O(1)$$

In the limit $\epsilon \to 0$ the integral in (5.3) has a leading asymptotic contribution expressed by

$\sqrt{2\epsilon\pi}\ \exp(-\Omega^2/2)\cos(\Omega\mu_i/\sqrt{\epsilon})$

so that the explicit solution becomes

$$x(\tau)\simeq x(\mu_i)\exp[(\tau^2-\mu_i^2)/2\epsilon]+\epsilon^{p-1/2}\sqrt{2\pi}\ \exp[(\tau^2-k^2-\sigma^2)/2\epsilon]\cos(\sigma\mu_i/\epsilon) \qquad (5.5)$$

The difference with the case 5b is that if $\mu_i^2>k^2+\sigma^2$ the dynamical bifurcation occurs at $\tau\sim\sqrt{\sigma^2+k^2}$ which therefore becomes a function of the frequency and the exponentially small amplitude of the imperfection.

5d. As σ becomes an O(1) quantity, the first exponential in (5.5) will be the first to diverge because $(\sigma^2+k^2)>\mu_i^2$ in which case the delayed bifurcation point is made independent of all properties of the periodic imperfection. This is easy to understand if we recognize the existence of multiple time scales: the perfect system has a characteristic time scale measured by ϵ whereas the imperfection varies on a characteristic time scale σ which here is O(1). Thus only the average imperfection is felt by the system and this average is zero.

6. TIME-DEPENDENT BIFURCATION; STOCHASTIC IMPERFECTION.

Another and most common source of imperfection is of course the stochastic influence of the environment on the system. In the laser case two sources of noise have to be distinguished:

(i) internal sources such as the spontaneous field;

(ii) external sources which usually reflect the unavoidable technological imperfections of lasers.

A general study of the Langevin equations associated with Eqs.(3.1) is still an open question. Here we shall consider the much more modest question of the bifurcation equation (3.3) in the presence of an additive white noise:

$$x_t=\mu(t)x+\xi(t), \qquad \langle\xi(t)\rangle=0, \qquad \langle\xi(t)\xi(t')\rangle=2D\delta(t-t') \qquad (6.1)$$

This problem has to be treated in two steps. First the system is <u>prepared</u> with a fixed value of $\mu=\mu_i<0$ and left to relax towards the stable state corresponding to μ_i. This state is the initial state, x(0), for the experiment in which the control parameter is swept slowly and linearly in time. The corresponding average solution of Eq.(6.1) is [Van Den Broeck and Mandel, 1987]

$$\overline{x(t)}=x(0)\exp[G(t)/2]$$

$$\overline{x^2(t)}=x^2(0)\exp[G(t)]+D\sqrt{\pi/\epsilon}\exp[G(t)+\mu_i^2/\epsilon]\{\mathrm{Erf}[\mu(t)/\sqrt{\epsilon}]-\mathrm{Erf}[\mu(0)/\sqrt{\epsilon}]\} \tag{6.2}$$

where

$$G(t)=\epsilon t^2+2\mu_i t \text{ and } \mathrm{Erf}(x)=(2/\sqrt{\pi})\int_0^x \exp(-y^2)dy$$

This experiment is then reproduced a large number of times leading to an ensemble average since x(0) is still a stochastic variable [see, e.g., Broggi et al., 1986 and Mannella et al., 1987 for investigations of this ensemble average using digital and analog numerical simulations]. The average of $x^2(0)$ is obtained by setting $\epsilon=0$ in (6.2) which gives $D/|\mu_i|$. In the case of an additive white noise we can perform the two averaging procedures separately since the correlation time vanishes. This leads to the result [Zeghlache et al., 1988]

$$\langle x^2(t)\rangle=\exp[G(t)]D/|\mu_i|+D\sqrt{\pi/\epsilon}\exp[G(t)+\mu_i^2/\epsilon]\{\mathrm{Erf}[\mu(t)/\sqrt{\epsilon}]-\mathrm{Erf}[\mu(0)/\sqrt{\epsilon}]\} \tag{6.3}$$

In terms of the parameters

$$y=(\mu_i+\epsilon t)/\sqrt{\epsilon}=[A(t)-1]/\sqrt{\epsilon}, \quad a=y(0) \tag{6.4}$$

Eq.(6.3) can be written as

$$|\mu_i|\langle x^2(t)\rangle/D=\exp(y^2-a^2)+\sqrt{\pi}|a|\exp(y^2)[\mathrm{Erf}(y)-\mathrm{Erf}(a)] \tag{6.5}$$

Let us stress that in the laser case the lhs of this last equation is nothing but the ratio of the average intensity at time t to the average initial intensity. We now define a critical time t* by the condition that the average intensity $\langle x^2(t)\rangle$ reaches a predefined level $\langle x^2(t^*)\rangle = x^2_{th}$. With the definition $z \equiv y(t^*)$ the implicit equation for the dynamical bifurcation time t^* becomes

$$\beta\exp(-z^2) - \sqrt{\pi}\mathrm{Erf}(z) = \exp(-a^2)/|a| - \sqrt{\pi}\mathrm{Erf}(a) \tag{6.6}$$

with β being defined as $\sqrt{\epsilon}x^2_{th}/D$. In the deterministic case we recover from (6.6) the solution $z=-a$ with $x^2_{th}=\langle x^2(0)\rangle$. In the general case, Eq.(6.6) has to be solved numerically [Zeghlache et al., 1988].

The main feature which we want to stress here is the occurrence of a noise-induced saturation of the delay suffered by the bifurcation. Indeed when the parameter a (which is related to the initial condition) tends to large negative values, the deterministic equations (3.8) predict an equivalently large delay of the bifurcation whereas in the stochastic case Eq.(6.6) becomes

$$\beta\exp(-z^2_{max}) - \sqrt{\pi}\mathrm{Erf}(z_{max}) = \sqrt{\pi} \tag{6.7}$$

Thus z_{max} appears to be independent of the initial state. This conclusion still holds if the system happens to have an initial condition which is deterministic, i.e., with zero dispersion $[\overline{x(0)}=x(0)]$. There are two remarkable values of β for which an analytic estimate of z_{max} can be derived from (6.7):

(i) When z_{max} is small, an expansion of (6.7) in powers of z_{max} gives $z_{max} \sim (\beta-\sqrt{\pi})/2$. This indicates that the bifurcation will be significantly delayed only when β is sufficiently larger than $\sqrt{\pi}$. Otherwise the delay becomes negligible.

(ii) When z_{max} is large, the asymptotic expansion of (6.7) leads to the result: $z_{max} \sim [\log(\beta/2\sqrt{\pi})]^{1/2}$.

7. TIME-DEPENDENT BIFURCATION: NONLINEAR THRESHOLD.

In the previous section, we have had to introduce a threshold intensity to determine the time at which the dynamical bifurcation takes place. This problem was not discussed explictly in the deterministic case. The implicit assumption made in sections 3 to 5 was that the dynamical bifurcation point was defined by the fact that the initial condition was again reached after the initial decay of x(t):

$$x(t^*)=x(0) \text{ for } t^*>0 \tag{7.1}$$

In the stochastic analysis, a threshold intensity had to be introduced since, by definition of the problem, the initial condition is useless as a discriminator.

In both cases it was still possible to solve the problem after performing a linearized stability analysis. A recent experiment, however, offers an alternative measure of the delayed bifurcation and some unexpected results as well [Scharpf et al. 1987]. In a commercial Ar^+ laser the laser cavity losses are swept in time by use of an acousto-optic modulator. The output intensity is recorded with a detector whose lower level is still above the average intensity noise level. Therefore the initial condition is undetectable and an alternative procedure has to be used. The method which was proposed is the following. After the intensity emerges from the noisy background, a jump transition occurs which brings the system in the vicinity of the dynamical solution A(t)-1. During this jump, the intensity has an inflexion point, i.e., its time derivative goes through a maximum. This point of maximum slope has been recorded and defined as the dynamical bifurcation point. This is an interesting definition which is neither (7.1) nor the arbitrary threshold condition used in the stochastic analysis of section 6. Quite the contrary, it is an implicit threshold condition which falls in the domain where the linearized analysis around the zero solution is no longer valid. Indeed with an equation of the type (3.3) it takes an infinite time to reach the point of maximum slope. Thus with this definition of the delayed bifurcation, we are probing the nonlinear nature of the system. In the present case it turns out that the range of parameters used in the experiment made it sufficient to consider Eq.(3.2) expanded up to the cubic term only:

$$E_t \simeq E[-1+(1-E^2)A(t)] \tag{7.2}$$

If $I=E^2$ is the laser intensity, the point of maximum slope takes place at the time T defined as

$$d^2I/dt^2|_{t=T}=0 \tag{7.3}$$

Although it has not been possible to solve (7.3) analytically, a numerical study [Mandel, 1987] revealed the following facts. The time T corresponds in a very good approximation to the time at which the dynamical intensity I(t) is half the time-dependent "intensity" A(t)-1. This clearly shows how critically T depends on the nonlinearity of the problem. Another difference between t^* and T is the way in which they scale as a function of the sweep rate. From (3.8) it is clear that $t^*\sim 1/v$. The experimental report from Scharpf et al. presented the result $T\sim 1/\sqrt{v}$. It was shown numerically that there is no contradiction between these two results. More precisely the exact scaling law of T is a complex function of the sweep rate v which has two limits: the inverse sweep rate in the asymptotic limit of very small sweep rates (which was not reached experimentally) and the square root of the inverse sweep rate at somewhat larger values of the sweep rate.

8. NONLINEAR ANALYSIS OF THE IMPERFECT BIFURCATION PROBLEM.

Other aspects of the delayed bifurcation problem require that nonlinear terms be retained [Haberman, 1979; Erneux & Mandel, 1986] rather than limiting the expansions to linear terms as in (3.3). The fact that Eqs.(3.1) have steady solutions I=0 and I=A-1 for the intensity suggest that we model the intensity bifurcation diagram by the equation

$$x_t=x(A-1-x)+\delta\equiv x(\mu-x)+\delta \tag{8.1}$$

In the static case δ may account for the intensity whose origin is in the spontaneous emission. The physical steady solution of (8.1) is

$$x_s=(1/2)[\mu+\sqrt{\mu^2+4\delta}] \tag{8.2}$$

which reduces to

$$\begin{cases} x=0 & \mu<0 \\ x=\mu & \mu>0 \end{cases}$$

when $\delta=0$. In the dynamical case we define a new time variable by $\tau=\mu(t)$ so that (8.1) becomes

$$\epsilon x_\tau = x(\tau - x) + \delta \tag{8.3}$$

In the limit $\epsilon \to 0$ we seek solutions of ((8.3) in powers of ϵ:

$$x(\epsilon;\tau,\delta)=x(0;\tau,\delta)+\epsilon x(1;\tau,\delta)+\dots$$

which yields a sequence of simple equations for the $x(n;\tau,\delta)$. In particular we easily find

$$x(\epsilon;\tau,\delta)=x_s(\tau,\delta)[1-\frac{\epsilon}{\tau^2+4\delta}+O(\epsilon^2)] \tag{8.4}$$

This expression remains regular as long as $4\delta \gg O(\epsilon)$. Thus when δ is O(1), we have to a good approximation

$$x(\epsilon;\tau,\delta)=x_s(\tau,\delta)=x_s[\mu(t),\delta] \tag{8.5}$$

which states that the time-dependent solution follows "adiabatically" the steady solution (8.2). In this case there is neither a significant delay... nor a well-defined bifurcation point.

The expression (8.4) indicates that the ϵ-expansion will no longer converges when

$$\delta \leq O(\epsilon) \tag{8.6}$$

$$\text{and} \quad \tau^2 \leq O(\epsilon) \tag{8.7}$$

To have a measure of the delay in this case, we consider specifically $\delta=\epsilon$ and $\tau=0$. At this point the solution (8.2) leads to

$$x_s(0,\epsilon)=\epsilon^{1/2}$$

whereas the solution of (8.3) becomes [Erneux & Mandel, 1986]

$x(\epsilon;0,\epsilon)\simeq\sqrt{2\epsilon/\pi}$

Hence in the domain $\delta=O(\epsilon)$ the solution follows adiabatically the steady solution except in the vicinity of the point $\tau=0$ corresponding to A=1 where a small deviation of order $\sqrt{\epsilon}$ from the "adiabatic" result (8.5) appears. Thus the perfect bifurcation point begins to be felt though rather weakly. As the imperfection parameter δ gets further smaller, the delay will grow and eventually the domain analyzed in section 3 will be reached.

9. CONCLUSIONS.

We have tried to show in these notes that the properties of second order phase transitions associated with algebraic bifurcation problems may significantly be altered when the control parameter is made time-dependent.

In particular there was a great emphasis on the influence of an imperfect transition point in order to get closer to most experimental situations. In the case of a stochastic imperfection, we have on purpose followed a procedure which is much simpler than the full stochastic analysis via Fokker-Planck-like equations. This is motivated by the fact that in the case of the laser, the intensity is the second moment of the distribution function and as shown in section 6 much information can already be derived by solving exactly the linearized Langevin equation directly. The limited amount of informations which can be obtained in this way is sufficient to characterize the dynamical steady bifurcation.

In sections 7 and 8 we studied problems which can no longer be described by linearized theories. This is obvious for the nonlinear threshold experiments [Sharpf et al., 1987] in which the switching time depends in a fairly complicated way on the sweep rate. This is also quite clear in section 8 where the domain of divergence of the ϵ-expansion depends on the type of nonlinearity. In the case of a quadratic nonlinearity as in Eq.(8.1) we obtain the conditions (8.6) and (8.7). For a cubic nonlinearity, however, the same type of behaviour is observed [Erneux & Mandel, 1986] but for $\delta=O(\epsilon^{3/4})$ and the same condition (8.7) for the reduced time τ.

Another aspect of the swept control parameter problem which is worth stressing is its use as a model discriminator. Indeed in the static case where the control parameter is kept fixed, only the steady states are usually studied and compared to the experimental results. On the contrary when the control parameter is swept, time-dependent properties of the phase transition are probed. Such properties are much more sensitive to all parameters of the problem than the steady states which usually depend on only a restricted number of the system parameters. This is apparent in (3.5) which is a nonlocal relation which requires a knowledge of $\mu(t)$ during the whole sweep whereas the corresponding static instability condition is A=1 which only depends on only one value of μ. Experimentally this richness of the dynamical problem was used recently to compare two laser models and therefore check the underlying physical mechanisms [Glorieux & Dangoisse, 1985; Arecchi et al., 1988]. In both cases the question was to determine to what extent the two-level model of laser physics is a reliable model for some classes of lasers. It was shown by both sets of authors that in some cases a more complete description of the atomic or molecular spectrum is necessary to expain the observed dynamics.

ACKNOWLEDGMENTS.

This work was supported in part by grants from the IAP program of the belgian government, the FNRS (Belgium), the European Community, the algerian government and a NSF grant DMS-8701302.

REFERENCES.

Arecchi F.T., Gadomski W., Meucci R. & Roversi J.A. 1988, Opt. Commun. **65**, 47.

Broggi G., Colombo A., Lugiato L.A. & Mandel P. 1986, Phys. Rev. **A33**, 3635.

DeGiorgio V. & Scully M.O., 1970, Phys. Rev. **2A**, 1170.

Erneux T. & Mandel P. 1986, SIAM J. Appl. Math. **46**,1.

Glorieux P. & Dangoisse D. 1985, IEEE J. Quantum Electron. **QE-21**, 1486.

Graham R. & Haken H. 1968, Z. Phys. **213**, 240.

——————————— 1970a, Z. Phys. **235**, 166.

——————————— 1970b, Z. Phys. **237**, 31.

Haberman R. 1979, SIAM J. Appl. Math.**37**, 69.

Haken H. 1975, Phys. Lett. **53A**, 77.

Haken H. 1985, Light, vol.2, North-Holland (Amsterdam).

Kazantsev A.P., Rautian S.Q. & Surdutovich, G.J. 1968, Sov. Phys. J.E.T.P. **27**, 756.

Mandel P. 1986, in "Frontiers in quantum optics" p.430, E.R.Pike and S.Sarkar eds (Adam Hilger, Bristol).

Mandel P. 1987, Opt. Commun. **64**, 549.

Mandel P. & Erneux T. 1984, Phys. Rev. Letters **53**, 1818.

Mandel P. & Erneux,T. 1987, J. Stat. Phys. **48**,1059.

Mannella R., Moss F. & McClintock P.V.E. 1987, Phys. Rev. **A35**, 2560.

Ruschin R. & Bauer S.H. 1979, Chem. Phys. Lett. **66**, 100.

Scharpf W.,Squicciarini M.,Bromley D., Green C., Tredicce J.R. & Narducci L.M. 1987, Opt. Commun. **63**, 344

Van Den Broeck C. & Mandel P. 1987, Phys. Lett. **122B**, 36.

Zeghlache H., Mandel P. & Van Den Broeck C. 1988, "Influence of colored noise on a delayed bifurcation", to appear.

QUANTUM TREATMENT OF DISPERSIVE OPTICAL BISTABILITY

H. Risken and K. Vogel
Abteilung für Theoretische Physik
Universität Ulm
D-7900 Ulm
Federal Republic of Germany

First the model of Drummond and Walls (DW) is introduced by which dispersive optical bistability is treated in a fully quantum mechanical way. The basic equation of the DW model is an equation for the density operator of the light mode inside the cavity. By using the Glauber-Sudarshan P-function or the Q-function the equation of motion for the density operator is transformed into a pseudo Fokker-Planck equation (FPE), i.e. a FPE where the diffusion matrix is not positive definite or semidefinite. For the Wigner-function the diffusion matrix is positive definite but additional third-order derivatives occur. It is shown that the pseudo FPE and the equation for the Wigner-function can be solved in terms of matrix continued fractions. Explicit results are given for the stationary Q- and Wigner-function as well as for the lowest nonzero eigenvalue. For vanishing thermal fluctuations this eigenvalue determines the tunneling rate between the two 'bistable' states.

1 Introduction

In an optical bistable device a coherent continuous wave (cw) laser beam is injected into an optical cavity being nearly in resonance with the incident light field. The cavity is filled with a nonlinear material, i.e. a material where some of its properties such as 'dielectric constant' or 'absorbtion constant' depend on the electric field. As it was first shown theoretically by Szöke et al. [1] and later demonstrated experimentally by Gibbs et al. [2] such a device can show bistability. This means that the system can be put into a low transmission regime or into a high transmission regime depending on the initial conditions. When the incident field is small the system is in the low transmission branch, whereas for very high incident field the system is in the high transmission branch. For intermediate incident fields the system can be in either one of the different states. Thermal and quantum fluctuations can cause transitions between these two states, similar to first-order phase transitions. The early investigations [1,2] initiated many further investigations on this effect, see [3-6] for reviews.

Usually one distinguishes between absorptive and dispersive optical bistability. In the former case mainly the absorption coefficient depends on the electric field. If the field is large enough, it can bring half of the atoms of the nonlinear material into the upper state. Because of this bleaching the absorption is small, i.e. it leads to high transmission of the incident light. For dispersive optical bistability the dielectric constant mainly depends on the electric field leading to different resonant frequencies of the cavity. If one starts with a low intensity and if the parameters are such that one has a mismatch between the frequency of the incident light and the cavity frequency this leads to a low transmission coefficient. If, on the other hand, the incident light has a large amplitude and the dielectric constant depends on the amplitude in such a way that the cavity frequency is now nearly in resonance with the incident field one gets a high transmission coefficient. A simple mathematical model describing both effects is a driven nonlinear Duffing equation [7,8] with an additional nonlinear absorption coefficient. In most materials one does not need to consider absorptive and dispersive effects simultaneously. One has either absorptive bistability (dispersive effects are neglected) or dispersive bistability (absorptive effects are neglected).

Without any fluctuations the bistable system stays in one of its two stable states. If fluctuations are taken into account, one has a certain probability that the system jumps from one state to the other. Obviously an appreciable jump rate would destroy the usefulness of such an optical bistable device. Therefore one tries to eliminate these fluctuations. In principle this is possible for external as well as for thermal fluctuations. Quantum fluctuations, however, cannot be eliminated. The transition rates between the two stable states due to quantum fluctuations are therefore very important because they determine the ultimate limits of working conditions for such a device.

In a fully quantum mechanical theory one starts with the system Hamilton operator. Damping is described by a coupling to a heat bath. By eliminating the heat bath variables one arrives at an equation for the density operator of the system. Because of damping irreversible terms occur in this equation. For solving the density operator equation one usually introduces continuous quasi-distribution functions like the Glauber-Sudarshan P-function [9,10], the Q-function [11], (i.e. the expectation value of the density operator in a coherent state), or the Wigner-function [12]. For absorptive bistability the equations for the quasi-distribution functions are partial differential equations with arbitrarily high order derivatives. Truncating these equations after the second term leads to a Fokker-Planck equation (FPE). As was demonstrated by Drummond [13], the transition rates for different representations, i.e. for P- and Q-functions, may then differ by orders of magnitude because the approximations made are different for different continuous quasi-distributions. For dispersive bistability the equation of motion for the quasi-distribution function also leads to partial differential equations. For the DW model, however, only derivatives up to the second order occur for the P- and Q-function, whereas derivatives up to the third order occur for the Wigner-function. Thus the equations for the P- and Q-function are Fokker-Planck like equations. These equations have, however, a nonpositive definite diffusion matrix and therefore cannot be interpreted as describing the Brownian motion of a particle under the influence of a suitable force. These equations have been termed pseudo FPEs (PFPE). A simulation of this PFPE is not possible. Nevertheless we show, that the stationary solution of this PFPE for Q-function can be obtained by the matrix continued fraction method (MCF) which was used nearly a decade ago by Vollmer and one of us [14] (see also [15] for a summary) for solving twodimensional FPEs describing Brownian motion in inclined periodic potentials. Furthermore, we show that the stationary solution for the Wigner-function containing derivatives up to third order can also be obtained by the MCF method. (Due to squeezing [16] the P-function does not exist.) The quantum transition rates are obtained for the DW model by calculating the lowest nonzero eigenvalue of the equations for the P- Q- and Wigner-function by the MCF method. It turned out that these eigenvalues for the P- Q- and the Wigner-function agree within numerical accuracy. Although, due to squeezing, the P-function does not exist in general its equation of motion can be used to obtain this eigenvalue as will be explained later on.

It should be mentioned that by doubling the phase space and by introducing the positive P-function it is possible to derive a FPE with a positive definite or positive semidefinite diffusion matrix for the DW model. (For a discussion of generalized representations of the density operator the reader is refered to [17-20].) The Langevin equation corresponding to the Fokker-Planck equation for the P- and Q-function can then be simulated [21] but one has to handle twice as many variables as in the original problem. Furthermore, we want to mention that the complex P-function can also be used to solve the DW model [22]. Here one has also to handle 2 complex = 4 real variables. Stationary expectation values have been obtained analytically for the Drummond and Walls model [22] for vanishing thermal fluctuations. Transition rates have not been obtained by the complex P-function. Transition rates, however, have been calculated for the DW model in the low cavity damping limit by solving a suitably truncated density matrix equation [23,24]. In this lecture we obtain the transition rate by solving the equations for the P- Q- and Wigner-function for the DW model with the MCF method.

These lecture notes are organized as follows. First in Sect. 2 we explain the DW model and write down the basic equation of motion for the density operator. Furthermore the quasi-classical equations are also derived in this section. Next in Sect. 3 the equations for the P- Q- and Wigner-functions are given. In Sect. 4 we make an expansion of the various quasi-distribution functions and derive the equation of motion for the expansion coefficients. Furthermore it is shown that these equations can be cast into a tridiagonal vector recurrence relation. In Sect. 5 we explain the MCF method for solving the tridiagonal vector recurrence relation. Finally in Sect. 6

the results for the stationary solution as well as for the eigenvalues are presented. We would like to mention that the main idea of the procedure of these lecture notes is outlined in [25]. The method is further developed in [24,26-29].

2 Model and Basic Equations

In their model for dispersive optical bistability Drummond and Walls [22] consider a single-mode field inside a cavity. The cavity is filled with a dispersive material, the nonlinearity of which is described by a third-order polarization. Making the rotating-wave approximation they obtained a single-mode Hamiltonian of the form

$$H_c/\hbar = \omega_c a^\dagger a + \chi a^{\dagger 2} a^2 \,, \tag{2.1}$$

where $a^\dagger$ and a are the creation and annihilation operators of the light mode, ω_c is the frequency of the cavity mode and χ is the anharmonicity parameter. The incident external field is assumed to be a classical coherent driving field with frequency ω_L. The interaction Hamilton operator reads

$$H_L/\hbar = F\left(a^\dagger e^{i\omega_L t} + a\, e^{-i\omega_L t}\right) . \tag{2.2}$$

As already mentioned in the introduction the loss mechanism is described by a coupling to a heat bath. In a reference frame rotating at frequency ω_L the equation of motion for the density operator takes the form

$$\dot{\rho} = -i[H/\hbar, \rho] + \kappa L_{ir}[\rho] \,, \tag{2.3}$$

where $H/\hbar$ and $L_{ir}[\rho]$ are given by

$$H/\hbar = -\Omega a^\dagger a + \chi a^{\dagger 2} a^2 - F(a + a^\dagger) \,, \tag{2.4}$$

$$L_{ir}[\rho] = 2a\rho a^\dagger - \rho a^\dagger a - a^\dagger a \rho + 2n_{th}[[a,\rho],a^\dagger] \,. \tag{2.5}$$

Here $\Omega = \omega_L - \omega_c$ is the difference between the frequency of the driving field and the frequency of the cavity mode. The irreversible term $L_{ir}[\rho]$ is well known in quantum optics [30,31] to describe the cavity damping, where κ is the damping constant. In (2.5) n_{th} is the number of thermal quanta inside the cavity due to the coupling to a heat bath. In the optical region n_{th} is very small and may be safely neglected.

Classical Limit

Before we try to solve the quantum mechanical master equation (2.3) it seems to be appropriate to look for some classical limit. For the expectation values of the operators a and $a^{\dagger}$

$$\alpha = Tr(a\rho), \qquad \alpha^{*} = Tr(a^{\dagger}\rho) \tag{2.6}$$

we obtain from (2.3-2.5) the classical equation of motion

$$\dot{\alpha} = \{i\Omega - \kappa - 2i\chi|\alpha|^{2}\}\alpha + iF \ . \tag{2.7}$$

In the derivation we have factored the terms of the form $Tr(a^{\dagger}a^{2}\rho)$ according to

$$Tr(a^{\dagger}a^{2}\rho) = Tr(a^{\dagger}\rho)\cdot[Tr(a\rho)]^{2} = \alpha^{*}\alpha^{2} \ . \tag{2.8}$$

One can show that the factorization in (2.8) is a good approximation in the limit $\Omega/\chi \to \infty$. Because the photon number inside the cavity is of the order Ω/χ (see below) this means that we have a large number of photons inside the cavity, i.e. that (2.8) corresponds to the classical limit.

For further considerations it is convenient to introduce normalized quantities $\tilde{\alpha}$, $\tilde{t}$, $\tilde{\kappa}$, $\tilde{F}$ in such a way that in (2.7) the parameters Ω and χ are normalized to one, i.e

$$\frac{d\tilde{\alpha}}{d\tilde{t}} = \{i(1 - 2|\tilde{\alpha}|^{2}) - \tilde{\kappa}\}\tilde{\alpha} + i\tilde{F} \ . \tag{2.9}$$

This is achieved by the transformation (Ω>0)

$$\tilde{t} = \Omega t, \quad \tilde{\alpha} = \sqrt{\chi/\Omega}\ \alpha, \quad \tilde{\kappa} = \kappa/\Omega, \quad \tilde{F} = \frac{1}{\Omega}\sqrt{\chi/\Omega}\ F \ . \tag{2.10}$$

The stationary solution for the normalized amplitude $|\tilde{\alpha}|$ as a function of the driving field $|\tilde{F}|$ is shown in Fig. 1. The broken lines are unstable stationary solutions of (2.9) which cannot be observed. It is clearly seen that between curve (1) and (4) bistable operation occurs. The lower part is the low transmission regime, the upper one is the high transmission regime. For an explicit expression of the stationary region see [22,28]. Because the normalized amplitude $|\tilde{\alpha}|$ and also the normalized intensity $\tilde{I} = \tilde{\alpha}^{*}\tilde{\alpha}$ is of order one, the unnormalized intensity $I = \alpha^{*}\alpha$, i.e. the number of photons inside the cavity, is of the order Ω/χ. It can be shown that (2.7) is the rotating-wave approximation to the driven Duffing-oscillator equation [7,8], see [28].

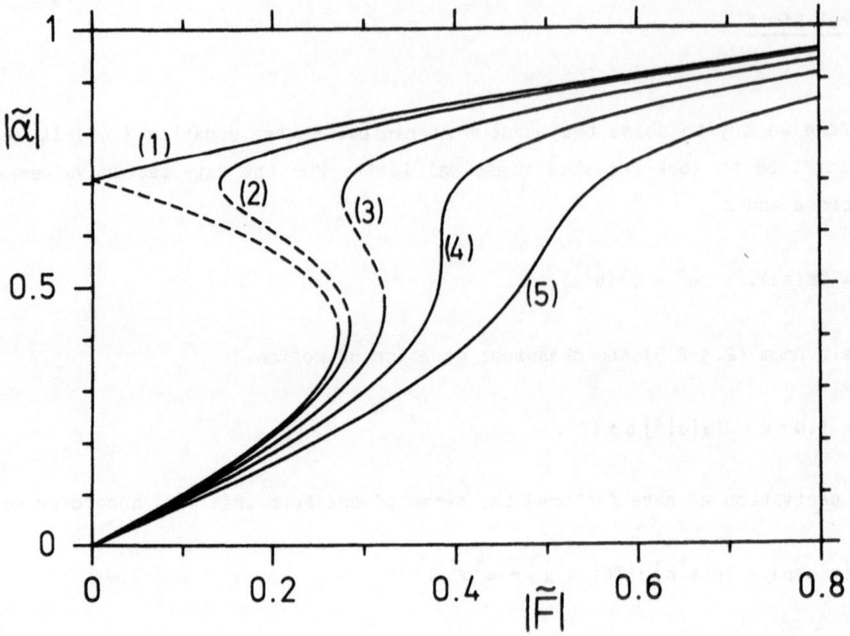

Fig. 1: Stable (solid lines) and unstable (broken lines) stationary solutions of (2.9) as a function of the scaled driving field $|\tilde{F}|$ for $\tilde{\kappa}=0$ (1), $\tilde{\kappa}=0.2$ (2), $\tilde{\kappa}=0.4$ (3), $\tilde{\kappa}=1/\sqrt{3}$ (4) and $\tilde{\kappa}=0.8$ (5).

3 Pseudo Fokker-Planck Equation and Fokker-Planck Equation with an Additional Third-Order-Derivative Term

In quantum optics the Glauber-Sudarshan P-function, the Q-function and the Wigner-function [9-12] are well known. These are quasi-distribution functions in the complex α plane, by which normally ordered, antinormally ordered or symmetrically ordered moments can be obtained according to

$$\langle a^{\dagger n} a^m \rangle = \int \alpha^{*n} \alpha^m P(\alpha)\, d^2\alpha \tag{3.1}$$

$$\langle a^m a^{\dagger n} \rangle = \int \alpha^{*n} \alpha^m Q(\alpha)\, d^2\alpha \tag{3.2}$$

$$\langle (a^{\dagger n} a^m)_{sym} \rangle = \int \alpha^{*n} \alpha^m W(\alpha)\, d^2\alpha\ . \tag{3.3}$$

In (3.3) $(a^{\dagger n} a^m)_{sym}$ means the symmetrically ordered form of the operators [32]. We have for instance

$$(a^\dagger a)_{sym} = \frac{1}{2}(a^\dagger a + aa^\dagger) \ ; \qquad (a^\dagger a^2)_{sym} = \frac{1}{3}(a^\dagger a^2 + aa^\dagger a + a^2 a^\dagger). \tag{3.4}$$

The Q-function can be expressed in terms of the density operator according to

$$Q(\alpha) = \langle\alpha|\rho|\alpha\rangle/\pi \ ; \qquad |\alpha\rangle = \text{coherent state} \ , \tag{3.5}$$

whereas the density operator itself can be expressed by the P-function according to

$$\rho = \int |\alpha\rangle\langle\alpha| \, P(\alpha) \, d^2\alpha \ . \tag{3.6}$$

Obviously the Q-function always exists and is positive. The Wigner-function also exists but it may have negative values. If squeezing occurs - as it does for the DW model - the P-function does not exist in general. (For a discussion of squeezed states see for instance [16].) Nevertheless the expansion coefficients of the P-function into a complete set of functions do exist because they are connected to the expectation values of $a^\dagger$ and a. Therefore the equations of motion for the expansion coefficients of the P-function may also be used for calculating expectation values and eigenvalues, see Sects. 4 and 5. The equation of motion for the density operator (2.3-2.5) transforms to equations of motion for the quasi-distribution functions. Using standard techniques [19,30,31] the equations are usually derived by first looking for the equations of the characteristic functions of P, Q and W. Employing operator relations of the form

$$[a, e^{a^\dagger\beta}] = \beta e^{a^\dagger\beta} \ , \qquad [e^{a\beta^*}, a^\dagger] = \beta^* e^{a\beta^*} \ ,$$

$$e^{a\beta^* + a^\dagger\beta} = e^{a\beta^*} e^{a^\dagger\beta} e^{-|\beta|^2/2} = e^{a^\dagger\beta} e^{a\beta^*} e^{|\beta|^2/2} \tag{3.7}$$

and performing a Fourier transform one obtains these equations. For the P- and Q-function they are given by

$$\begin{aligned}\frac{\partial}{\partial t}\begin{Bmatrix}P\\Q\end{Bmatrix} = &-\frac{\partial}{\partial\alpha}[-\kappa\alpha + i\Omega\alpha + 2i(1\mp1)\chi\alpha - 2i\chi\alpha^2\alpha^* + iF]\begin{Bmatrix}P\\Q\end{Bmatrix} \\ &-\frac{\partial}{\partial\alpha^*}[-\kappa\alpha^* - i\Omega\alpha^* - 2i(1\mp1)\chi\alpha^* + 2i\chi\alpha^{*2}\alpha - iF]\begin{Bmatrix}P\\Q\end{Bmatrix} \\ &\mp i\chi\frac{\partial^2}{\partial\alpha^2}\alpha^2\begin{Bmatrix}P\\Q\end{Bmatrix} + 2\kappa\left(n_{th} + \frac{1}{2} \mp \frac{1}{2}\right)\frac{\partial^2}{\partial\alpha^*\partial\alpha}\begin{Bmatrix}P\\Q\end{Bmatrix} \pm i\chi\frac{\partial^2}{\partial\alpha^{*2}}\alpha^{*2}\begin{Bmatrix}P\\Q\end{Bmatrix},\end{aligned} \tag{3.8}$$

whereas the equation of the Wigner-function takes the form

$$\frac{\partial W}{\partial t} = -\frac{\partial}{\partial\alpha}[-\kappa\alpha + i\Omega\alpha + 2i\chi\alpha - 2i\chi\alpha^2\alpha^* + iF]W$$

$$-\frac{\partial}{\partial\alpha^*}[-\kappa\alpha^* - i\Omega\alpha^* - 2i\chi\alpha^* + 2i\chi\alpha^{*2}\alpha - iF]W$$

$$+ 2\kappa(n_{th}+\frac{1}{2})\frac{\partial^2 W}{\partial\alpha^*\partial\alpha} + \frac{i}{2}\chi\left[\frac{\partial^3}{\partial\alpha\partial\alpha^{*2}}\alpha^* - \frac{\partial^3}{\partial\alpha^*\partial\alpha^2}\alpha\right]W \,. \qquad (3.9)$$

In (3.8) only derivatives up to second order occur. Thus (3.8) is of Fokker-Planck form. The diffusion matrix, however, is not positive definite or semidefinite for large intensitities, i.e. for

$$\alpha^*\alpha = \alpha_1^2 + \alpha_2^2 > \frac{\kappa}{\chi}(n_{th}+\frac{1}{2}\mp\frac{1}{2}) \,. \qquad (3.10)$$

where again the upper sign is valid for the P-function and the lower sign for the Q-function. Therefore (3.8) has been termed pseudo Fokker-Planck equation (PFPE). In [28] we have shown for a simple example that solutions of a PFPE may exist even if the diffusion matrix is not positive definite or semidefinite. Thus nonpositivity of the diffusion matrix should not be taken as an objection to the usefulness of (3.8). We will show in Sect. 6 that the stationary Q-function as well as eigenfunctions do exist and can be calculated from (3.8). The equation of motion for the Wigner-function has a positive definite diffusion matrix, it contains, however, beyond the ordinary Fokker-Planck terms a third-order derivative-term. As was shown by Pawula [33] for the onedimensional case an equation with a third-order derivative term (generally for a truncated Kramers-Moyal expansion truncated at a finite derivative of order larger than 2) is in contradiction to the positivity of the transition probability for small times. In the present case the Wigner-function needs not to be positive. Therefore the inclusion of a third-order term is of no objection either. (It is shown in [34] for a simple example that Kramers-Moyal expansions truncated at a finite derivative of order larger than 2 may be quite useful.) As explained in Sects. 4, 5 our MCF method is also applicable to the partial differential equation (3.9) having higher than second order derivatives.

If one nevertheless neglects the third-order derivative terms in (3.9) one obtains an ordinary FPE for two real variables. It is also worth mentioning that this FPE can be obtained from the equation (3.8) for the P- and Q-function by neglecting the $\mp$ or $\pm$ terms, i.e. those terms by which the equation for the P- and Q-function differ from each other. This equation is then just the FPE which is obtained from the classical equation (2.7) with an additional drift term $2i\chi\alpha$ by adding complex Gaussian white noise $\Gamma(t)$ of the form

$$\langle\Gamma(t)\rangle = \langle\Gamma(t)\Gamma(t')\rangle = 0 \; ; \qquad \langle\Gamma(t)\Gamma^*(t')\rangle = (2n_{th}+1)\kappa\delta(t-t') \; , \tag{3.11}$$

i.e. from the Langevin equation

$$\dot{\alpha} = \{i\Omega + 2i\chi - \kappa - 2i\chi|\alpha|^2\}\alpha + iF + \Gamma(t) \; . \tag{3.12}$$

Graham and Schenzle [35] and Haug et al. [36] already obtained a similar FPE by starting from classical equations and adding white noise.

4 Expansion into Complete Sets and Derivation of Vector Recurrence Relations

In order to apply our MCF method for solving the pseudo FPE (3.8) or equation (3.9) with derivative terms up to third order we have to expand the quasi-distribution functions into complete sets satisfying the boundary conditions. In our problem the functions are defined in the whole α plane. The only condition is that integrals of the form (3.1-3.3) should exist. One choice of sets fulfilling these requirements are Hermite functions for the real part ($\mathrm{Re}\,\alpha$) and the imaginary part ($\mathrm{Im}\,\alpha$) of α. In the present case, however, it seems to be more appropriate to use Laguerre functions for the expansion into the intensity variable I and Fourier expansions for the phase variable ϕ where I and ϕ are defined by

$$\alpha = \sqrt{I}\, e^{i\phi} \; . \tag{4.1}$$

The expansion is written in the form

$$\left\{\begin{matrix} P(I,\phi,t) \\ W(I,\phi,t) \\ Q(I,\phi,t) \end{matrix}\right\} = \sum_{m=0}^{\infty} \{a_m^0 e^{-I/I_0} L_m^0(I/I_0)$$

$$+ \sum_{n=1}^{\infty} \frac{2}{\sqrt{n!}} (a_m^n \cos n\phi - b_m^n \sin n\phi) e^{-I/I_0} (I/I_0)^{n/2} L_m^n(I/I_0)\} \; , \tag{4.2}$$

where L_m^n are the generalized Laguerre polynomials [37] and I_0 is an arbitrary scaling intensity which is chosen such that good numerical convergence is achieved. The factor $1/\sqrt{n!}$ was added for numerical reasons. It reduces the numerical errors for the matrix inversions which occur in the MCFs in Sect. 5. An expansion of this type was already used in [36,38]. By inserting the expansion (4.2) into (3.8,3.9) we obtain an equation of motion for the expansion coefficients. To derive these equations we have first transformed (3.8,3.9) to I and ϕ coordinates. Inserting (4.2) and multiplying the resulting equation by tje generalized Laguerre polynomials and $\sin n\phi$, $\cos n\phi$, using orthogonality relations and recurrence relations for the Laguerre polynomials [37] we finally arrive at [28,29] ($n \geq 1$, $m \geq 0$, coefficients with a negative lower index formally occuring for $m = 0$ can be omitted because of the prefactor m):

$$\dot{a}_m^0 = \frac{2Fm}{\sqrt{I_0}} b_{m-1}^1 - \frac{2\kappa}{I_0}\left(n_{th} + (1-s)/2 - I_0\right) m\, a_{m-1}^0 - 2\kappa m\, a_m^0$$

$$\begin{aligned}\dot{a}_m^n = {} & \frac{Fm}{\sqrt{(n+1)I_0}} b_{m-1}^{n+1} + F\sqrt{n/I_0}\, b_m^{n-1} - \frac{2\kappa}{I_0}\left(n_{th} + (1-s)/2 - I_0\right) m a_{m-1}^n \\ & - \kappa(2m+n) a_m^n - n\{(2I_0+s)(2m+n+1)\chi - 2\chi - \Omega\}\, b_m^n \\ & + 2nm\chi\{(I_0+s) - (1-s^2)/(4I_0)\}\, b_{m-1}^n + 2\chi I_0 n(n+m+1)\, b_{m+1}^n\end{aligned}$$

$$\begin{aligned}\dot{b}_m^n = {} & -\frac{Fm}{\sqrt{(n+1)I_0}} a_{m-1}^{n+1} - F\sqrt{n/I_0}\, a_m^{n-1} - \frac{2\kappa}{I_0}\left(n_{th} + (1-s)/2 - I_0\right) m b_{m-1}^n \\ & - \kappa(2m+n) b_m^n + n\{(2I_0+s)(2m+n+1)\chi - 2\chi - \Omega\}\, a_m^n \\ & - 2nm\chi\{(I_0+s) - (1-s^2)/(4I_0)\}\, a_{m-1}^n - 2\chi I_0 n(n+m+1)\, a_{m+1}^n \,, \end{aligned} \tag{4.3}$$

where the parameter s is 1 for the P-function, 0 for the Wigner-function and -1 for the Q-function. (Cahill and Glauber [39] have introduced quasi-distribution functions where s is a continuous parameter. For these quasi-distribution functions (4.3) is still valid, see [29].) The normalization

$$\int P(\alpha,t)\, d^2\alpha = \int Q(\alpha,t)\, d^2\alpha = \int W(\alpha,t)\, d^2\alpha = 1 \tag{4.4}$$

requires

$$a_0^0(t) = 1/(\pi I_0) \,. \tag{4.5}$$

(The volume element $d^2\alpha = d\mathrm{Re}\alpha\, d\mathrm{Im}\alpha$ transforms to $dI\, d\phi/2$.) Higher expansion coefficients are connected with the moments. We have for instance

$$\langle a \rangle = \pi I_0^{3/2}(a_0^1 - i b_0^1) \,, \quad \langle a^\dagger a \rangle = \pi I_0^2 (a_0^0 - a_1^0) + (s-1)/2 \,, \tag{4.6}$$

where s is the parameter defined above. Because the expansion coefficients are connected with the moments, the equations of motion for the expansion coefficients (4.3) also exist for the P-function even if squeezing occurs. In principle, the P-function and the corresponding PFPE (3.8) for the P-function can be avoided in the derivation of (4.3) by only using the master equation (2.3-2.5) and the connection between moments and expansion coefficients for the derivation of (4.3).

The equation of motion for the expansion coefficients can be cast into a vector recurrence relation. Introducing the vectors $\mathbf{c}_m$ by

$$\mathbf{c}_m = (a_m^0, a_m^1, b_m^1, a_m^2, b_m^2, \dots) \tag{4.7}$$

we may write (4.2) in the form

$$\dot{\mathbf{c}}_m = \mathbf{Q}_m^+ \mathbf{c}_{m+1} + \mathbf{Q}_m \mathbf{c}_m + \mathbf{Q}_m^- \mathbf{c}_{m-1} \; . \tag{4.8}$$

The matrices $\mathbf{Q}_m^+$, $\mathbf{Q}_m$, and $\mathbf{Q}_m^-$ follow from (4.3). The explicit forms for the P- and Q-functions have been given in [28]. For the Wigner-function they will be given in [29].

For solving the time-dependent equation (4.8) we make the ansatz

$$\mathbf{c}_m(t) = \hat{\mathbf{c}}_m e^{-\lambda t} \tag{4.9}$$

by which (4.8) is transformed to the eigenvalue equation

$$\mathbf{Q}_m^+ \hat{\mathbf{c}}_{m+1} + (\mathbf{Q}_m + \lambda \mathbf{I}) \hat{\mathbf{c}}_m + \mathbf{Q}_m^- \hat{\mathbf{c}}_{m-1} = \mathbf{0} \; , \tag{4.10}$$

where $\mathbf{I}$ denotes the identity matrix.

5 Matrix Continued Fraction Method

The tridiagonal recurrence relation (4.10) can be solved by matrix continued fractions. For pedagogical reasons let us first explain the main idea for solving (4.10) for the tridiagonal scalar recurrence relation

$$Q_m^+ \hat{c}_{m+1} + (Q_m + \lambda) \hat{c}_m + Q_m^- \hat{c}_{m-1} = 0 \; . \tag{5.1}$$

If we divide (5.1) by $\hat{c}_m$ and if we introduce the ratio or transfer coefficients

$$S_m = \hat{c}_m / \hat{c}_{m-1} \tag{5.2}$$

(5.1) takes the form

$$Q^+_m S_{m+1} + (Q_m + \lambda) + Q^-_m / S_m = 0 \; . \qquad (5.3)$$

Thus, instead of the tridiagonal recurrence relation (5.1), we now get a two term recurrence relation for the transfer coefficients S_m. We can therefore express S_m by S_{m+1} according to

$$S_m(\lambda) = - \frac{Q^-_m}{Q_m + \lambda + Q^+_m S_{m+1}(\lambda)} \; , \qquad (5.4)$$

or, if we use (5.4) again and again, we obtain the following infinite continued fraction

$$S_m(\lambda) = - \cfrac{1}{Q_m + \lambda - Q^+_m \cfrac{1}{Q_{m+1} + \lambda - Q^+_{m+1} \cfrac{1}{Q_{m+2} + \lambda - \ldots} Q^-_{m+2}} Q^-_{m+1}} Q^-_m \; . \qquad (5.5)$$

If (5.1) is a onesided tridiagonal recurrence relation ($\hat{c}_{-1} = \hat{c}_{-2} = \ldots = 0$) eq. (5.1) reads for m=0

$$\left(Q^+_0 S_1(\lambda) + Q_0 + \lambda\right) \hat{c}_0 = 0 \; . \qquad (5.6)$$

For nontrivial solutions we thus have

$$Q^+_0 S_1(\lambda) + Q_0 + \lambda = 0 \; . \qquad (5.7)$$

This equation determines the eigenvalues λ of (5.1).

A similar procedure as above can be used for the tridiagonal vector recurrence relation (4.10). In order to obtain matrices of finite dimensions the expansion in n is truncated at n=N. Then the matrices $\mathbf{Q}^{\pm}_m$ and $\mathbf{Q}_m$ are matrices of dimensions (2N+1)×(2N+1). Introducing the "ratio" or transfer matrices $\mathbf{S}_m$ defined by

$$\hat{\mathbf{c}}_m = \mathbf{S}_m \hat{\mathbf{c}}_{m-1} \qquad (5.8)$$

we obtain similar to (5.3)

$$\mathbf{Q}^+_m \mathbf{S}_{m+1} + (\mathbf{Q}_m + \lambda \mathbf{I}) + \mathbf{Q}^-_m \mathbf{S}^{-1}_m = \mathbf{0} \; . \qquad (5.9)$$

Solving for $\mathbf{S}_m$ yields

$$S_m(\lambda) = -[Q_m + \lambda I + Q_m^+ S_{m+1}(\lambda)]^{-1} Q_m^- \qquad (5.10)$$

in analogy to (5.4). In contrast to (5.4) the order of the matrices in (5.10) is now important. Iteration of (5.10) leads to an infinite matrix continued fraction. If the matrix inversions are written by fraction lines $\mathbf{A}^{-1} = \frac{I}{\mathbf{A}}$ this MCF reads explicitly for m=1

$$S_1(\lambda) = -\cfrac{I}{Q_1+\lambda I - Q_1^+ \cfrac{I}{Q_2+\lambda I - Q_2^+ \cfrac{I}{Q_3+\lambda I - \dots} Q_3^-} Q_2^-} Q_1^- . \qquad (5.11)$$

For $m = 0$ and for a onesided recurrence relation (4.10) reduces to

$$K(\lambda)\hat{c}_0 = [Q_0^+ S_1(\lambda) + Q_0 + \lambda I]\hat{c}_0 = 0 . \qquad (5.12)$$

Equation (5.12) has only a nontrivial solution if the determinant of $K(\lambda)$ vanishes, i.e.

$$D(\lambda) = \mathrm{Det}[K(\lambda)] = 0 . \qquad (5.13)$$

This condition determines the eigenvalues of (4.10) or equivalently of the equation of motion for the quasi-distribution functions (3.8,3.9) or equivalently of the operator equation (2.3-2.5).

In the actual calculation of the infinite matrix continued fraction (5.11) the MCFs have to be truncated at the M'th iteration. This is equivalent to truncating the expansion (4.2) in m at m=M. (This truncation index M and the truncation index N must be determined in such a way that increasing M and N does not alter the results beyond a given accuracy.) By iterative use of (5.10) the matrix $K(\lambda)$ and its determinant $D(\lambda)$ can thus be calculated. By a proper root finding technique we may then find one of the eigenvalues λ. The eigenvalue $\lambda = 0$ is already known and need not be determined in this way. In order to obtain eigenfunctions we then calculate the matrices S_M, S_{M-1}, ..., S_2, S_1 according to (5.10) and solve (5.12) for $\hat{c}_0$. The vectors $\hat{c}_1$, $\hat{c}_2$, ...$\hat{c}_M$ are obtained by repeatedly using (5.8) i.e. by

$$\hat{c}_1 = S_1\hat{c}_0, \quad \hat{c}_2 = S_2\hat{c}_1, \quad \hat{c}_3 = S_3\hat{c}_2 \dots . \qquad (5.14)$$

The eigenfunctions and the stationary solution finally follow by using (4.7) and performing by the summation in (4.2) truncated at $m = M$ and $n = N$. The iteration according to (5.14) is numerically stable whereas the iteration of (4.10) starting with $\hat{c}_0$ and $\hat{c}_1 = S_1\hat{c}_0$ is numerically unstable [15].

6 Results

First we discuss the stationary mean value $\langle a\rangle$ for $n_{th}=0$. According to (4.6) it is connected with the expansion coefficients a_0^1 and b_0^1. For $n_{th}=0$ one can also use the analytic results derived in [22] with the help of the complex P-function. Because n_{th} is very small for optical photons we confine ouselves to the case $n_{th}=0$ (pure quantum noise). In Fig. 2 the real and the imaginary part are shown as a function of the normalized driving field for various Ω/χ values together with the stationary solution of the "classical" equation (2.9). As discussed at the end of Sect. 2 the number of photons inside the cavity scales with the Ω/χ ratio. As is seen in Fig. 2 for small Ω/χ one has a rather smooth transition between the low intensity branch and the high intensity branch. For higher Ω/χ values this transition becomes sharper. For $\Omega/\chi\to\infty$ one expects a straight vertical line well approximated by the $\Omega/\chi=50$ curve. This sharp transition is similar to a first-order phase transition in e.g. a liquid→gas transition. The straight line then corresponds to the Maxwell line in such a first-order transition though Maxwell's area rule theorem seems not to be valid in the present case. For smaller κ values one obtains similar plots. The following interesting feature, however, occurs for these smaller damping constants. The curves now depend critically on the ratio Ω/χ in such a way that for integer or nearly integer values of Ω/χ the transition region is shifted to lower fields, see Fig. 3. For Ω/χ ratios not integer or not nearly integer no such drastic changes occur. In Fig. 4 we have plotted the transition field, i.e. that field where $\mathrm{Re}\langle a\rangle=0$, as a function of Ω/χ. The oscillating variation is clearly visible in this figure. As explained in [24] this rapid variation for integer Ω/χ values is due to the fact that for integer Ω/χ values some of the lower eigenvalues of the Hamiltonian (2.4) are exactly degenerate for F=0 and remain almost degenerate for finite fields.

Next we discuss the stationary Q-function and the Wigner-function. In Fig. 5 and Fig. 6 we have plotted Q-functions for $\tilde{\kappa}=0.001$ and for $\tilde{\kappa}=0.1$ for various driving fields and for $\Omega/\chi=10$. For $\Omega/\chi=9.5$ (not shown here) the situation is quite similar except that the transition field for small κ is not as low as it is for $\Omega/\chi=10$, see also the discussion above. For small driving fields in Figs. 5a and 6a we have only one maximum and no bistability occurs. The lines $Q=\text{const}$ have a circular shape. In Figs. 5b and 6b the driving field is in the transition region so that two maxima of the Q-function occur. If the driving field is increased further the left maximum of the Q-functions disapperars (see Figs. 5c and 6c) and only one ear-shaped distribution remains indicating that the system is no longer bistable. Comparing Figs. 5 and Figs. 6 one realizes that the influence of $\tilde{\kappa}$ is twofold. For small cavity damping ($\tilde{\kappa}=0.001$) the Q-functions are almost symmetric with respect to the real axis. For higher cavity damping the Q-functions seems to be rotated. It is interesting to notice that our ear-shaped Q-functions in Fig. 5c, 6c and the right

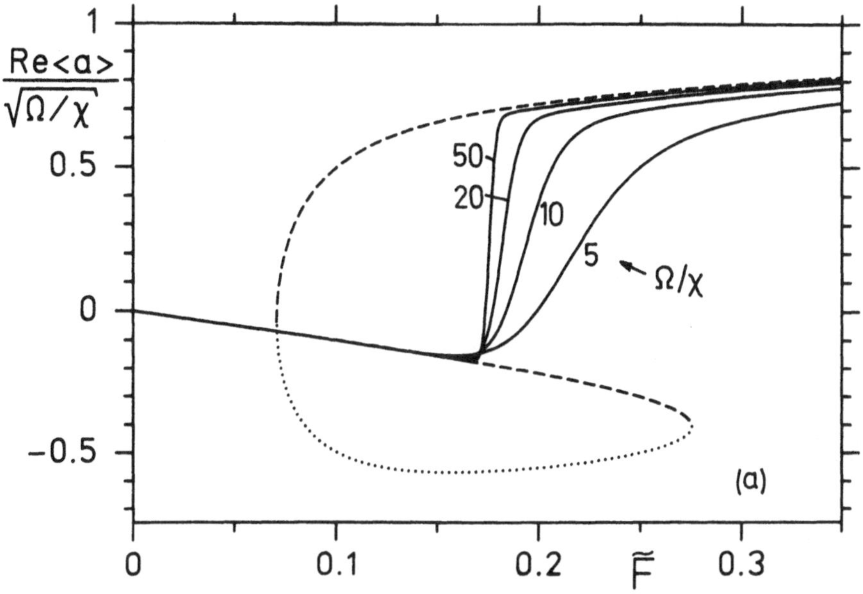

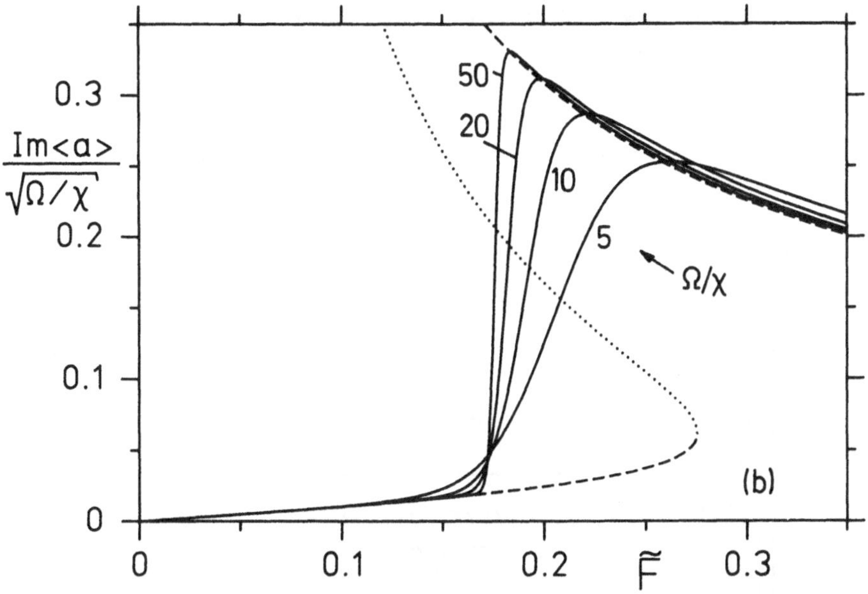

Fig. 2: The real part (a) and the imaginary part (b) of the normalized expectation value $\langle a\rangle/\sqrt{\Omega/\chi}$ for various Ω/χ values for $\tilde{\kappa} = 0.1$ and for $n_{th} = 0$ (solid lines). The stationary solution of (2.9) is also shown (stable = broken line, unstable = dotted line).

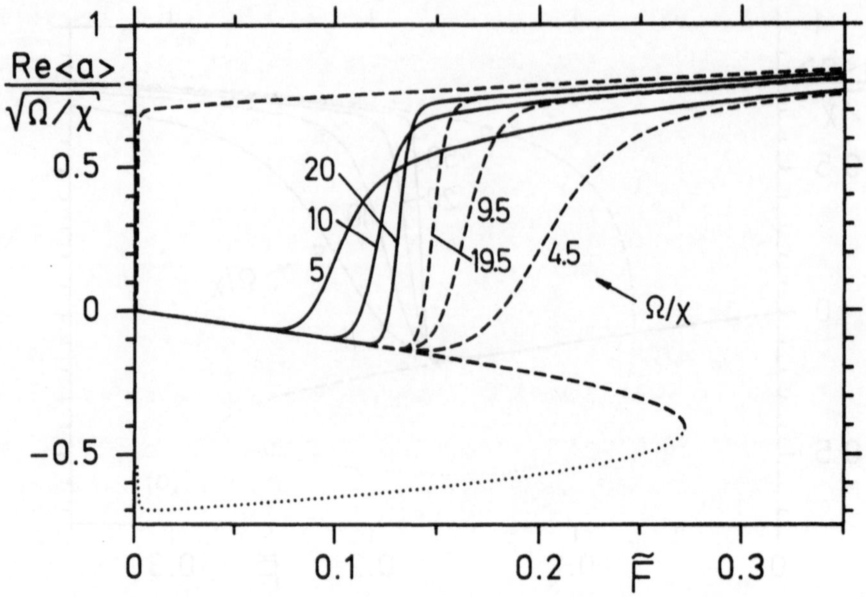

Fig. 3: The real part of the normalized expectation value $\langle a\rangle/\sqrt{\Omega/c}$ for integer Ω/χ (solid lines) and halfinteger Ω/χ (broken lines) for $\tilde{k} = 0.001$ and for $n_{th} = 0$. The stationary solution of (2.9) is also shown (stable = broken line, unstable = dotted line).

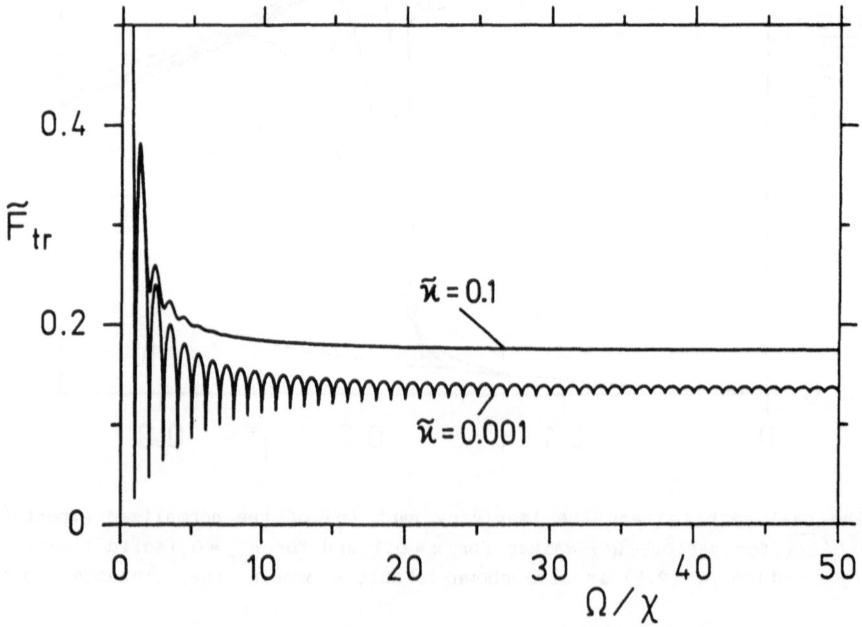

Fig. 4: The field $\tilde{F}_{trans}$, i.e. the field for which $\mathrm{Re}\langle a\rangle = 0$, as a function of Ω/χ for $\tilde{\kappa} = 0.1$ and $\tilde{\kappa} = 0.001$.

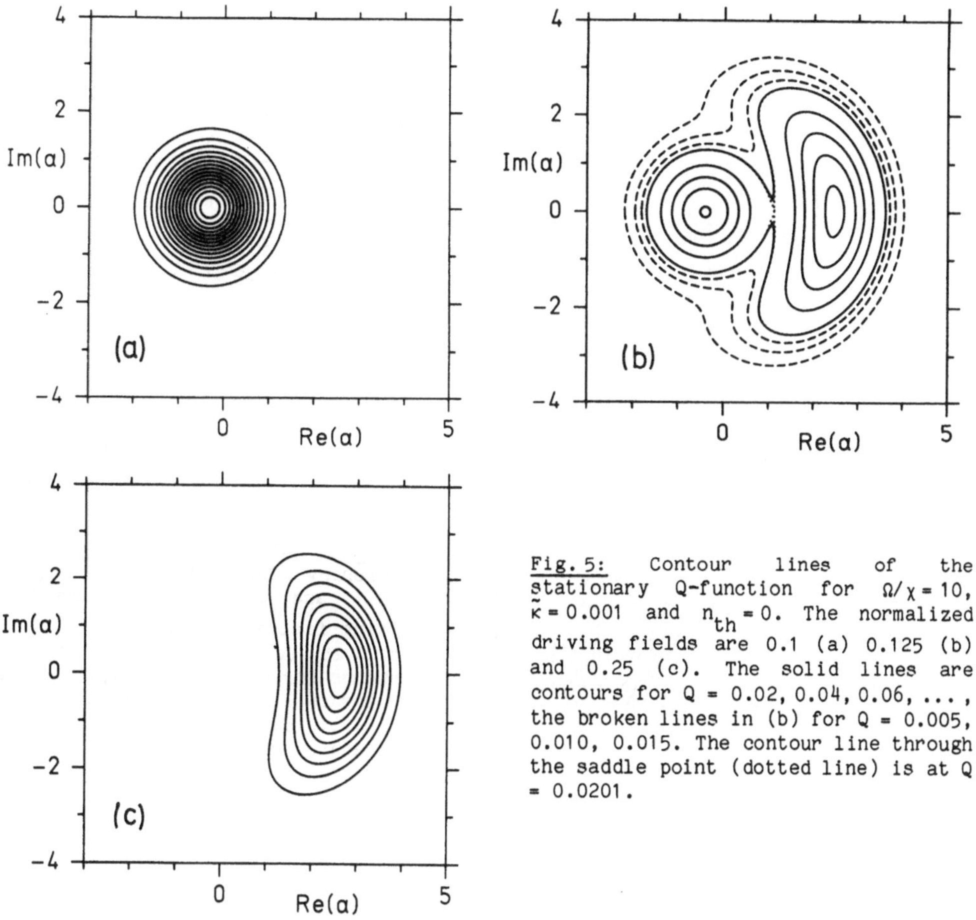

Fig. 5: Contour lines of the stationary Q-function for $\Omega/\chi = 10$, $\tilde{\kappa} = 0.001$ and $n_{th} = 0$. The normalized driving fields are 0.1 (a) 0.125 (b) and 0.25 (c). The solid lines are contours for Q = 0.02, 0.04, 0.06, ... , the broken lines in (b) for Q = 0.005, 0.010, 0.015. The contour line through the saddle point (dotted line) is at Q = 0.0201.

part of our Q-functions in Figs. 5b, 6b are similar to the Q-functions generated by a nonlinear Mach-Zehnder interferometer [40] which have been called crescent-shaped. In [40] it is shown that these Q-functions describe number-phase minimum-uncertainty states. The nonlinear Kerr medium inside the interferometer is described by the operator $a^{\dagger 2}a^2$ also occuring in our Hamiltonian (2.4). In Fig. 7 we have plotted the Wigner-distribution for a driving field in the transition region. The curves are similar to Fig. 6b except that in between the two maxima a wavy structure emerges, see in particular Fig. 7b. As it was already discussed the P-function does not exist if squeezing occurs. It can be shown [26] that squeezing exists for a driving field in the transition region. Therefore the P-function does not exist for the parameter of Figs. 5b and 6b. If one nevertheless tries to sum up the expansion (4.2) for the P-function, the results oscillate and depend on the scaling intensity I_0 as well as on the truncation indices N and M. In Fig. 4 of [28] the convergence respectively divergence of the sum (4.2) for the Q- and P-function is demonstrated.

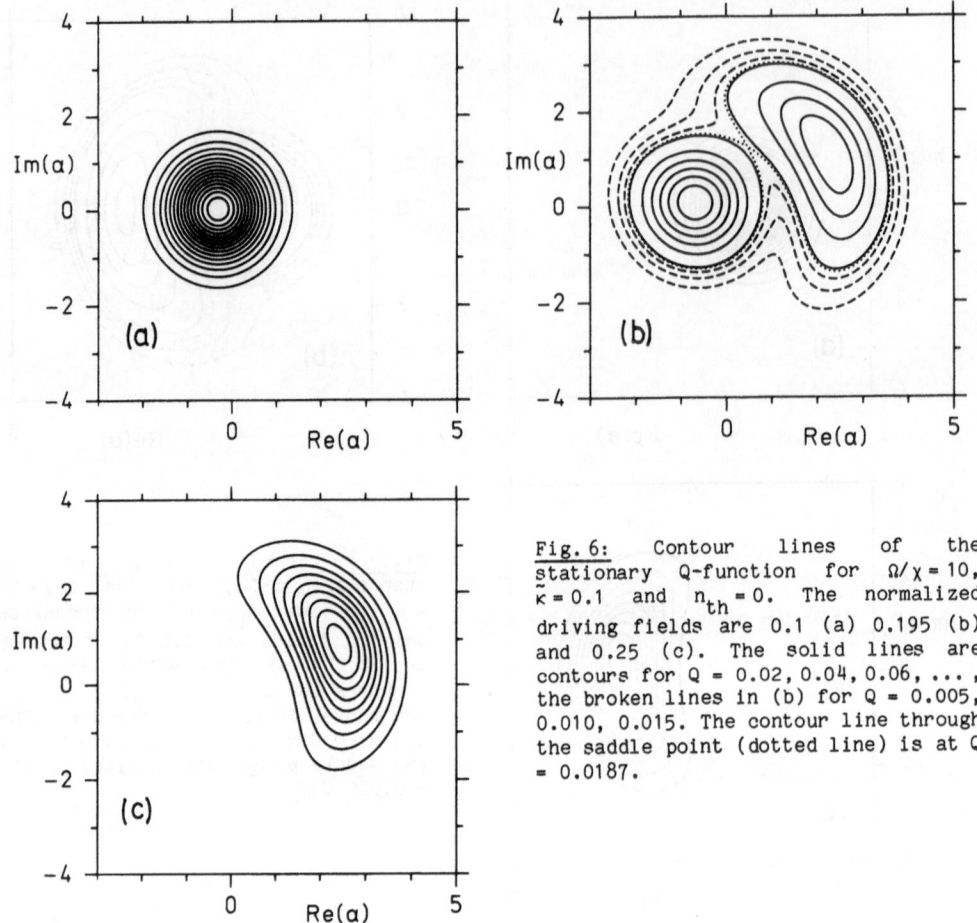

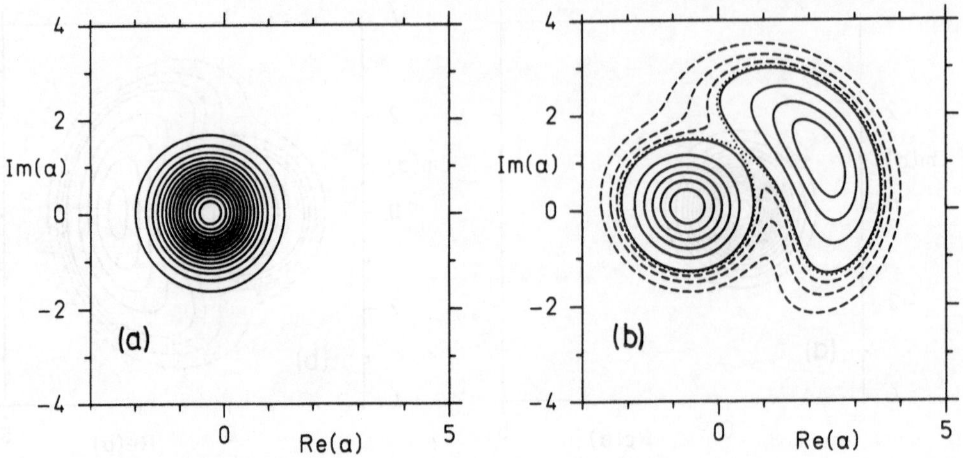

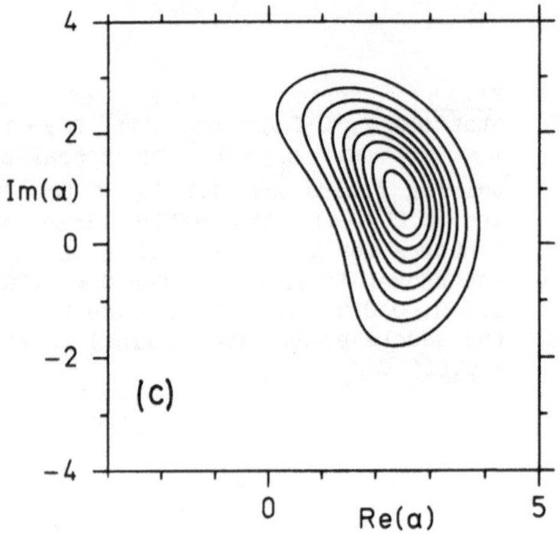

Fig. 6: Contour lines of the stationary Q-function for $\Omega/\chi = 10$, $\tilde{\kappa} = 0.1$ and $n_{th} = 0$. The normalized driving fields are 0.1 (a) 0.195 (b) and 0.25 (c). The solid lines are contours for Q = 0.02, 0.04, 0.06, ... , the broken lines in (b) for Q = 0.005, 0.010, 0.015. The contour line through the saddle point (dotted line) is at Q = 0.0187.

Finally we discuss the the results for the eigenvalues of equation (3.8,3.9), or equivalently of (4.8), i.e. we look for nonzero λ satisfying (5.13). Some of the lowest real eigenvalues for small ($\tilde{\kappa} = 0.001$) and larger ($\tilde{\kappa} = 0.1$) damping constants are shown in Fig. 8 and Fig. 9 for $\Omega/\chi = 9.5$ (a) and $\Omega/\chi = 10$ (b). It can be shown that for zero driving fields the eigenvalues of (3.8,3.9) are 0, 2κ, 4κ, 6κ,... . Therefore we have plotted the eigenvalue divided by 2κ. Furthermore, we want to emphasize that the eigenvalues obtained from the Q-function, the P-function and the Wigner-function agree within the numerical accuracy. Thus the problem of representation-dependent results as discussed by Drummond [13] for absorptive optical bistability does not arise in our calculations. For $\Omega/\chi = 9.5$ and $\tilde{\kappa} = 0.001$ our results agree with the results obtained from the procedure used in [23] for the small-damping limit. For integer Ω/χ values, however, this procedure does not work. Here, as discussed in [24], one has to include appropriate nondiagonal elements of the density operator in addition to the diagonal elements only taken into account in [23].

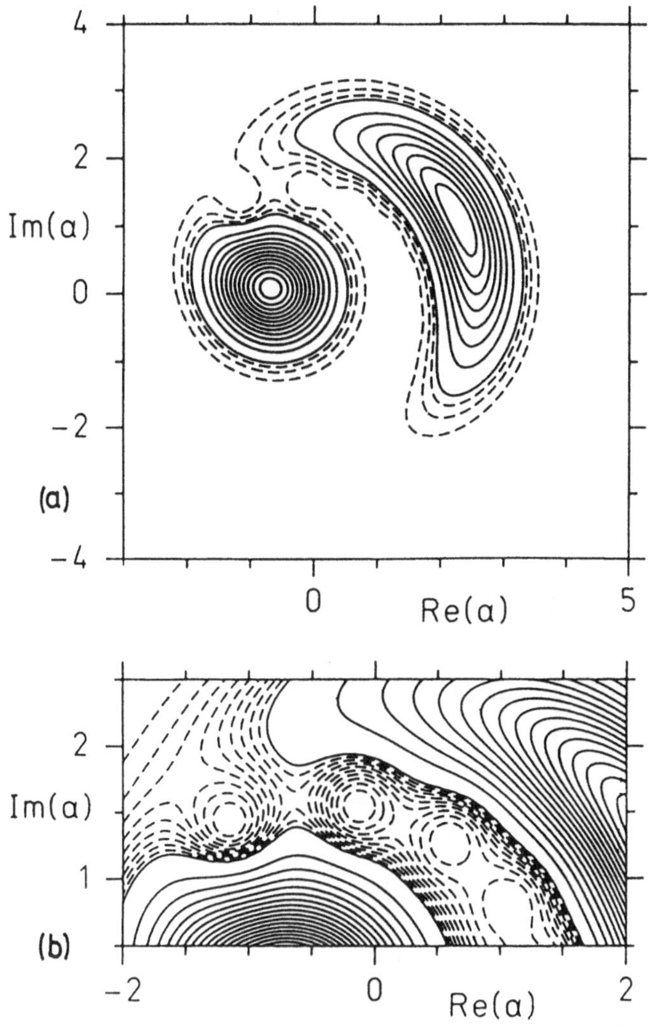

Fig. 7: Contour lines of the Wigner-function for $\tilde{F} = 0.195$, $\tilde{\kappa} = 0.1$, $\Omega/\chi = 10$ and $n_{th} = 0$. The contour lines in Fig. 7a are for Q = 0.02, 0.04, 0.06, ... (solid lines) and for Q = 0.005 0.010 0.015 (broken lines). In Fig. 7b a section in between the two maxima is shown for the contour lines Q = 0.01, 0.02, 0.03, ... (solid lines) and for Q = 0.001, 0.002, 0.003, ..., 0.009 (broken lines).

For driving fields allowing bistability the lowest nonzero eigenvalue is well separated from the higher eigenvalues and therefore determines the long-time behavior of the system. An eigenfunction (Q-function) corresponding to the lowest nonzero eigenvalue for a driving field which allows bistability is shown in Fig. 10. By comparing this figure to Figs. 5b and 6b (note the positive sign for the stationary Q-function and the different signs of the eigenfunction of Fig. 10) we conclude that the lowest nonzero eigenvalue and the corresponding eigenfunction describe

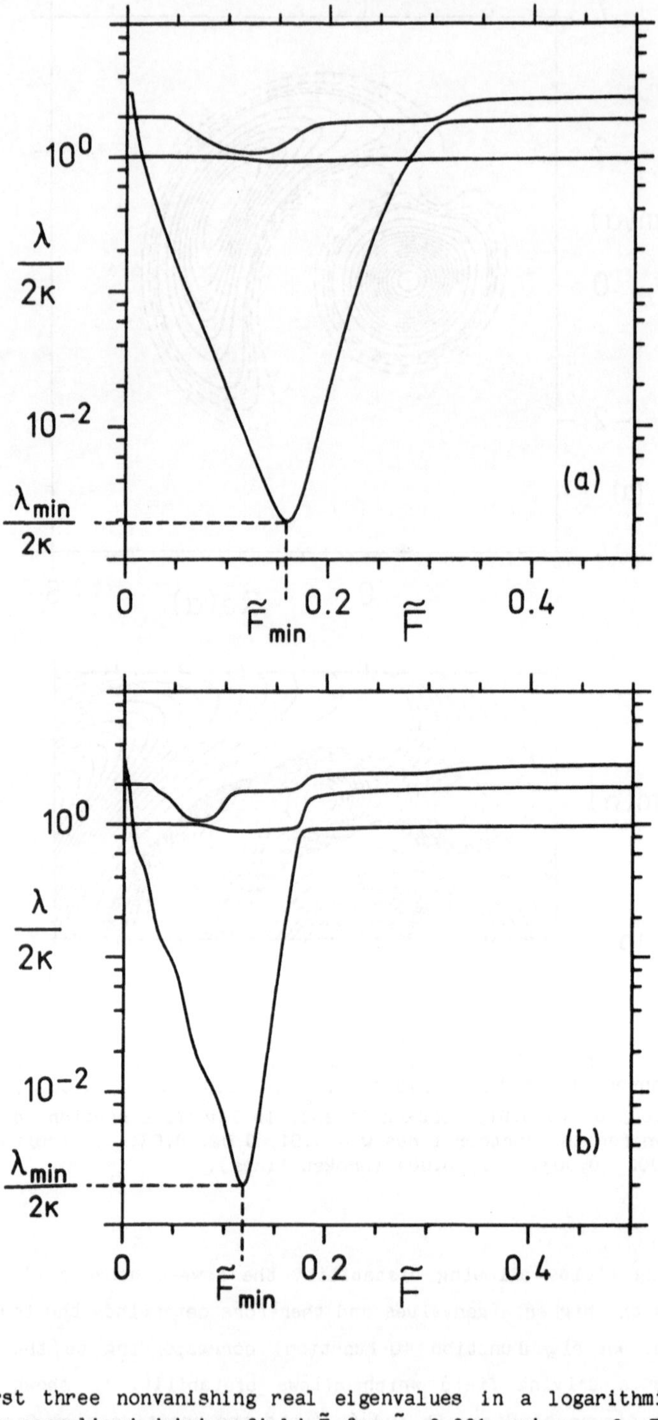

Fig. 8: The first three nonvanishing real eigenvalues in a logarithmic scale as a function of the normalized driving field $\tilde{F}$ for $\tilde{\kappa} = 0.001$ and $n_{th} = 0$. The values for Ω/χ are 9.5 (a) and 10 (b).

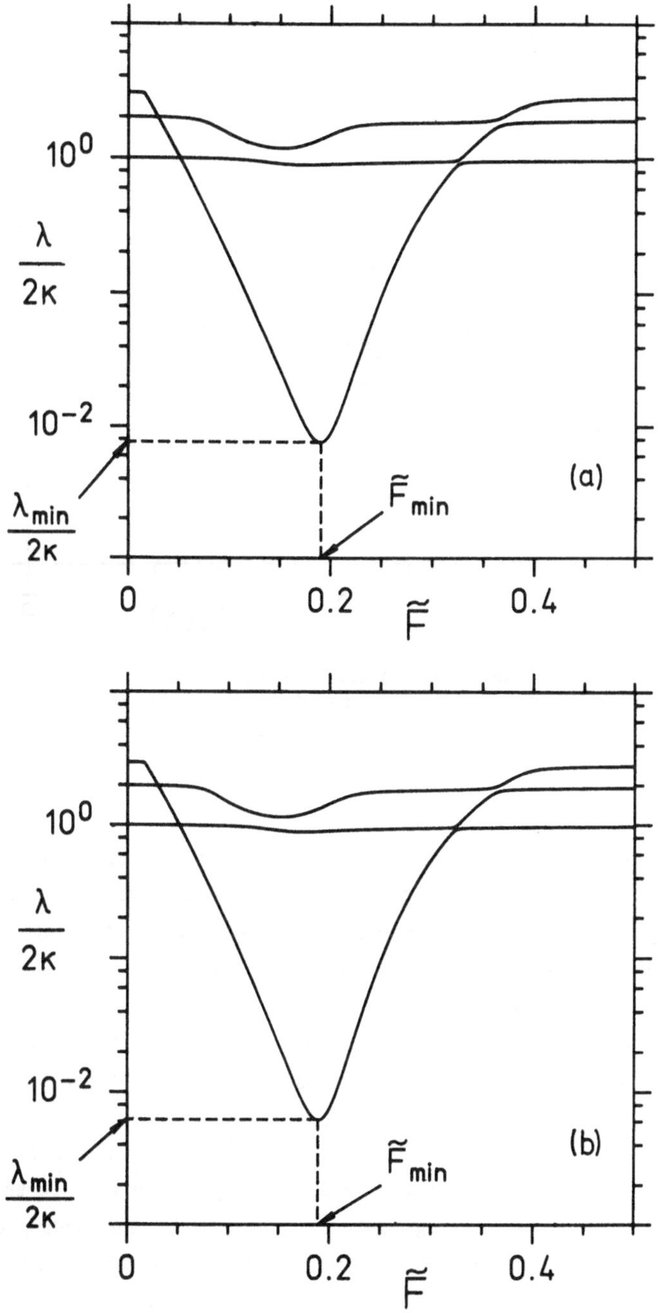

<u>Fig. 9:</u> same as Fig. 8 but for $\tilde{\kappa} = 0.1$.

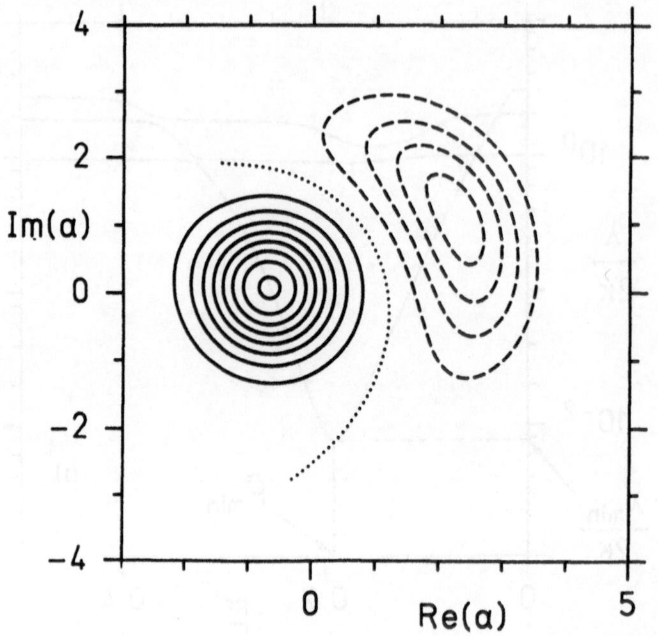

Fig. 10: Contour lines of the Q-eigenfunction corresponding to the lowest nonvanishing eigenvalue for the parameters of Fig. 8b. The solid lines and the broken lines indicate the different sign of the eigenfunction. At the dotted line the eigenfunction is zero.

transitions between the two almost stable states of our bistable system, i.e. the tunneling rate is determined by the lowest nonzero eigenvalue. Because we have neglected thermal fluctuations this eigenvalue determines the ultimate stability of our system as already mentioned in the introduction.

For $\tilde{\kappa} = 0.1$ the results for $\Omega/\chi = 9.5$ and $\Omega/\chi = 10$ do not differ very much (see Fig. 9), for $\tilde{\kappa} = 0.001$, however, the driving field $\tilde{F}_{min}$, where the lowest nonzero eigenvalue has its minimum λ_{min}, changes appreciably if the parameter Ω/χ is slightly increased from 9.5 to 10.0. Furthermore, ripples occur for integer Ω/χ values whereas for noninteger Ω/χ values these ripples disappear. In order to show the shift of $\tilde{F}_{min}$ more clearly we ploted the driving field $\tilde{F}_{min}$ and the corresponding eigenvalue λ_{min} as a function of the parameter Ω/χ. The results are shown in Fig. 11. For small cavity damping ($\tilde{\kappa} = 0.001$) one can clearly see that the field $\tilde{F}_{min}$ reaches very low values at integer values of Ω/χ and increases very sharply when moving away from integer values. For large Ω/χ values this oscillating behavior is less pronounced. It is similar to the oscillating behavior of $\tilde{F}_{trans}$ shown in Fig. 4. The minimum eigenvalue corresponding to $\tilde{F}_{min}$ decreases roughly exponentially if Ω/χ is increased. Because the number of photons inside the cavity scales with Ω/χ this means that for large photon numbers the tunneling rate becomes extremely small and vanishes in the classical limit. In addition to this exponential dependence on Ω/χ, the eigenvalue

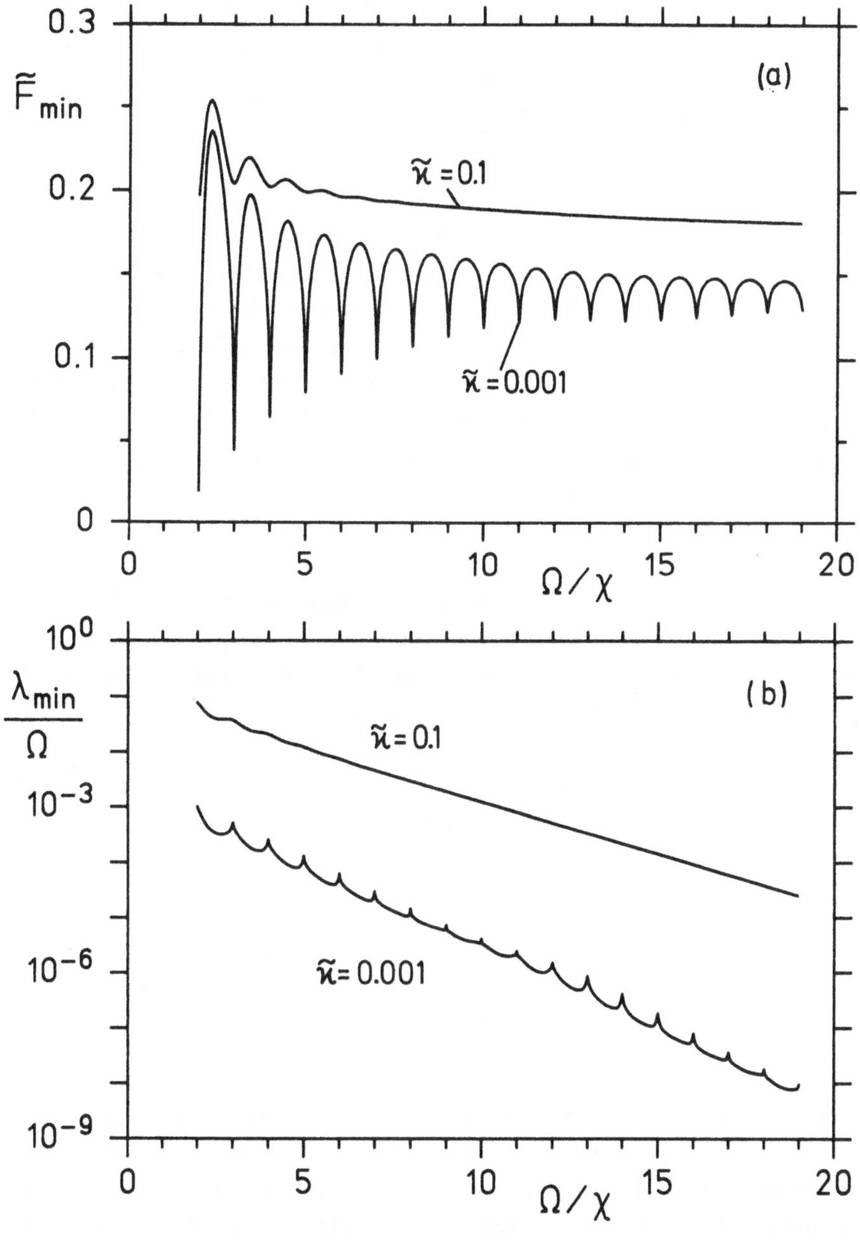

Fig. 11: The normalized driving field $\tilde{F}$, where the lowest nonzero eigenvalue has its minimum, (a) and the corresponding mimimum eigenvalue (b) (see Figs. 8 and 9) as a function of Ω/χ for $\tilde{\kappa} = 0.001$ and $\tilde{\kappa} = 0.1$.

has peaks leading to somewhat larger values at and near integer values for Ω/χ. (For larger damping constants the results for integer Ω/χ do no longer differ appreciably from the results for noninteger Ω/χ values.) For small cavity damping and Ω/χ values which are not too large we have therefore the following result: At integer Ω/χ values the bistable region is appreciably shifted towards smaller driving fields and the tunneling rate itself increases, i.e. the system becomes less stable. It can be shown [24] that this remarkable result for integer Ω/χ values at small damping constants is due to the fact that for integer Ω/χ values some of the lower eigenvalues of the Hamiltonian (2.4) are exactly degenerate for $F = 0$ and remain almost degenerate for finite driving fields. Using an expansion of the density operator in eigenstates of the Hamiltonian (2.4) the lowest nonzero eigenvalue was obtained in [24] by diagonalizing an appropriate matrix. This procedure is similar to the procedure used in [23]. In addition to the procedure in [23], where only diagonal elements of the density operator have been taken into account, some appropriate nondiagonal elements must be included for integer or nearly integer values of Ω/χ. For small cavity damping the results of [24] agree very accurately with the results by the MCF method. Because the MCF method and the method used in [24] are completely different, the excellent agreement of the results is a very strong indication that both methods are correct.

Concerning the 'nonclassical' form of the pseudo FPE (3.8) or of (3.9) with the additional third-order derivative term the following remark seems to be worth mentioning: As discussed above both equations for the quasidistributions (3.8) and (3.9) lead to the oscillating behavior of $\tilde{F}_{min}$ and $\lambda_{min}/(2\kappa)$. If the third-order term in (3.9) is neglected one obtains an ordinary FPE. As shown in [29] such a 'classical' FPE does not show this oscillating behavior. It also leads to a tunneling rate which may, for the parameters of Fig. 9, differ appreciable from the correct value.

7 Conclusion

We have calculated quantum tunneling rates, the stationary Q-functions and the Wigner-function for the model of Drummond and Walls describing dispersive optical bistability by solving a pseudo Fokker-Planck equation and a partial differential equation with derivatives up to third order with the matrix continued fraction method. We have shown that the stationary Q-function is well-behaved although the diffusion matrix of the pseudo Fokker-Planck equation is not positive definite. The Wigner-function behaves similar to the Q-function. An additional wavy structure, however, appears in between the two maxima. Generally the P-function does not exist. Nevertheless, we have demonstrated that its equation of motion can also be used for calculating expectation values and eigenvalues. The expectation values and eigenvalues obtained from the P-function, from the Wigner-function and from the Q-function all agree within the numerical accuracy. The investigation of tunneling rates was one important feature of our calculations. In the bistable region the lowest nonzero eigenvalue is well-separated from the higher ones and therefore describes the decay rate of the two almost stable states. Although the lowest nonzero eigenvalue may become very small it can be calculated with high precision by the matrix continued fraction method. We have found that for small cavity damping constants the driving field, where the tunneling rate has its minimum, varies appreciably in an oscillating fashion as a function of the parameter Ω/χ. Thus, due to the quantum nature of the problem, the system behaves differently for integer and noninteger Ω/χ values if the cavity damping is small enough.

References

[1] A. Szöke, V. Daneu, J. Goldhar and N. A. Kurnit: Appl. Phys. Lett. **15**, 376 (1969)
[2] H. M. Gibbs, S. L. McCall and T. N. C. Venkatesan: Phys. Rev. Lett. **36**, 113 (1976)
[3] C. M. Bowden, M. Cliftan, H. R. Robl (Eds.): **Optical Bistability** Plenum Press, New York (1981)
[4] R. Bonifacio (Ed.): **Dissipative Systems in Quantum Optics**, Topics in Current Physics, Vol. **27**,Springer, Berlin (1982)
[5] L. A. Lugiato, **Progress in Optics XXI**, Ed. E. Wolf, page 69, North-Holland, Amsterdam (1984)
[6] J. C. Englund, R. R. Snapp, W. C. Schieve: **Progressin Optics XXI**, Ed. E. Wolf, page 355, North-Holland, Amsterdam (1984)
[7] G. Duffing: **Erzwungene Schwingungen bei veränderlicher Eigenfrequenz und ihre technische Bedeutung**, Vieweg, Braunschweig (1918)
[8] A. H. Nayfeh and D. T. Mook: **Nonlinear Oscillations**, page 162ff, Wiley, New York (1979)
[9] R. J. Glauber: Phys. Rev. Lett. **10**, 84 (1963); R. J. Glauber: Phys. Rev. **131**, 2766 (1963)
[10] E. C. G. Sudarshan: Phys. Rev. Lett. **10**, 277 (1963)
[11] M. Hillery, R. F. O'Connell, M. O. Scully and E. P. Wigner: Physics Reports **106**, 121 (1984)
[12] E. Wigner: Phys. Rev. **40**, 749 (1932)
[13] P. D. Drummond: Phys. Rev. **A33**, 4462 (1986)
[14] H. Risken and H. D. Vollmer: Z. Phys. **B33**, 297 (1979); H. D. Vollmer and H. Risken: Z. Phys. **B34**, 313 (1979); H. D. Vollmer and H. Risken: Physica **110A**, 106 (1982); H. Risken and H. D. Vollmer: Mol. Phys. **46**, 555 (1982)
[15] H. Risken: **The Fokker-Planck Equation**, Springer Series in Synergetics, Vol. **18**, Springer, Berlin (1984)
[16] D. F. Walls: Nature **306**, 141 (1983)
[17] P. D. Drummond and C. W. Gardiner: J. Phys. **A13**, 2353 (1980)
[18] P. D. Drummond , C. W. Gardiner and D. W. Walls: Phys. Rev. **A24**, 914 (1981)
[19] C. W. Gardiner: **Handbook of Stochastic Methods**, Springer Series in Synergetics, Vol. **13**, 2nd Ed., Springer, Berlin (1985)
[20] H. J. Carmichael and M. Wolinsky: In **Quantum Optics IV**, page 208, Ed. J. D. Harvey and D. F. Walls, Springer, Berlin (1986)
[21] M. Dörfle and A. Schenzle: Z. Phys. **B65**, 113 (1986)
[22] P. D. Drummond and D. F. Walls: J. Phys. **A13**, 725 (1980)
[23] H. Risken, C. Savage, F. Haake and D. F. Walls: Phys. Rev. **A35**, 1729 (1987)
[24] H. Risken and K. Vogel: Quantum Tunneling Rates in Dispersive Optical Bistability for Low Cavity Damping, submitted to Phys. Rev. A
[25] K. Vogel and H. Risken: Optics Comm. **62**, 45 (1987)
[26] H. Risken and K. Vogel: In: **Fundamentals of Quantum Optics II**, page 225,

Ed. F. Ehlotzky, Springer, Berlin (1987)
[27] H. Risken and K. Vogel: **Eigenvalues of the Quantum Fokker-Planck Equation for Dispersive Optical Bistability**, Talk given at the **Topical Conference on Optical Bistability, Instability and Optical Computing**, Peking University, Beijing, Aug. 1987
[28] K. Vogel and H. Risken: Quantum Tunneling Rates and Stationary Solutions in Dispersive Optical Bistability, accepted for publication in Phys. Rev. A
[29] K. Vogel and H. Risken: work in preparation
[30] H. Haken: **Laser Theory, In: Encyclopedia of Physics, Vol. XXV/2c**, Ed. S. Flügge, Springer, Berlin (1970)
[31] W. L. Louisell: **Quantum Statistical Properties of Radiation**, Wiley, New York (1973)
[32] K. E. Cahill and R. J. Glauber: Phys. Rev. **A177**, 1857 (1969)
[33] R. F. Pawula: Phys. Rev. **162**, 186 (1967)
[34] H. Risken and H. D. Vollmer: Z. Phys. **B35**, 313 (1979) and **B66**, 257 (1987)
[35] R. Graham and A. Schenzle: Phys. Rev. **A23**, 1302 (1981)
[36] H. Haug, S. W. Koch, R. Neumann and H. E. Schmidt: Z. Phys. **B49**, 79 (1982)
[37] W. Magnus, F. Oberhettinger and R. P. Soni: **Formulas and Theorems for the Special Functions of Mathematical Physics**, Springer, Berlin (1966)
[38] H. Risken, H. D. Vollmer: Z. Phys. **B39**, 89 and 339 (1980)
[39] K. E. Cahill and R. J. Glauber: Phys. Rev. **A177**, 1882 (1969)
[40] M. Kitagawa and Y. Yamamoto: Phys. Rev. **A34**, 3974 (1986)

Spontaneous symmetry breaking and spatial structures in optical systems

L.A. Lugiato, C. Oldano, L. Sartirana, Wang Kaige*
Dipartimento di Fisica del Politecnico, Corso Duca degli Abruzzi 24
10129 Torino, Italy

L.M. Narducci, G.-L. Oppo, M.A. Pernigo, J.R. Tredicce
Physics Department, Drexel University
Philadelphia, Pa. 19104, USA

F. Prati
Dipartimento di Fisica dell'Universita', Via Celoria 16
20133 Milano, Italy

and

G. Broggi
Physik Institut der Universität Zurich, Schönberggasse 9
8001 Zürich, Switzerland

1. Introduction

The spontaneous formation of stationary spatial structures in homogeneous systems far from thermal equilibrium has been the object of extensive investigations in such fields as nonlinear chemical reactions and developmental biology [1-3]. Here the instabilities that are responsible for the emergence of spatial patterns arise from a diffusive mechanism and are usually referred to as Turing instabilities [4].

Optical systems are much more widely known for their propensity to produce temporal structures in the form of spontaneous oscillations of the regular or chaotic type [5-6]. Only very recently has a Turing instability been discovered [7-11] in an optical model. Here the resulting stationary pattern is produced by the interplay between diffraction and the nonlinear coupling among several transverse modes, and not by a diffusion process.

The arrangement that is predicted to yield these new effects consists of a resonant medium and an optical cavity that allows stationary solutions which are uniform in a plane transverse to the direction of propagation. When the input intensity, in the case of a passive driven system, or the pump parameter for a laser exceed a certain threshold level, diffraction may cause an instability that produces, spontaneously, a transverse stationary spatial pattern. A typical condition for the emergence of this instability is that the frequency spacing between the resonant longitudinal mode and the nearest transverse resonances be of the order of the cavity linewidth. This situation creates a competition between longitudinal modes whose end result is the loss of stability of the spatially homogeneous stationary solution.

This optical Turing instability [7] was linked recently [12] to the phenomenon of spatial soliton formation [13], and to the modulational instabilities analyzed in Ref. 14. As a natural extension of the plane Cartesian geometry discussed in Refs. 7-11, we have investigated the more realistic model of a cavity with spherical mirrors, in which the transverse modes have a Gauss-Laguerre structure, instead of the cosine shape produced by the cavity with plane mirrors. In this case, we found that the transverse profile of the steady-state electric field intensity undergoes significant variations as one changes appropriate control parameters. In this paper we review these recent advances and, in addition, discuss a new phenomenon that produces a spontaneous breaking of the cylindrical symmetry: with a laser operating in a symmetrical single-mode Gauss-Laguerre configuration, an increase of the pump parameter produces a spatial instability that causes the appearance of a radially asymmetric steady state pattern.

2. The Kerr model with diffraction: passive, externally driven systems

We begin our discussion of spatial patterns in optical systems with the help of an especially simple model. We consider a Fabry-Perot cavity of length L (see Fig. 1) filled with a Kerr medium and driven by an external coherent field.

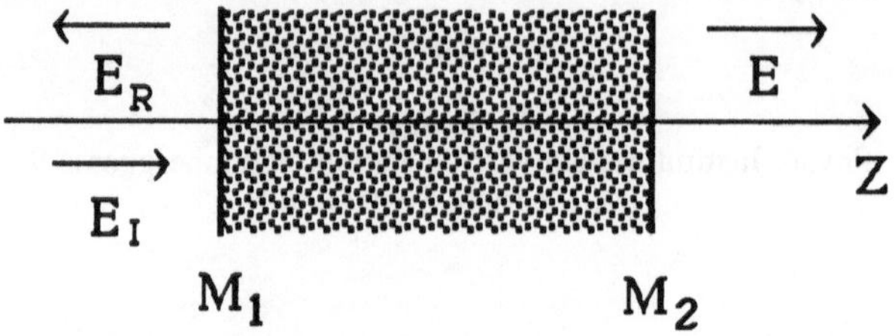

Fig. 1 Fabry-Perot cavity filled with nonlinear material. The mirrors M_1 and M_2 have transmittivity coefficient T. E_I, E, and E_R are the input, output and reflected fields, respectively.

The field internal to the cavity has the structure

$$\mathbb{E} = E\cos(k_z z)\, e^{-i\omega_0 t} + c.c. \tag{1}$$

where ω_0 is the frequency of the injected field; $k_z = \pi n_z/L$ is the longitudinal wave number, and n_z is an integer. The envelope function E obeys the time-evolution equation

$$\frac{dE}{dt} = -\kappa\left\{E - E_I - iE\,(|E|^2 - \Theta)\right\} \tag{2}$$

where $\kappa = cT/2L$ is the cavity damping constant, or cavity linewidth, and E_I is the amplitude of the input field which we assume to be real and positive for definiteness. The detuning parameter Θ accounts for the mismatch between the frequency of the input field and the cavity and for the linear, intensity independent part of the refractive index of the medium. The nonlinear contribution to the refractive index of the material is represented by the cubic term of Eq. (2). This model is valid in the uniform field limit, which assumes that the transmittivity coefficient T of the mirrors is much smaller than unity [15].

From Eq. (2) we can obtain immediately the steady state equation linking the input intensity E_I^2 to the transmitted intensity $|E|^2$, i.e.

$$E_I^2 = |E|^2\left\{1 + (|E|^2 - \Theta)^2\right\} \tag{3}$$

This equation, well known from the earliest days of optical bistability, was proposed for the first time in Ref. 16. According to Eq. (3) the steady state curve linking $|E|^2$ with E_I^2 is single-valued for

$\Theta<\sqrt{3}$ and S-shaped (i.e. bistable) for $\Theta>\sqrt{3}$.

Equation (2) is based on the plane-wave approximation and excludes the possibility of diffractive effects from the outset. In order to include diffraction we complete Eq. (2) as follows

$$\frac{\partial E(x,y,t)}{\partial t} = -\kappa\left\{E(x,y,t) - E_I - i\,E(x,y,t)\left[|E(x,y,t)|^2 - \Theta\right]\right\} +$$

$$+\, i\,\frac{c}{2k_z}\left(\frac{\partial^2}{\partial x^2} + \frac{\partial^2}{\partial y^2}\right) E(x,y,t) \qquad (4)$$

where x and y are the transverse coordinates. Because we assume that the input field E_I has a flat transverse profile, Eq. (4) allows stationary solutions that are homogeneous along the x and y directions and, in fact, coincide with those calculated using Eq. (2) as the starting point [see Eq. (3)].

In general, when E_I is a plane wave, it is also common to assume that the cavity field is independent of x and y, so that Eq. (4) automatically reduces to the form of Eq. (2). This assumption, however, is not always correct, as one can verify from the study of the linear stability properties of the homogeneous stationary solutions. The main problem with Eq. (2) is that it allows only those fluctuations that are uniform along the transverse directions. Random perturbations, on the other hand, do not have to be uniform at all, so that a more general linear stability analysis is needed using Eq. (4) as the starting point. The importance of diffraction in connection with the stability properties of plane-wave stationary solutions was already emphasized in Ref. 14.

The presence of second order space derivatives in Eq. (4) requires the specification of additional boundary conditions for our problem. For this purpose, we consider a cavity defined by four mirrors with transmittivity coefficient $T<<1$; two mirrors are orthogonal to the longitudinal z-axis and are separated by a distance L from one another; the other two are orthogonal to the x-axis and are spaced a distance b (see Fig. 2). The cavity is open along the y direction.

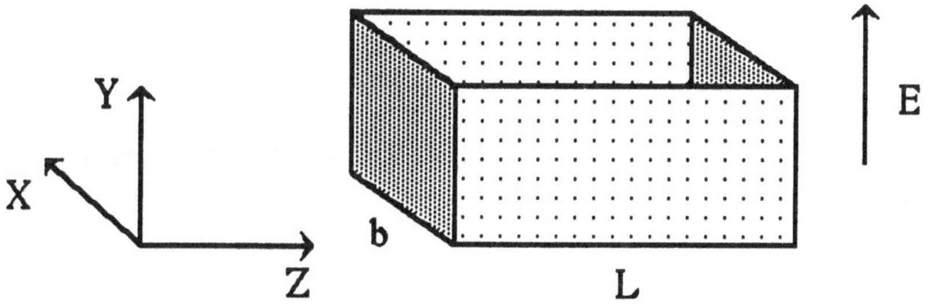

Fig. 2 Schematic representation of the cavity configuration discussed in this article. The four mirrors have reflection coefficients $T<<1$. The cavity is open along the y direction which is also the direction of polarization of the electric field.

We assume that both the input and the cavity field are linearly polarized along the y direction (other choices of polarization, leading again to a transverse modal configuration of the type $\cos k_x x$ as in Eq. (6a) are described in Ref. 7). From this requirement and from the transversality condition, it follows that E is independent of y. In this case, Eq. (4) reduces to

$$\frac{\partial E(x,t)}{\partial t} = -\kappa\left\{E(x,t) - E_I - i\,E(x,t)\left[|E(x,t)|^2 - \Theta\right]\right\} + i\frac{c}{2k_z}\frac{\partial^2 E(x,t)}{\partial x^2} \qquad (5)$$

Furthermore, we assume that the cavity can support modes of the type

$$\cos(k_x x)\cos(k_z z) \qquad (6a)$$

where

$$k_x = \frac{\pi n}{b}, \qquad k_z = \frac{\pi n_z}{L} \qquad (6b)$$

and n, n_z are nonnegative integer numbers.

Note that Eq. (5) fixes the value of n_z, while the integer n remains free to vary over the range 0,1,2,... As a result, Eq. (5) is a single longitudinal mode model, but it can account for an infinite number of transverse modes. The stationary solution, Eq. (3), corresponds to selecting n=0; however, as we shall show in the next section, solutions of this type can become unstable under appropriate conditions and allow the growth of transverse modes with $n \neq 0$.

3. Stationary spatial patterns

In order to describe this instability, it is useful to focus on the cavity frequencies which are given by the usual expression

$$\omega = c\sqrt{k_x^2 + k_z^2} \qquad (7)$$

The model (5) assumes the paraxial approximation $k_x \ll k_z$ so that Eq. (7) can be approximated by

$$\omega \approx c k_z + \frac{c}{2k_z} k_x^2 \qquad (8)$$

or, with the help of Eq. (6b), by

$$\omega \approx \pi n_z \frac{c}{L} + \pi^2 n^2 \frac{c}{2k_z b^2} \qquad (9)$$

Finally, if we take into account that $k_z = 2\pi/\lambda$, and if we introduce the Fresnel number

$$\mathbb{F} = \frac{b^2}{\lambda L} \qquad (10)$$

we obtain the expression

$$\omega = \pi n_z \frac{c}{L} + a(n)\kappa \qquad (11)$$

In Eq. (11) we have defined

$$a(n) = \frac{\pi n}{2T\,\mathbb{F}} \qquad n = 0,1,2,\ldots.. \qquad (12)$$

The parameter a(n) represents the frequency difference between the n-th transverse mode and the resonant longitudinal mode, measured in units of the modal linewidth κ. Note that, on the frequency axis, the transverse modes lie to the right of the longitudinal mode. Because we assume that $T \ll 1$ and we want a(1) to be of the order of unity, we must require that $\mathbb{F} \gg 1$.

We are now in a position to understand the origin of the transverse mode instability. If we consider first only the longitudinal cavity resonances, we note that they are very narrow (for $T<<1$) and well removed from one another because the modal width is much smaller than the mode spacing $2\pi c/2L$. As a result, the frequency ω_0 of the input field interacts only with the resonant cavity mode and the other longitudinal resonances do not affect the stationary state. If we now take into account also the transverse modes, we see from Eq. (11) that, when a(1) is of the order of unity, the nearest transverse resonances overlap the longitudinal resonant mode. This situation triggers a mode-mode competition which is at the origin of the destabilization of the homogeneous stationary solution.

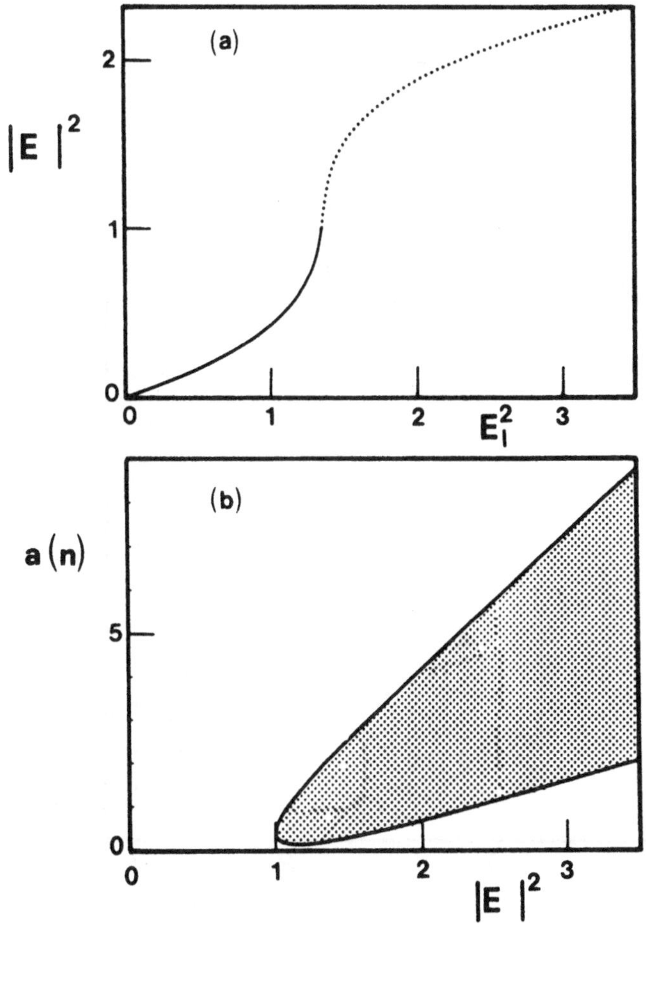

Fig. 3 (a) Steady state curve for $\Theta=1.6$. The broken part indicates the stationary solution that may become unstable. (b) The shaded region corresponds to the unstable domain in the plane of the variables a(n) [see Eq. (12)] and $|E|^2$.

Typical results of the formal linear stability analysis are illustrated in Fig. 3 for Θ=1.6. Consider, first, the steady state curve for the transmitted versus incident intensity (Fig. 3a): each point on the curve such that $|E|^2 \geq 1$ is unstable provided that for the corresponding value of $|E|^2$ at least one of the numbers a(n), n=1,2,... lies in the shaded region of Fig. 3b. This region, in other words, identifies the domain of the unstable transverse modes.

The special and interesting feature of this robust instability is that it does not lead to oscillatory behavior or to other dynamical effects, but it produces a new stationary state. To be more precise, the electric field in the new stationary configuration is still characterized by a single oscillation frequency that coincides with the incident frequency ω_0. Hence the input beam imposes its frequency to all the cavity modes that appear in the new stationary state.

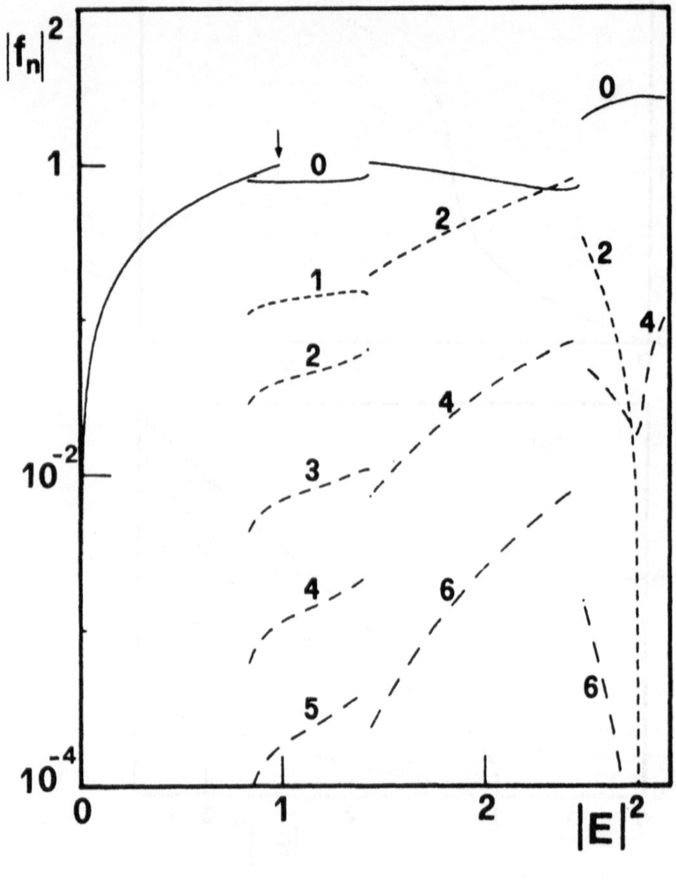

Fig. 4 Passive Kerr medium, Cartesian cavity configuration, Θ = 1.6. The intensities of the modes with n = 0, 1, ..., 6 are plotted as functions of the normalized output intensity of the homogeneous stationary solution. The arrow indicates the critical point.

In the neighborhood of the critical point we calculated the new stationary solution analytically using bifurcation theory [17], and found that the bifurcation is subcritical (i.e. the spatial structure emerges discontinuously at the critical point) for $\Theta > 41/30 \approx 1.4$. Figure 4 shows the result of a numerical calculation of the modal intensities $|f_n|^2$ as a function of the output intensity $|E|^2$ of the homogeneous stationary solution which is linked to the input intensity by Eq. (3). The modal amplitudes are defined by the equation

$$E(x',t) = \sum_{n=0}^{\infty} f_n(t) \cos(\pi n x') \; ; \qquad x' = x/b \; ; \tag{13}$$

in steady state the amplitudes f_n are time independent.

At the critical point there is a discontinuous jump in the intensity of the homogeneous mode, n=0, and the abrupt appearance of the transverse modes 1, 2,...,5. On decreasing the input intensity below the critical point, the system remains in the inhomogeneous state producing hysteresis, as is typical of subcritical bifurcations. If instead one increases the input power above the critical point over the interval of $|E|^2$ shown in Fig. 4, one finds other two values of the input intensity such that the transverse profile undergoes a discontinuous transition.

The analysis of the spatial instability formulated in Ref. 7 for a Kerr medium has been extended in Ref. 10 to the case of a two-level passive system. In this situation the model displays not only spatial but also temporal instabilities.

4. Passive systems, cavity with spherical mirrors

If we now insert the expansion (13) into Eq. (5) and take into account the orthogonality of the functions $\cos(\pi n x')$ over the interval $0<x'<1$, we obtain a set of ordinary differential equations for the modal amplitudes f_n:

$$\kappa^{-1} \frac{df_n}{dt} = - f_n + E_I \delta_{n,0} - i a(1) n^2 f_n + i \frac{1}{4(1+\delta_{n,0})} \Big(\sum_{n'n''n'''}^{*} f_{n'} f_{n''} f^{*}_{n'''} - \Theta f_n \Big) \tag{14}$$

where the asterisk implies that the sum is restricted to terms with positive n''' and with n''' = ±n±n'±n''; note that all the combinations of upper and lower signs must be included. The parameter a(1) is given by

$$a(1) = \frac{\omega_1 - \omega_0}{\kappa} \tag{15}$$

where ω_0 and ω_1 are the frequencies of the modes n=0 and n=1, respectively; for the cavity shown schematically in Fig. 2, a(1) coincides with the value given by Eq. (12) for n=1.

The Cartesian waveguide geometry of Fig. 2 is not very common in quantum optical studies, with the possible exception of the microwave frequency range. For this reason, we consider also a more standard cavity configuration with spherical mirrors in which the transverse modes are still discrete in spite of the absence of the lateral mirrors. For definiteness, we consider the ring cavity configuration of Fig. 5,

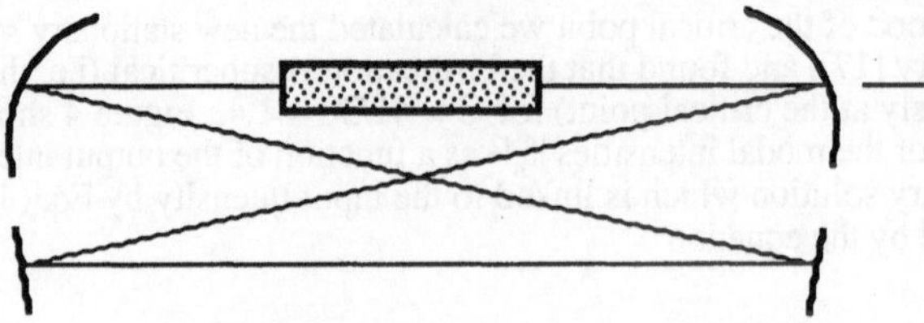

Fig. 5 Schematic representation of a ring cavity. The spherical mirrors have transmittivity coefficient T, the plane mirrors are ideal reflectors. E denotes the output field.

The cavity modes have the Gauss-Laguerre transverse structure [18, 19]

$$A_{pl}(r,\varphi) = \left(\frac{r}{w_0}\right)^l L_p^l\left(\frac{2r^2}{w_0}\right) \exp\left(-\frac{r^2}{w_0^2}\right) \begin{Bmatrix} \sin l\varphi \\ \cos l\varphi \end{Bmatrix} \tag{16}$$

where $r = (x^2+y^2)^{1/2}$ and φ are the radial and angular transverse variables, respectively, w_0 is the beam waist parameter and L_p^l denotes the Laguerre polynomials, with p,l = 0, 1,..... The cavity frequencies are given by

$$\omega = 2\pi n_z \frac{c}{L} + \frac{1}{2} a(1)\,(2p+l+1) \tag{17}$$

where n_z has the same meaning as in Eq. (9), L is the cavity length and the constant a(1) depends on the exact geometrical details of the cavity [19]. For the moment, we assume cylindrical symmetry and therefore we expand the electric field in the following way

$$E(r',t) = \sum_p f_p(t) \exp(-r'^2)\, L_p(2r'^2) \tag{18}$$

with $r' = r/w_0$. Assuming, as in the Cartesian case, that the Fresnel number $\mathbb{F} = w_0^2/\lambda L$ is much greater than unity in such a way that $\mathbb{F}T = O(1)$, one obtains [20] a set of dynamical equations for the amplitudes f_p of the form

$$\kappa^{-1}\frac{df_p}{dt} = -f_p + E_I\delta_{p,0} - ia(1)pf_p + i\Big(\sum_{p'p''p'''} f_{p'}f_{p''}f^*_{p'''}\, c_{pp'p''p'''} - \Theta f_p\Big) \tag{19}$$

where a(1) is still given by Eq. (15) and ω_0 and ω_1 are the frequencies of the modes p=0 and p=1, respectively. If we compare Eqs. (14) and (19) we note two main differences: (1) the term containing a(1), which arises from diffraction, is proportional to n^2 in Eq. (14) and to p in Eq. (19); (2) the coefficients $c_{pp'p''p'''}$ of the nonlinear term in Eq. (19) couple all the amplitudes f_p. As a consequence, we have been unable to calculate analytically any stationary solutions of Eq. (19).

If, however, in Eq. (19), one neglects every amplitude except for f_0 and takes into account that $c_{0000} = 1$, one finds that $|f_0|^2$ obeys a steady state equation identical to Eq. (3). We shall call the solutions of this equation the "single-mode stationary solutions"; we must keep in mind, however, that these are not exact stationary solutions, although they play a useful role in analyzing the actual stationary configurations and in studying the influence and behavior of the various transverse modes. Figure 6 illustrates the steady state behavior of the system for $\Theta = 4$; these results were obtained by

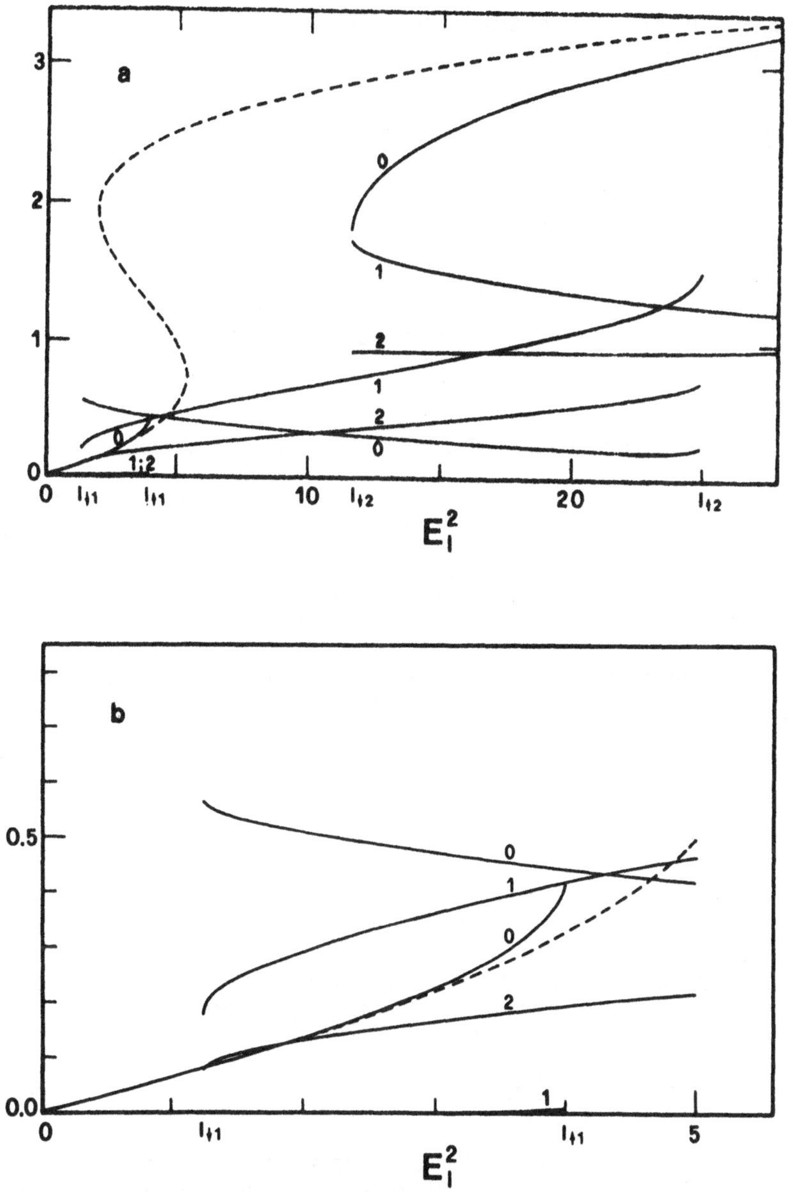

Fig. 6 Passive Kerr medium, cavity with spherical mirrors, $\Theta = 4$. (a) The broken line corresponds to the single-mode steady state curve (see text); the solid lines display the intensities of the modes $p = 0, 1, 2$ as functions of the input intensity. (b) Expanded version of part of figure (a).

solving numerically the time-dependent equations (19) with the three modes n = 0, 1, and 2. The broken curve traces the single-mode steady state; the solid lines show the intensities $|f_p|^2$ (for p=0,1,2) in the long-time regime.

For small values of the input intensity E_I^2 the steady state solution is practically of the single-mode type, the amplitudes f_1 and f_2 being negligible. For $E_I^2 = I_{\uparrow 1}$ one observes a discontinuous transition in which the modes p=1,2 acquire a sizable intensity. If at this point one decreases E_I^2 one finds hysteretic behavior, as expected. If, on the other hand, one increases E_I^2 beyond the value $I_{\uparrow 1}$, the intensity of the transverse modes p=1,2 becomes larger than that of the fundamental mode p=0 until one reaches the point $I_{\uparrow 2}$ where a second discontinuous transition brings the fundamental mode to dominate again. Figure 6 displays two hysteresis cycles for $I_{\downarrow i} < E_I^2 < I_{\uparrow i}$ (i=1,2); this picture is very different from the standard hysteresis cycle of the single-mode solution because the bistability obtained here arises from discontinuous changes of the transverse profile.

5. The spatial instability in the laser: cooperative frequency locking

We consider again the same cavity configuration described in Fig. 2, but now we assume that the medium is prepared in a state of inversion. Initially we imagine the parameters of the problem to be fixed in such a way that the atomic line is in exact resonance with the longitudinal cavity mode corresponding to the integer value n_z. Under these conditions the laser approaches a stationary state in which the field oscillates with the frequency of the selected mode. Now, we detune the laser from the resonant condition and look for an instability of the stationary state.

We describe this system [11] with the two-level model in the presence of diffraction

$$\frac{\partial E}{\partial t} = -\kappa\left\{(1+i\theta)E - 2CP\right\} + i\frac{c}{2k_z}\frac{\partial^2 E}{\partial x^2} \tag{20a}$$

$$\frac{\partial P}{\partial t} = -\gamma_\perp\left\{(1+i\Delta)P - ED\right\} \tag{20b}$$

$$\frac{\partial D}{\partial t} = -\gamma_\parallel\left\{D - 1 + \tfrac{1}{2}\left(E^*P + EP^*\right)\right\} \tag{20c}$$

Here the normalized variables P and D describe the atomic polarization and population inversion, respectively, $\gamma_\perp$ and $\gamma_\parallel$ are the respective relaxation rates ($\gamma_\perp$ coincides with the atomic linewidth). The cavity detuning parameter θ and the atomic detuning parameter Δ are given by

$$\theta = \frac{\omega_C - \omega_R}{\kappa}, \qquad \Delta = \frac{\omega_A - \omega_R}{\gamma_\perp} \tag{21}$$

where ω_C is the frequency of the longitudinal mode that is closest to resonance with the atomic line, ω_A is the transition frequency of the two-level atoms and the reference frequency ω_R is given by the mode pulling formula

$$\omega_R = \frac{\kappa\omega_A + \gamma_\perp\omega_C}{\kappa + \gamma_\perp} \tag{22}$$

which is equivalent to the relation

$$\Delta = -\theta\,; \tag{23}$$

C is the pump parameter. If one ignores the diffraction term in Eq. (20a), the set of equations (20) reduces to the well known Haken-Lorenz model [21, 22]. As they are, Eqs. (20) are more appropriate for the description of a ring cavity with lateral mirrors of the type described in Ref. 10, rather than the Fabry-Perot cavity of Fig. 2. However, when the diffusional motion of the atoms washes out the standing wave effects, this model can describe also the case of a Fabry-Perot cavity.

The steady state of this system with the field amplitude oscillating at a frequency ω_R is spatially homogeneous along the transverse direction, i.e. it corresponds to the choice n=0 in Eq. (6). More precisely, in steady state we have

$$|E|^2 = 2C - 1 - \theta^2 \tag{24}$$

with an arbitrary phase for the field E. The linear stability analysis of this stationary solution shows [11] the existence of a spatial instability, that arises for $\theta < 0$ when, for a given value of C, at least one transverse mode with n>0 satisfies the condition

$$a(n) \leq - \frac{\theta}{C} (2C - 1 - \theta^2) \tag{25}$$

Note that,in the case of ring cavity, a(n) is one half of the value given by Eq. (12). When the condition (23) is satisfied the system enters a new regime. In order to explore the various possibilities, we look for other stationary intensity solutions of the type

$$E(x,t) = e^{-i\delta t} E_S(x) = e^{-i\delta t} \sum_n f_n \cos(\pi n x') \tag{26}$$

where, as usual, we have set $x' = x/b$. A simple calculation leads to the predicted value of δ acording to the formula [11]

$$\delta = \kappa \frac{\sum_n a(n) |f_n|^2}{\sum_{n=1}^{\infty} |f_n|^2 + 2 |f_0|^2} \tag{27}$$

These new types of stationary intensity solutions can be classified in two distinct groups:

1. Consider a fixed integer $k \geq 0$. This group of solutions is characterized by modal amplitudes $f_n \neq 0$ only for $n = (2m+1)k$, with m=0,1,2,...;

2. Consider a fixed integer $k \geq 0$. These solutions are characterized by modal amplitudes $f_n \neq 0$ only for n=mk, with m=0,1,2,...

Note that a special case of solutions of type (1), for k=0, is the spatially homogeneous solution for which the only non vanishing component of the field is f_0 and $\delta=0$. In the case of the other solutions corresponding to k>0, the only non-vanishing components of the field are f_k and its odd harmonics. It turns out that the harmonic components are very small so that these solutions are essentially of the single-mode type and have a spatial configuration determined by the transverse mode n=k.

The most interesting feature of this problem is brought about by the solutions of type (2) in which the transverse modes coexist. Just as in the passive case discussed in section 3, these solutions are representative of nontrivial spatial structures. In addition, they signal the appearance of a new regime of operation of the laser. Traditionally, we identify two distinct types of operation for an active device of the laser type:

i. single-mode operation, where the emergence of one mode leads to the suppression of all the others, and

ii. multimode operation, in which different modes coexist.

In the usual multimode behavior each mode oscillates with a different frequency and the total electric field takes the form

$$\mathbb{E}(x,t) = \sum_n f_n\, g_n(x)\, e^{-i\omega_n t} + c.c. \tag{28}$$

where ω_n is the mode-pulled frequency of the n-th mode, and the factor $g_n(x)$ describes its spatial structure. The associated intensity, of course, has a time-dependent behavior as a result of the interference among the different components. In our case, instead, the situation is quite different because from Eqs (1), and (26) we have

$$\mathbb{E}(x,t) = \sum_n f_n \cos(\pi n x') \cos(k_z z)\, e^{-i(\omega_0+\delta)t} + c.c. \tag{29}$$

so that the intensity is actually stationary in time. In this type of operation, which is not to be confused with the traditional mode-locking, whose end result is the generation of output pulses, the modes compete with one another but, eventually, find a common oscillation frequency which they all adopt. For this reason we call this phenomenon "cooperative frequency locking". This behavior eliminates the beat notes and produces a stationary but spatially inhomogeneous output.

For $\kappa/\gamma_\perp = 0.1$, $\gamma_{||}/\gamma_\perp = 1$, $\theta = -0.6$, and $2C = 3.36$ we investigated Eqs. (20) numerically in the unstable region (25). For a(1) = 0.65 the mode n = 1 builds up initially and the system appears to approach a multimode stationary solution ehich, however, turns out to be unstable in this case; as a result, the system eventually develops a single-mode structure of type (1) with k=2. Figure 7 shows the intensities $|f_n|^2$ of the modal amplitudes n = 0 and 1 in the interval $0.2 < a(1) < 0.55$, where the system approaches a multimode stationary solution of type (2) with k=1, characterized by

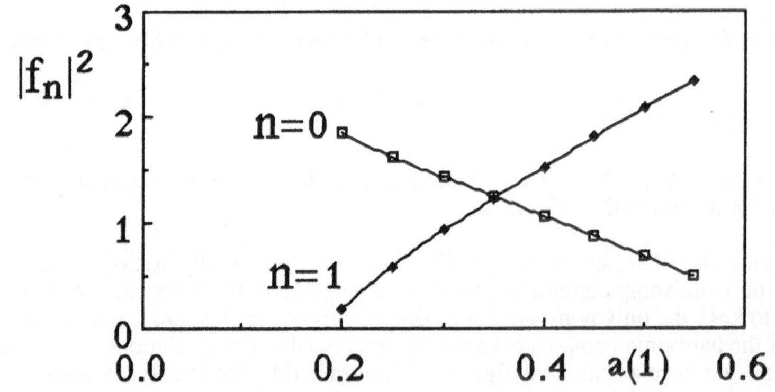

Fig. 7 The Cartesian cavity configuration. The laser parameters are $\kappa/\gamma_\perp = 0.1$, $\gamma_{||}/\gamma_\perp = 1.0$, $\theta = -0.6$, and $2C = 3.36$. Plot of the stationary modal intensities for n = 0 and n = 1 as functions of a(1).

cooperative frequency locking. The intensities of modes n = 2, 3, ... are much smaller than those of modes n = 0 and 1.

Figure 8 shows the spatial profile of the steady state intensity for a(1) = 0.05, a case in which the modes n = 1, 2, 3 are simultaneously unstable. The long term situation is a multimode stationary state with odd harmonics that are stronger than their even neighbors.

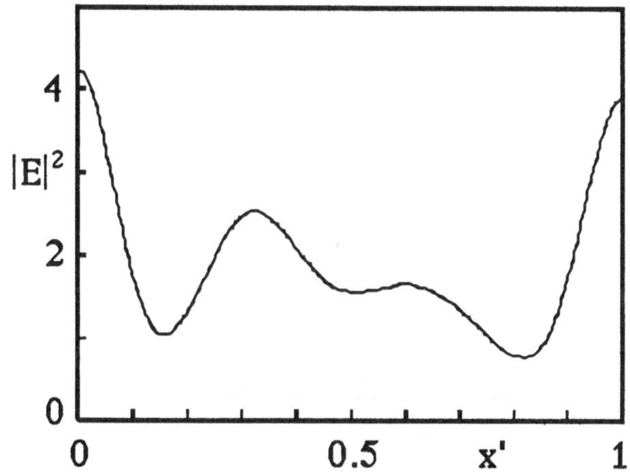

Fig. 8 Transverse profile of the stationary output intensity for a(1) = 0.05. The remaining parameters are the same as in Fig. 7.

6. Ring laser with spherical mirrors

We turn now to the case of a laser with the ring cavity described by Fig. 5. The transverse structure of the cavity modes is given by Eq. (16); again, we assume cylindrical symmetry and consider only the modes with l=0. We assume a pump profile which is uniform along the transverse directions. In order to limit the number of modes that play a significant dynamical role, we assume that the loss parameter κ is mode-dependent (as one expects in a real laser system because of the confinement of the side walls) and we choose, somewhat arbitrarily,

$$\kappa_p = \kappa\,(1+\beta p^4) \tag{30}$$

where β is a constant. The selection $\beta = 0.05$ leads to two dominant modes, p = 0 and 1.

An overview of the results [23] is given in Fig. 9 where the solid line traces the behavior of the steady state modulus of the mode p = 0. The system displays hysteretic behavior as indicated by the arrows. Upon decreasing the value of a(1) [see Eq. (15)] beyond a certain threshold value an instability develops that leads to an oscillating time dependence of the output and which appears to be induced by a Hopf bifurcation. The frequency of the spontaneous oscillations in the output intensity is very close to the beat between the pulled frequencies of modes p = 0 and p = 1 whose strong interference leads to a reduced average value of the power output, a feature that may be undesirable for certain applications.

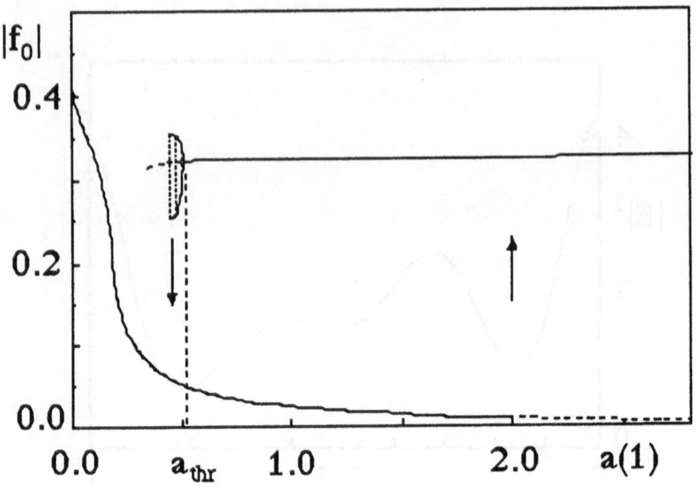

Fig. 9 Ring laser with spherical mirrors, 2C = 1.2, θ = 0., $\kappa/\gamma_\perp$ = 0.1, and $\gamma_\parallel/\gamma_\perp$ = 1. The steady state value of the modulus $|f_0|$ [see Eq. (18)] is plotted as a function of the mode spacing a(1). Beginning at a(1) = a_{crit}, and continuing for smaller values on the upper branch, one has the appearance of oscillations in the output intensity.

The growth of the oscillations for decreasing values of a(1), perhaps because of the presence of a saddle node, eventually causes a sudden disappearance of the oscillatory instability whose range of existence is sensitive to the value of $\gamma_\parallel/\gamma_\perp$ (in these simulations the instability range is rather small, but it increases significantly for smaller values of $\gamma_\parallel/\gamma_\perp$). A further decrease of the transverse mode spacing yields the development of long term stationary states which are characterized by modal intensities $|f_0|^2$ and $|f_1|^2$ of comparable magnitude (Fig. 10). Note that we find again a crossover of the two intensities as we also observed in the plane-wave case (Fig. 7).

In these steady state multimode configurations, it is clear that the natural oscillation frequencies of the transverse modes have been pulled together, or frequency locked, into a common value. Typical stationary solutions in this range of parameters displays a significant distortion of its transverse pattern which should be easily recorded by a suitable scan of the radial profile (Fig. 11).

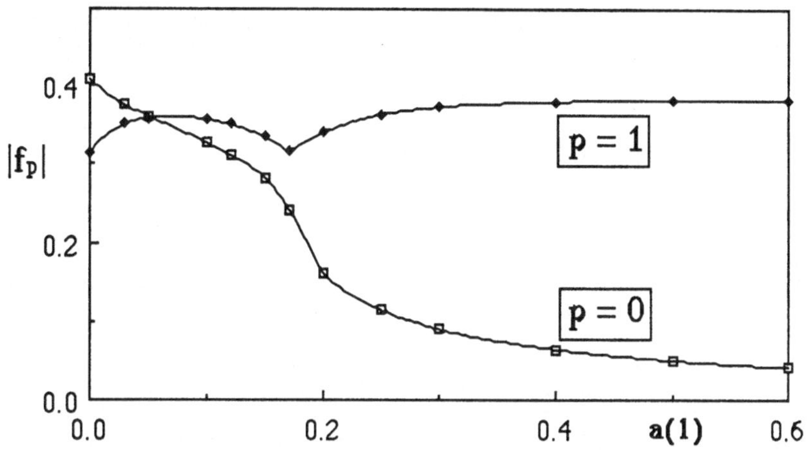

Fig. 10 The steady state intensities of modes p = 0 and 1 are plotted as functions of the a(1). The values of the parameters are the same as in Fig. 9.

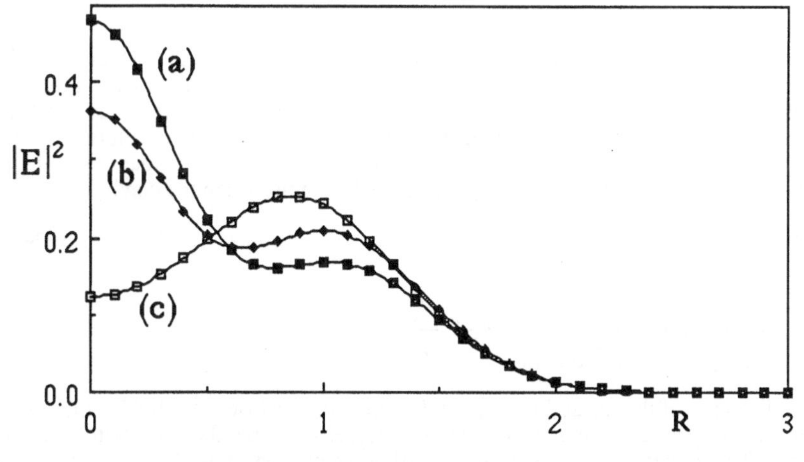

Fig. 11 Transverse profile of the stationary output intensity for (a) a(1) = 0.1, (b) a(1) = 0.05, (c) a(1) = 0. The values of the other parameters are the same as in Fig. 9.

We expect that the power output should be higher in the presence of cooperative frequency locking because a synchronized multimode operation has a better chance to use the available gain in the active region.

7. Spontaneous breaking of the cylindrical symmetry

In the case of the ring cavity with spherical mirrors we found transitions in the transverse spatial structure both in the active and in the passive case. Strictly speaking. however, these transitions are not accompanied by a symmetry breaking. In this section we demonstrate that this type of cavity configuration can also generate a real symmetry breaking phenomenon: this is associated with the breaking of the cylindrical symmetry.

As shown in Eq. (17), different transverse modes with the same value of 2p+l are degenerate in frequency. We consider a ring laser with spherical mirrors in which the atomic line is exactly resonant with the three degenerate modes $p = 1, l = 0$ and $p = 0, l = 2$ [sine and cosine; see Eq. (16)]. We assume that all the other modes are quite removed in frequency from the atomic resonance, so that their dynamical contribution can be ignored and the problem can be reduced to the interaction of three modes. We assume that the pump profile has the Gaussian form

$$\exp\left(-\frac{r^2}{2R_p^2}\right) \tag{31}$$

wher R_p is the pump waist. The system admits three distinct single-mode stationary solutions governed by the steady state equations

$$2C\int_0^{2\pi} d\varphi \int_0^{\infty} dr'\, r' \frac{A_{pl}^2(r',\varphi)}{1+A_{pl}^2(r',\varphi)\, f_{pl}^2} \exp\left(-2\frac{r'^2}{\psi^2}\right) = 1 \tag{32}$$

where $r' = r/w_0$, A_{pl} is given by Eq. (16), f_{pl} is the amplitude of the mode (p,l) and $\psi = 2R_p/w_0$. The cylindrically symmetric stationary state $p = 1, l = 0$ (Fig. 12a) has the lowest threshold. Hence, if we increase the pump parameter gradually, the laser starts oscillating with the symmetric mode. We performed analytically [24] the linear stability analysis of this stationary solution obtaining two eigenvalue equations, one for the amplitude and the other for the phase fluctuations of the electric field. The latter equation predicts the emergence of an instability when the condition

$$2CI > 1 \tag{33a}$$

$$I = \int_0^{2\pi} d\varphi \int_0^{\infty} dr'\, r' \frac{A_{02}^2(r',\varphi)}{1+A_{10}^2(r')\, f_{10}^2} \exp\left(-2\frac{r'^2}{\psi^2}\right) \tag{33b}$$

is satisfied. The ratio of the instability threshold to the ordinary laser threshold is close to unity for $\psi >> 1$. When the symmetric stationary solution becomes unstable, the asymmetric solutions are also unstable. The numerical solution of the equations of motion shows [24] that, if the system is initially in the symmetric state (Fig. 12a) and the pump parameter satisfies the instability condition (33), the system evolves into a new stationary state in which the three modes coexist and the cylindrical symmetry is broken (Fig. 12b). There is in fact an infinite number of asymmetric stationary solutions that can be obtained from one another by a simple rotation around the axis of the system; an initial fluctuation determines which of these solutions is approached by the system.

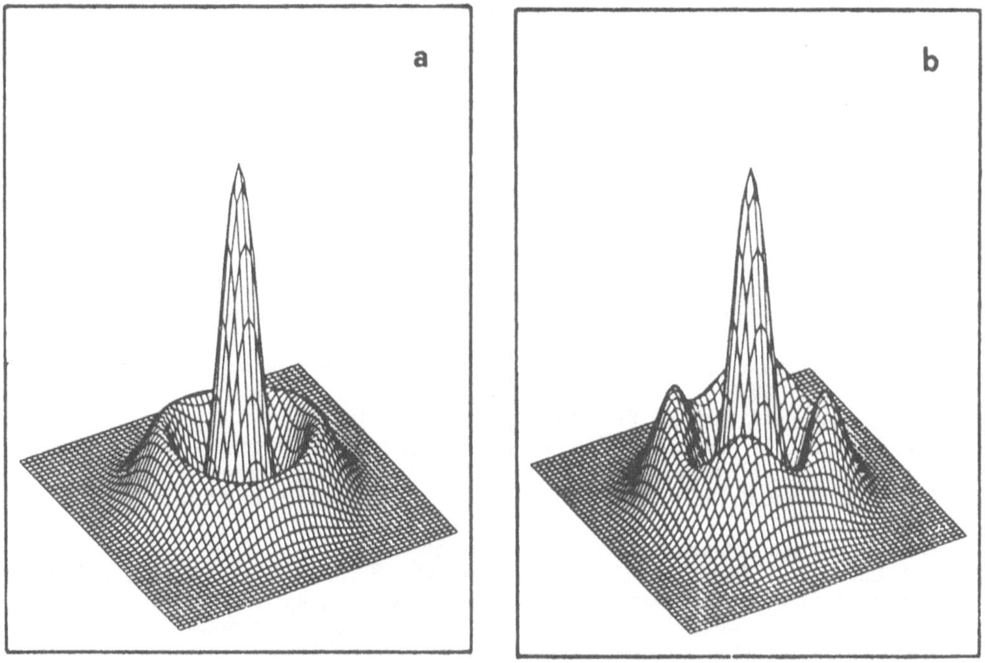

Fig. 12 Three-mode ring laser with spherical mirrors. Evolution from a cylindrically symmetric (a) to an asymmetric (b) steady state transverse configuration of the output intensity. The laser is 50% above threshold, $\psi = 1.5$ and $\kappa = \gamma_{\perp} = \gamma_{\parallel}$.

Similar phenomena involving the spontaneous breaking of the cylindrical symmetry have been discovered by numerical studies in Ref. 25 and have also been observed experimentally in four-wave mixing experiments [26] without a resonator and in a configuration involving a single mirror [27].

Acknowledgements

This work has been performed with the partial support of a NATO Collaborative Research Grant and in the framework of the EEC twinning project on "Dynamics of Nonlinear Optical Systems".

References

* Permanent address: Department of Physics, Beijing Normal University, Beijing, People's Republic of China.

1. H. Haken, Synergetics - An Introduction, Springer Verlag, Berlin, 1977.
2. G. Nicolis and I. Prigogine, Self-Organization in Nonequilibrium Systems, Wiley, New York, 1977.
3. J.D. Murray, J. Theor. Bio. 88, 161 (1981).
4. A.M. Turing, Phil. Trans. Roy. Soc. London B237, 37 (1952).
5. N.B. Abraham, L.A. Lugiato, and L.M. Narducci, Eds., Feature Issue on Instabilities in Active Optical Media, J. Opt. Soc. Am. B2, January 1985; D.K. Bandy, J.R. Tredicce and A. Oraevsky, Eds., Feature Issue on Nonlinear Dynamics of Lasers, J. Opt. Soc. Am., May 1988 (in press).
6. F.T. Arecchi, and R. Harrison, Eds., Instabilities and Chaos in Quantum Optics, Springer Verlag, Berlin, 1987.
7. L.A. Lugiato and R. Lefever, Phys. Rev. Lett. 58, 2209 (1987).
8. L.A. Lugiato, L.M. Narducci, and R. Lefever, in "Lasers and Synergetics - a volume in honor of the 60th birthday of Hermann Haken", Springer Verlag, Berlin 1987.
9. L.A. Lugiato, and R. Lefever, volume in honor of the 70th birthday of Adriano Gozzini, in press.
10 L.A. Lugiato and C. Oldano, submitted for publication.
11. L.A. Lugiato, C. Oldano and L.M. Narducci, in J. Opt. Soc. Am. B, Feature issue on Laser Dynamics, edited by D.K. Bandy, J.R. Tredicce and A. Oraevsky, to appear.
12. A. Ouazzardini, H. Adachihara and J.V. Moloney, private communications.
13. J.V. Moloney and H.M. Gibbs, Phys. Rev. Lett. 48, 1607 (1982).
14. D.W. McLaughlin, J.V. Moloney and A.C. Newell, Phys. Rev. Lett. 54, 681 (1985).
15. L.A. Lugiato, "Theory of Optical Bistability", in Progress in Optics, Vol. XXI, edited by E. Wolf, North Holland, Amsterdam, 1984.
16. H.M. Gibbs, S.L. McCall and T.N.C. Venkatesan, Phys. Rev. Lett. 36, 113 (1976).
17. D. Sattinger, Topics in Stability and Bifurcation Theory, Springer Verlag, Berlin, 1973.
18. A. Yariv, Optical Electronics , 3rd Edition, Holt, Rinehart and Winston, New York, 1985.
19. P. Ru, L.M. Narducci, J.R. Tredicce, D.K. Bandy and L.A. Lugiato, Opt. Comm. 63, 310 (1987).
20. L.A. Lugiato, L.M. Narducci and Wang Kaige, in preparation.
21. H. Haken, Phys. Lett. 53A, 77 (1975).
22. H. Haken, Light 2 - Laser Dynamics, North Holland, Amsterdam, 1985.
23. L.A. Lugiato, G.-L. Oppo, M.A. Pernigo, J.R. Tredicce, L.M. Narducci and D.K. Bandy, submitted for publication.
24 L.A. Lugiato, F. Prati, L.M. Narducci and G.-L. Oppo, in preparation.
25. J.V. Moloney, H. Adachihara, D.W. McLaughlin and A.C. Newell, in Chaos, Noise and Fractals, E.R. Pike and L.A. Lugiato, Eds., Adam Hilger, Bristol, 1987.
26. G. Grinberg, Opt. Comm. to be published.
27. G. Giusfredi, J.F. Valley, R. Pon, G. Khitrova and H.M. Gibbs, J. Opt. Soc. Am. B, Feature Issue on Laser Dynamics, D.K. Bandy, J.R. Tredicce and A. Oraevsky, Eds., May 1988.

SCALING FOR AN INTERFACIAL INSTABILITY

by

David Jasnow

Department of Physics and Astronomy

University of Pittsburgh

Pittsburgh, Pennsylvania 15260 U.S.A.

1 introduction

In these lectures I will review recent work exploring the dynamics of some interfacial growth processes. Some of the work is in initial stages and empirical in nature, and these lectures are, in part, a progress report.

Considerable progress has been made over the last few years in the field of interfacial pattern formation. There is a great deal of conceptual as well as practical interest in studies of two-phase interfaces under non-equilibrium conditions. On the theoretical side one must deal with the complex, nonlinear behavior of moving boundaries in which nonlocality plays an essential role. On the more practical side, questions of solidification, kinetics, hydrodynamic instabilities and so on have a variety of technological connections. Examples of specific systems that have attracted attention include (i) directional solidification in binary systems [1,2], (ii) viscous fingering in a Hele - Shaw cell, [3,4] and (iii) dendritic growth of a solid from a melt. [5] Additional references and background are contained in, for example, the review by Langer. [6]

A great deal of the attention in these problems has concentrated on the question of the selection and characterization of steady state patterns which emerge reproducibly in experiments. For example, in the case of viscous fingering in a rectangular Hele - Shaw cell, one would like to predict the finger shape, in particular, the width of the finger in the channel, as a function of the velocity of advance. Experimentally one finds that a unique width is selected; theoretical analysis typically yields a family of steady state solutions. A mechanism, which focuses on certain "essential" aspects of the system, has been proposed to explain the selection. [7] The general idea is that in some cases a continuous family of steady state solutions is already broken down to a countable number (by, in the viscous fingering case, for example, the existence of nonzero surface tension), and only a small number, perhaps a single one, of these solutions is linearly stable. A detailed numerical check of this scheme has been made in the case of viscous fingering, [8] which stands out as a prototype case since the starting equations and system-dependent parameters are well known, and in the case of directional solidification. [9,10,11,12]

Not as much attention has been paid as yet to to the dynamics of interfacial growth (as opposed to the particular steady state characteristics). Some systematics of cluster growth and growth from an initially planar boundary have been studied using one or another variant of diffusion-limited aggregation first studied in detail by Witten and Sander. [13] In such simulations random walkers are introduced into the system far from the planar boundary (or"seed", depending on the system under consideration), and they walk until contact is made. The original idea was to model mass or concentration diffusion and show that complex patterns could evolve from rather simple attachment kinetics. Numerous studies, for example Refs. [14,15,16,17,18], have yielded a great deal of empirical and other information about these processes.

Initial DLA-based studies, however, were not designed to deal with local thermodynamics and relaxation of the growing structures, a feature central to the Mullins - Sekerka instability [19] which lies at the heart of a variety of interfacial growth instabilities. Later efforts did aim at introducing the major effects of surface tension into DLA-like processes. [20,17] It is also possible to study matter or concentration diffusion and an instability of the Mullins - Sekerka type by basing the simulations on lattice gas models, in which additional modeling of local thermodynamics and the *ad hoc* introduction of the effects of surface tension are not required. The Gibbs - Thomson effect, which may be understood microscopically in terms of condensation and evaporation of particles at the interface, can be taken care of automatically, without macroscopic or phenemenological considerations. Such simulations based on Ising lattice gas energetics and kinetics will be described below.

Once a lattice gas description is accepted for the simulation of an interfacial growth instability, one is struck by qualitative similarities with other growth or ordering processes which depend crucially on interfacial dynamics, such as phase separation or ordering in binary systems both with nonconserved and conserved order parameters. The ordering in such systems has revealed a scaling (self-similar) behavior discovered empirically. [21] There indeed has been a revitalization of theoretical interest and activity in studies of spinodal decomposition owing to the discovery of a possible scaling regime seen first in computer simulations at the microscopic level. [22]

In these lectures I will review work on simulating the growth dynamics of a system undergoing an interfacial instability from a microscopic point of view and will discuss the nature of the scaling which is seen to emerge. [23,24] Preliminary results of simulations based on macroscopic interfacial equations of motion will also be discussed. The original lattice-gas simulations forced further considerations on the nature of ordering phenomena in the presence of a gradient in the ordering

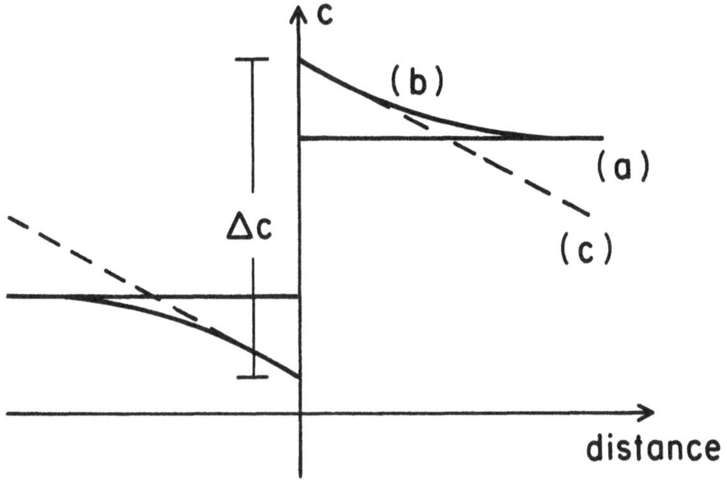

Fig. 1. Flat interface shown schematically on a macroscopic scale.

field; a digression on such dynamics is included.

An outline of these lectures is as follows. In the next section the lattice gas model and simulations are discussed. This is followed in Section 3 by a discussion of the results, while Section 4 is devoted to a proposed scaling law. A discussion of the macroscopic equations and the effect of a field gradient occupies the next section. Section 6 is devoted to interface equations and some preliminary simulations of the macroscopic equations. The final section is devoted to some concluding remarks.

2 model and simulations

The situation considered, one in which an instability of the Mullins - Sekerka type develops, is as follows. Imagine a two-phase system, say an A-B binary system, in equilibrium at some temperature below the phase separation critical point. Suppose the system were driven into a nonequilibrium state by, for example, quenching rapidly to a temperature deeper in the two-phase region. The situation is shown schematically in Fig. 1. The initial flat interface and the corresponding miscibility gap are shown on a macroscopic scale in Fig. 1(a). The compositions refer, for example,to the A-species in the binary system. A very long while after the quench the compositions reach the new equilibrium values determined by the final temperature; the A-rich phase is slightly richer in the A species, and so on. (Experimentally this may take a long while indeed, requiring diffusion over macroscopic distances of the sample.) On the other hand, very near the interface the compositions may adjust to their final equilibrium values quite rapidly, and the final (larger) miscibility gap Δc can be established at the interface. As shown schematically in Fig. 1(b) this leads to a concentration gradient and, hence, a concentration flux across the interface. Events are delayed far from the interface because of the finite time necessary for macroscopic diffusion of the composition. If one is interested in interfacial effects of such a nonequilibrium situation, one might as well consider a flat initial interface subjected to a steady flux, suggested by the dashed line in Fig. 1(c). For a symmetric binary system this configuration has been examined by Langer and Turski [25] and in the context of binary fluids by Jasnow et al. [26] A standard linear stability analysis (briefly reviewed below) shows that at sufficiently long wavelength, surface tension effects are insufficient to stabilize the flat interface, and small distortions grow exponentially.

An actual quench rather than a steady driving force is more difficult to treat. As matter is transported across the interface, the bulk phases approach their final equilibrium values. Local gradients in the vicinity of the interface are diminished, and the driving force for the instability is thereby reduced. The question of whether or not a time domain of exponential or nearly exponential growth exists is important from an experimental viewpoint and was addressed by Onuki et al. [27] These authors examined the evolution of the concentration profile forming right after the quench. In spite of the driving force for the instability evolving dynamically, a time domain was found with exponential growth of interfacial distortions. The simplification to the ramp profile suggested in Fig. 1(c), that is to a constant driving force for the instability, certainly makes microscopic simulations as well as theoretical treatments easier. This simplification is made in all of the following discussion.

One may simulate the evolution of a system in which the interface separating two coexisting phases of a two-phase system is driven unstable in the following manner. [For simplicity, all terminology is appropriate to two dimensions; generalization of the ideas to higher dimensions is straightforward.] An Ising system on a square lattice of $N_x \times N_z$ sites is initialized below T_c with an interface at $z = 0$, the spin-up phase occupying the top half ($z > 0$) of the system. Clearly one may translate the terminology to a binary lattice gas, where spin-up corresponds to A species, and so on. The initial equilibrium situation is established by pinning the top row ($z = z_{max}$) of spins "up" and the bottom row ($z = z_{min}$) of spins "down." Periodic boundary conditions are used in the x-direction. A field gradient, which will be discussed in more detail below, is introduced in such direction to stabilize the interface. In the presence of the field gradient alone, interface distortions set up initially relax back to the flat interface.

The energy of a spin configuration is thus

$$E = -J \sum_{<ij>} S_i S_j - \sum_i G z_i S_i \tag{1}$$

where, as usual, i denotes a site with coordinates (x_i, z_i). The first sum is over nearest neighbor pairs and each S_i may take the value +1 or -1. In the language of binary mixtures an up (down) spin in a cell corresponds to an A (B) particle. At the temperatures considered, $T < T_c$, the equilibrium interface is extremely flat. For (small) systems of width $N_x \sim 100$, the equilibrium interface is extremely flat even without the field gradient. The field gradient removes even the possibility of "interface wandering."[28] One must specify the kinetics describing the evolution of a spin configuration. To model concentration diffusion, conserving kinetics of the simplest type are adopted, namely Kawasaki spin exchange dynamics. [29] The system evolves under the exchange of a down spin with an up neighbor thus conserving the total magnetization of the system. The following rules are followed. (i) A down spin may exchange with an up neighbor; for each movable down spin a neighbor is randomly selected. (ii) If a neighboring up spin happens to be selected, the exchange is automatically effected if the energy computed according to Eq. (1) decreases. (iii) If the energy were to increase after the exchange, the exchange is made with probability determined by the Boltzmann factor for the energy increase, ΔE, i.e., the value of $\exp(-\Delta E/k_B T)$ is compared to a random number on the unit interval. In a pass all potentially mobile down spins are given an attempt to exchange.

Such dynamics allow the system to approach equilibrium. These same dynamics have been widely used in simulations of spinodal decomposition and in the study of spin dynamics with conserved order parameter. [21,30] To study spinodal decomposition, one begins with a high temperature equilibrium state and examines the phase separation process aiming at equilibrium at a temperature well below the critical temperature. By contrast, in the present simulations one begins with a low temperature state with two-phase coexistence. More recently there have been studies of Kawasaki dynamics in driven, open systems carrying a current; [31] such systems can exhibit steady state phase transitions and have been receiving considerable attention including preliminary linear stability studies of a flat interface. [32]

A nonequilibrium flux is established in these simulations by introducing a down (up) spin at a random position deep in the up (down) phase at regular intervals. This is simply accomplished by overturning a present spin. These introduced spins evolve under the same dynamics as the others and hence diffuse toward the interface. The field gradient is not essential in principle, but its presence induces spins to execute a biased random walk toward the interface. Note that the overturned spins will be relatively few in number, but even so, the phases are driven metastable by

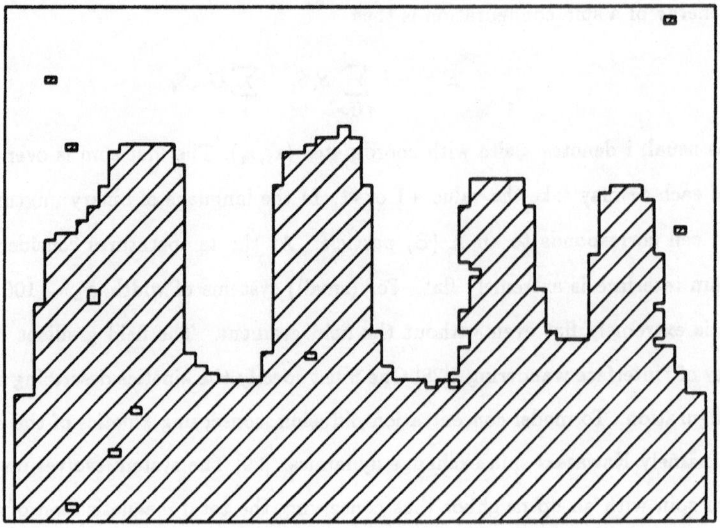

Fig. 2. Sample run exhibiting the developed pattern in the two-sided model after 64000 M. C. steps. System width 64.

the introduction of "wrong" spins, and the bias helps prevent clustering in the bulk. It is important to recognize that no special boundary conditions are required at the interface; all spins, whether on the interface or elsewhere in the system, move according to the same stochastic dynamics (which sample the local energetics) previously described. As noted above, these simulations represent an extension to interfacial nonequilibrium processes of techniques used to study nonequilibrium and near-equilibrium bulk processes.

In such simulations one hopes, of course, to make contact with macroscopic processes or macroscopic treatments; this is possible if the structures which develop are sufficiently larger than microscopic lengths which enter the problem. Relatively low temperatures are considered so that thermal correlation lengths are small, on the order of a lattice spacing. Hence if structures which develop are more than several lattice spacings in extent, one may hope to reach the beginnings of the macroscopic domain. The macroscopic treatment, at least at the level of linear stability analysis, will be considered below.

Two distinct types of simulations will be commented on here. The symmetric or two-sided problem has been discussed above. The second type is one-sided, in which diffusers are only introduced at one end. In this second case one of the phases grows in volume since the number of one of the species grows at a regular rate; hence the average interfacial position advances at a regular rate. In Fig. 2 a sample run for the symmetric, two-sided case is shown, while in Fig. 3 a

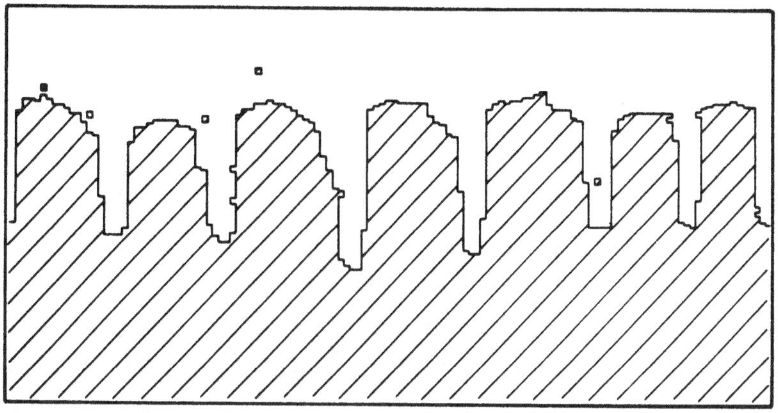

Fig. 3. Sample run for the one-sided model. System width 128.

sample run for the one-sided case is shown. All figures will correspond to $T = 0.5T_c$, $G = 1$ in units of the exchange energy. The initial condition for the one-sided case was a thin layer of one of the phases at the bottom. Note the strong asymmetrical pattern, the grooves being narrower, in the one-sided case.

A final comment is in order on the simulations. Although the concept is straightforward, it should be noted that exchange dynamics are relatively slow at the temperatures considered. Time must not be wasted on immobile spins. Data structure techniques along with dynamic memory allocation were employed initially; further details are discussed in Ref. [23]

3 results

The microscopic simulations are analyzed using the Fourier transform, or more specifically the "quasistatic power spectrum," $P(k,t)$, of the interface shape, where t is the time measured in passes through the system of spins. This is defined as follows. If $\varsigma(x,t)$ is the height deviation of the interface from the average position (assumed single valued; see below) at horizontal position x, then one may find the lattice transform in the usual fashion

$$\hat{\varsigma}(k,t) = \sum_{x} \varsigma(x,t) \exp(ikx), \tag{2}$$

where $k = 2\pi n/N_x$, with n an integer. Then the power spectrum is defined as $P(k,t) = \mid \hat{\varsigma}(k,t) \mid^2$. This quantity is closely related to the exchange, or short range energy associated with the interface (by analogy with discussion of capillary waves and roughening). The exchange energy associated with the interface can be thought of as proportional to the length of interface, which involves $P(k,t)$ as discussed briefly in the next section. It comes as no surprise that $P(k,t)$ tends to increase with time; exchange energy is being fed into the system at a regular rate through the overturned spins. This energy finds itself, through an "inverse cascade process," in interfacial energy. A crude argument based on this will be used in the next section to suggest a scaling law relating the exponents δ and z which are introduced now.

The conclusion of the microscopic simulations, for which some evidence is presented in this section, is that there appears to be a scaling regime at intermediate times for the evolution of a flat interface driven out of equilibrium by a flux as described in Section 2. One finds that the power spectrum described above takes the form

$$P(k,t) \sim t^{\delta} f(kt^{1/z}). \tag{3}$$

In addition there appears to be a degree of universality in the scaling form which may be investigated by varying (as much as reasonable within rather tight constraints) some of the microscopic parameters (e.g., temperature, field and driving flux). The scaling form introduces two exponents z and δ, the former of which can be interpreted as a 'dynamical exponent' governing the coarsening.

Two further points need to be made before discussing the results. The processes are stochastic; all future references to the power spectrum will refer to an average over 50–100 individual runs. The second point is that the interface shape is not single valued. At the low temperatures considered, the correlation length is on the order of a cell size so that one expects extremely short wavelength fluctuations which are not of interest here. They are merely averaged out. Other types of overhangs are possible as a consequence of the macroscopic nonequilibrium growth. This has been discussed elsewhere [23]; for the moment let it suffice to say that the runs are over intermediate times in which serious multivaluedness does not occur (in the simulations). The time domain for single valued interfaces (on the macroscopic scale) might be quite long. This situation is to be contrasted with the numerical solution of the interface equations in Section 6.

Some results for the two-sided case have appeared in an earlier publication, [23] and the reader is referred to it for further details. A sample power spectrum for a few times is shown in Fig. 4. It was shown that the power contained in the dominant mode (that is, the value $k = k_{max}(t)$ for which $P(k,t)$ is maximal at the time t) grows as a power law, i.e.,

$$P(k_{max}(t),t) \sim t^{\delta}. \tag{4}$$

Plotting $k_{max}(t) vs. t$ in a log-log fashion suggests $\delta \sim 2.3-2.4$ for the growth exponent independent of the driving flux. [23] (For lower flux the system takes longer to organize and get into the scaling

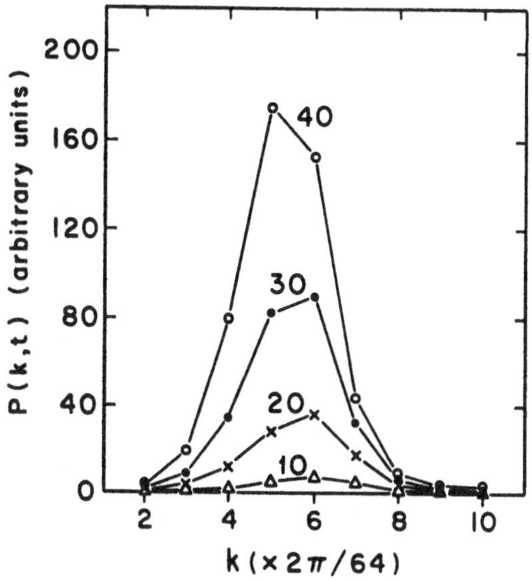

Fig. 4. Sample power spectrum for the two-sided model. The labels give the time in thousands of M. C. steps.

regime.) Independence with respect to limited changes in temperature and field gradient was also noted. This suggests some degree of *universality* for the growth exponent. Now, $k_{max}(t)$ provides a characteristic wavenumber for the interfacial structures. It is reasonable to expect coarsening of the structures; one has ample experience with other diffusive processes in which screening plays a role. As one finger lags it is deprived of new "material" and the relaxation processes embodied in the local energetics described by Eq. (1) act to flatten it. The coarsening is easily observed by plotting $P(k,t)$ and noting that longer wavelength modes dominate at later times. In Fig. 5 the situation is shown for the one-sided case. The same general behavior can be seen in Fig. 4. As described elsewhere, it is natural (from general considerations and by analogy with other coarsening processes) to attempt to fit the simulation data to a scaling form

$$P(k,t) \approx A(t-t_0)^{\delta} f(b(t-t_0)^{1/z}k). \tag{5}$$

This is consistent with the observed power-law time dependence of the dominant mode. One generally expects the constants A and b to be nonuniversal (i.e., to depend on the flux and field gradient, for example), but one hopes the scaling function $f(x)$ is independent of such details. The fitting parameter t_0 is needed because one cannot expect, nor does one observe, the scaling regime to set in at the earliest times. The scaling function $f(x)$ itself is shown in Fig. 6 for the two-sided case; note there are two different flux values shown. As described earlier [23,24] the fact that the

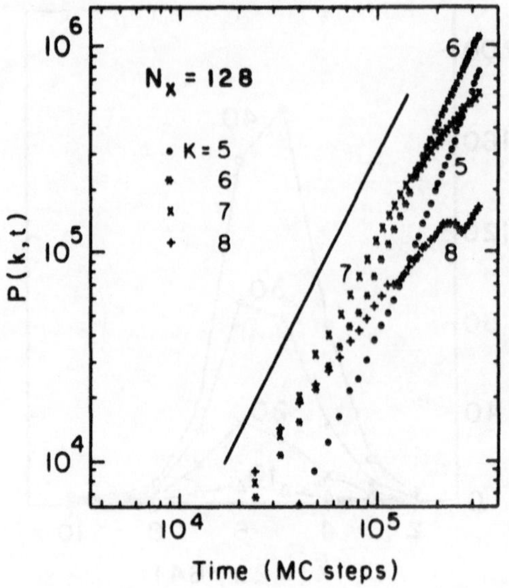

Fig. 5. Growth of several modes as a function of time for a sample of runs of the one-sided case. The straight line is drawn with slope 2. Note the crossing; curves are labeled by relative wavenumber.

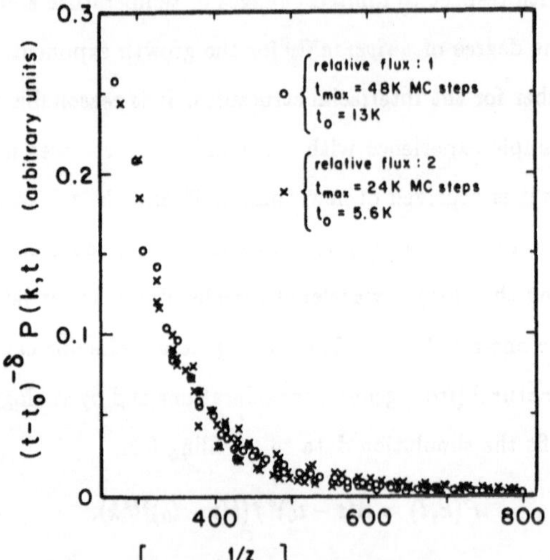

Fig. 6. Collapsed data showing the scaling function $f(x)$ for the two-sided case. The nonuniversal amplitudes A and b have been adjusted to bring the data for two different flux values together. Exponent values $\delta = 2.4$ and $z = 2.2$ have been used for both data sets.

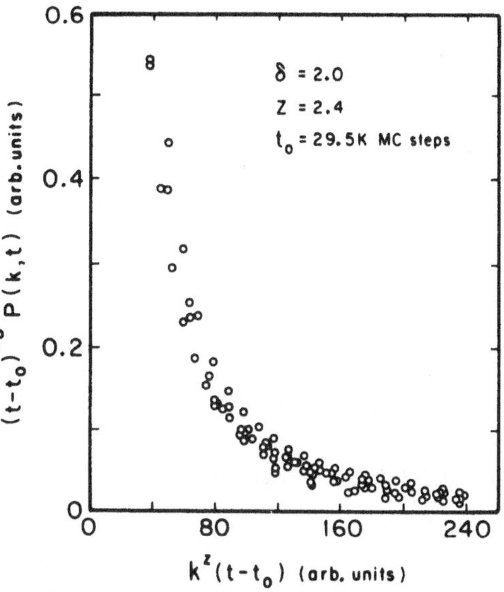

Fig. 7. Similar collapsed data for the one-sided case. Values $\delta = 2$ and $z = 2.2$ have been used.

data "collapse" and fall on a smooth curve is evidence for a scaling regime; furthermore the data collapse for different fluxes is evidence for universality of the scaling function. Preliminary data for the one- sided case are shown in Fig. 7; once again it appears that the data fit a scaling form at least over a limited range.

Unbiased, but as yet fairly crude, estimates for the exponents δ and z are $\delta \approx 2.4, z \approx 2.2$ for the two-sided case and $\delta \approx 2, z \approx 2.4$ for the one-sided case. These initial simulations suggest a confidence range of about 10% in the value of δ. While the precision can certainly stand improvement, it certainly appears that the one- and two - sided cases belong to different universality classes. This is consistent with the qualitative appreciation that systems behave differently when there is strong asymmetry in the diffusion coefficients of the two phases (as compared to the symmetric case). Furthermore, in the one-sided case there is a net advance of the average interface position, while in the symmetric case the average interface position is constrained by the conservation to remain fixed. Within the available precision both sets of exponents appear to satisfy the equality in the exponent inequality

$$z(\delta - 1) \leq d + 1, \tag{6}$$

where d is the spatial dimensionality. In the next section a rough argument is sketched, specific to the present simulations, which suggests such an exponent relation. In Section 6 interface equations and initial efforts toward their numerical solution are discussed. The intention is use such equations to develop greater precision and flexibility to reinforce the studies initiated by the Monte Carlo approach discussed here.

4 a proposed scaling law

In this section a "derivation" of the scaling law relating the exponents z and δ suggested in Eq. (6) is given. First consider the analogy of the phase separation occurring in spinodal decomposition. There are two exponents at play there as well, but the scaling of the structure factor takes the form

$$S(q,t) = L(t)^d F(qL(t)), \tag{7}$$

where q is the wave number and $L(t)$ is the characteristic domain size at time t. The size $L(t)$ grows as a power law, $L \sim t^{1/z}$ so that from Eq. (7), the analog of the exponent δ introduced above (that is, the power of the time in the prefactor) is not independent of the 'dynamical exponent' z. In the case of spinodal decomposition and, indeed, for other ordering processes, the integrated structure factor (properly normalized) is time independent. An argument is now presented suggesting that there is a relation between the two exponents z and δ describing the scaling associated with the interfacial growth instability under consideration. Such an argument is based on the particular microscopic processes in the simulations.

In Ising language the nonequilibrium flux involves an introduction of "up" ("down") spins in a "down" ("up") phase, and each such introduction adds short range exchange energy to the system. [In the language of AB mixtures, like particles attract, so introduction of an A particle into a B-rich phase raises the interaction energy.] The interface itself, with its ever increasing length, is a repository for the exchange or interaction energy continually introduced on the shortest length scales. Note that the kinetics of the model are particle-conserving but not energy-conserving, so that one cannot rigorously track the exchange energy. However the short-range energy is introduced at a regular rate, so that the total introduced exchange energy increases linearly in time. It is reasonable to assume that the excess exchange energy associated with the interface (over and above the energy of the flat interface) is bounded by the energy injected.

Now, at low temperatures the interface energy is proportional to its length (in the case of two-dimensional bulk systems) or more generally the "area", which, in a continuum approximation in d dimensions, is given by

$$A = \int [1 + (\nabla\varsigma(x,t))^2]^{1/2} d^{d-1}x. \tag{8}$$

For small amplitudes (or more precisely, small gradients) the incremental area and hence the incremental exchange energy is simply proportional to

$$\Delta E_{ex} \sim \int (\nabla \varsigma(x,t))^2 d^{d-1}x \sim \int k^2 P(k,t) d^{d-1}k, \tag{9}$$

where the identification of the Fourier transform of the interface shape with the power spectrum $P \sim |\hat{\varsigma}(k,t)|^2$ has been used. If the form of the scaling function introduced in Eq. (3) is used in Eq. (9), and one further assumes, as argued above, that the excess interface energy is bounded by the energy injected, one arrives at the scaling law $z(\delta - 1) \leq d+1$ suggested previously. In has been confirmed within the simulations themselves for the one-sided case that the interfacial length grows linearly in time essentially for the entire run. At least within the small amplitude approximation made above (*i.e.*, for the amplitude of the growth modes small compared to the wavelength), the linear growth of the length suggests equality should be satisfied in the above relation between z and δ. The relation, which involves the dimensionality d, between the phenomenologically introduced exponents, and its suggested origin, makes the proposed interfacial scaling even more analogous to the situation in spinodal decomposition and other ordering processes.

There is even a further extension suggested by the observed growth of the interfacial length. The full length (or 'area' more generally) involves not only $P(k,t)$, which might be thought of as a two-point function, but, according to Eq. (8), also involves higher functions of the interface displacement. Certain additional insights may be obtained by a study of these multipoint correlations of interfacial displacement, although the indications are for a more complicated spectrum of correlations than one has, for example, when hyperscaling holds in critical phenomena. [33] On the other hand both renormalization group treatments [34] and simulations on driven systems [31] offer some encouragement for this line of investigation.

5 macroscopic description and dynamics in a field gradient

The macroscopic description and linear stability analysis are briefly reviewed. Interestingly, these simulations involved a field gradient in a material way, and considerations on the macroscopic description led to developments in the treatment of phase separation dynamics, specifically spinodal decomposition, in the presence of such a field. At the end of this section modifications in the conventional treatment based on the Cahn - Hilliard equation [35] will be briefly discussed.

The problem considered involves isothermal diffusion in a binary system, and one expects, by analogy with Brownian motion in a force field, that the following equation must be satisfied in each of the bulk phases

$$\partial_t u = D[\nabla^2 u + Q_\gamma G \partial_z u]. \tag{10}$$

The subscript γ refers to the two phases (spin up or A-rich and spin- down or B-rich) denoted by α and β. Here u is the deviation of the concentration from a stationary profile described in this section below, and Q, which will also be identified below, depends on which of the two phases one is considering. G is a constant field gradient, so that the magnetic field, in Ising language, varies linearly through zero at the nominal interface center. Note that the product QG is dimensionally an inverse length, and so introduces a diffusion length into the problem. The incremental current (say of the A-species) consistent with Eq. (10) is given by

$$\delta\vec{j} = -D[\nabla u + QGu\hat{e}], \tag{11}$$

where $\hat{e}$ is a unit vector in the z-direction, perpendicular to the plane of the interface. The interface moves if there is a flux imbalance, so that

$$[\delta\vec{j_\alpha} - \delta\vec{j_\beta}] \cdot \hat{n} = (\Delta c)v_n, \tag{12}$$

where the normal velocity v_n is directed into the α phase, and Δc is the equilibrium miscibility gap. These equations must be supplemented by a condition of local equilibrium at the interface. The appropriate Gibbs - Thomson relation takes the form

$$u\,|_S = (j_0/D)\varsigma - \chi G\varsigma + \Gamma K(\varsigma), \tag{13}$$

where the interface shape is given by $\varsigma(x,t)$ and $K(\varsigma)$ is the local curvature of the interface, here taken as positive if the center of curvature lies within the α phase. (The subscript S indicates that the quantity is to be evaluated as the limit approaching the interface from one or the other of the bulk phases.) Eq. (13) is.standard, with the concentration deviations at the interface of the form $c - c^{eq} = \Gamma K$. It may appear unusual because of the reference state, described below, used in defining u. The physical content of Eq. (13) is the instantaneous adjustment of local compositions to maintain coexistence in the face of curvature, displacements and so on. The coefficient Γ contains thermodynamic information and is of the form $\Gamma = \sigma\chi/\Delta c$, where σ is the surface tension and χ is the ordering susceptibility $\chi = \partial c/\partial\mu$ in the equilibrium phases. Note for this symmetric model the susceptibility is equal for the two coexisting phases. There is some interesting physics in the interplay of quantities in the capillarity Γ which will be commented on below. Finally, note in Eq. (13) that j_0 is the driving force for the instability, the steady flux impinging on the interface. [25]

It is worth commenting on some features of the thermodynamic boundary condition Eq. (13). Surely in modeling a solid the surface tension should be anisotropic, and this is easily included in principle. In fact, it is a strength of the approach that for simulations based on the Ising model, one knows the anisotropic surface tension from equilibrium calculations (in two dimensions) and simulations (in two and three dimensions). [36] This point has been discussed by Guo and Jasnow [23] to which the interested reader may refer. Indications are that the simulations are carried out under conditions of about 10 percent anisotropy, which may be considered rather large from some perspectives. A second point also discussed further in the references is that, strictly speaking, the interface "stiffness" should appear in the thermodynamic boundary condition when comparing to results for a solid. [37] The stiffness diverges exponentially at low temperatures, but it should be remembered that the susceptibility goes even more rapidly to zero, yielding an interesting interplay.

A second point just noted briefly here is that in the macroscopic description the thermodynamic boundary condition might need modification in the presence of an advancing interface, and one should generally expect a term (anisotropic in general) proportional to the normal velocity to appear on the right hand side of Eq. (13). [6,38,39,40] Physically speaking, the advance of the interface is expected to pull the system away from local equilibrium. A rough calculation based on the Ginzburg - Landau equation for this system (one scalar field, isothermal conditions) indicates the effect is small; in any event such effects are not included here.

The linear stability analysis corresponding to Eqs. (10) to (13) is straightforward. One assumes a small interfacial distortion of the form

$$\varsigma = \hat{\varsigma} \exp(ikx + \Omega t) \tag{14}$$

and corresponding solutions to the bulk diffusion equations (10), $u_\alpha \sim \exp(ikx + \Omega t - p_\alpha z)$, with a similar form for the β phase. The parameters p_α, p_β hence become known functions of k, G, Q and signs are chosen so that the deviations u decay at large distances from the interface. In the quasistatic approximation (made for convenience) the time derivative is dropped in Eq. (10), and one finds the dispersion relation

$$(\Delta c)\Omega = D[p_\alpha + p_\beta - G(Q_\alpha - Q_\beta)][(j_0/D) - \chi G - \Gamma k^2], \tag{15}$$

where Ω is understood as $v_n/\hat{\varsigma}$. Note the main features. A positive real part of Ω corresponds to growth. For $G \to 0$, $p_\alpha, p_\beta \to k$, and one recovers the usual symmetric model dispersion relation. Interfacial distortions grow for $k^2 < k_c^2 = (j_0/D\Gamma)$ for *any* positive value of the driving flux, j_0, *i.e.*,flux from the poor to the rich phase. [This feature makes the model somewhat unusual.] In the other limit $k \to 0$ for fixed G, the field gradient generally acts to stabilize the interface, but the ultimate stability or instability depends on the relative strength of j_0. For $j_0 = 0$ the field gradient acts to stabilize interfacial fluctuations absolutely. Finally, note that the critical wavenumber is reduced by the competing effect of the field gradient. Hence, at least in the linear regime, the essential features of the symmetric model instability are not modified seriously by the presence of the field gradient. The field gradient, recall, was initially introduced as a device to make the simulations possible (to reduce run-times, to reduce the tendency of nucleating clusters in the bulk, and so on). For long enough wavelength and strong enough driving flux j_0, small interfacial distortions are expected to grow exponentially. The field gradient may have an important effect on the non- linear regime, namely the width of the scaling regime, as preliminary results to be discussed below, suggest.

The origin of Eq. (10) and the identification of coefficients is now discussed. The more microscopic language of the Ginzburg - Landau level will be used, but ultimately the limit of long wavelengths will be considered and nonlinear fluctuation terms will be neglected. In the standard treatment of a symmetric binary model with conserved order parameter one has order parameter ϕ, with positive (negative) values corresponding to A-rich (B- rich). The most widely studied model for the dynamics (say, phase separation during spinodal decomposition) is the Cahn - Hilliard equation

$$\partial_t \phi = L \nabla^2 \frac{\delta H}{\delta \phi}, \tag{16}$$

where L a kinetic coefficient. For simplicity a Gaussian white noise satisfying the fluctuation-dissipation theorem has been omitted. The content of Eq. (16) is that changes of ϕ occur because of a nonvanishing local divergence of a current, itself proportional to the gradient of an appropriate chemical potential, $\delta H/\delta\phi$. The usual ϕ^4 Hamiltonian is used

$$H = H_0 = \int d^d r [\frac{1}{2}(\nabla\phi)^2 - \frac{1}{2}r\phi^2 + g\phi^4] \tag{17}$$

with $r > 0$ essentially measuring the temperature distance below the phase separation critical point, and g is an additional phenomenological constant. When an ordering field having a constant gradient in, say, the z-direction exists, the Hamiltonian is modified to become

$$H = H_0 - \int Gz\phi d^d r, \tag{18}$$

where G is a phenomenological constant proportional to the field gradient. Since the extra term is linear in ϕ and in the spatial coordinate, it simply disappears from the equation of motion (16). The effect of the boundaries, say zero flux at the top and bottom walls, will involve G explicitly. Hence to obtain the equilibrium in the presence of the field gradient, one need only modify the boundary conditions. One reaches a paradoxical situation in that one can postpone the effects of the boundaries for as long as necessary by, for example, increasing the height of the system. On the other hand in modeling the dynamics of a binary system with conserved order parameter using a master equation and Kawasaki exchange, as discussed in Section 2, the effect of the field gradient on an initial condition sets in immediately. The resolution is that in the presence of a field gradient, the coarse-graining leading to the Ginzburg - Landau level description must be handled carefully. Further details of the brief summary to follow have been provided by Kitahara et al. [41]

In Kawasaki exchange dynamics the local flux of the order parameter is proportional to a binary collision probability. In the usual spin-1/2 Ising versions this is usually not discussed; a cell has an A or a B (spin-up or down). Coarse graining so that the order parameter in a cell can vary continuously from, say $\phi = \phi_m$ (pure A) to $\phi = -\phi_m$ (pure B), one expects the collision probability proportional to $\rho_A \rho_B \sim (\phi_m + \phi)(\phi_m - \phi) \sim (1 - \phi^2/\phi_m^2)$. Kitahara *et al.* follow this line of argument within the framework of a cell dynamical model developed by Oono and Puri, [42] but it is possible to incorporate these ideas into a more conventional master equation approach. [43] The result of this line of thinking is that at the level of the Ginzburg - Landau equation describing the dynamics of a conserved order parameter, an order-parameter-dependent mobility naturally enters the equations. The generalization of the Cahn - Hilliard equation takes the form

$$\partial_t \phi = \nabla \cdot L(1 - \phi^2/\phi_m^2)[\nabla \frac{\delta H_0}{\delta \phi} - G\hat{e}]. \tag{19}$$

In many situations, without a field gradient, the important configurations are such that ϕ may be treated as small, and the new term makes no essential difference. However, at late stages in spinodal decomposition there may be a significant slowing of bulk diffusion. [44] In addition the drift or bias term in Eq. (10) arises from the mobility.

For convenience we write, consistent with Eq. (19)

$$\vec{j} = -M(\phi)[\nabla \frac{\delta H_0}{\delta \phi} - G\hat{e}] \tag{20}$$

and linearize about each of the equilibrium ordered phases. In the macroscopic limit higher gradients and nonlinear fluctuation terms are discarded, so that at this level the current becomes $\vec{j} = -M(\phi)[\chi^{-1}\nabla(\phi - \phi^{eq}) - G\hat{e}]$. The stationary solution has in each phase $\phi = \phi^{eq} + \chi G z$, at least not too far from the nominal center $z = 0$. This is analogous to the 'barometric' distribution for a gas; the current vanishes in this state. These considerations allow the identification of D as the bulk diffusion constant for the ordered phases (equal in this symmetric model) and $Q_\alpha = D^{-1}(\partial M/\partial \phi)_\alpha G$, with a similar equation for the β phase, the derivatives evaluated in the respective ordered phases. Note the important fact that, by symmetry, $Q_\alpha = -Q = -Q_\beta$.

The identifications are completed by noting that the stationary distribution is imagined to be "distorted" accounting for a steady flux j_0 in the z-direction. Quite by analogy with the symmetric model in the absence of the field, we find near the interface

$$\phi = \phi^{eq} + \chi G z - (j_0/D)z + u \tag{21}$$

The term in j_0 is the limiting form (near the interface) of the solution of the biased diffusion equation with the current in Eq. (20) equal to $j_0\hat{e}$. Hence, for $u = 0$, one has a stationary distribution with steady flux on a flat interface. In the phenomenological discussion at the start of this section, the stability of such a flat interface was examined.

6 interface equations

In this section an alternative approach to studying the interfacial dynamics of the symmetric model evolving unstably from a perturbed flat configuration is discussed. The method involves the solving of the equations for the interfacial coordinates; bulk degrees of freedom are completely removed from the problem. For the present problem the equations were first derived by Langer and Turski, [25] to which the reader is referred for further discussion. The method also has been applied to a variety of situations including directional solidification [9] and the radial finger patterns evolving in a circular Hele - Shaw cell. [45,46]

Return to the basic diffusion equation, Eq. (10), and for the purposes of this discussion ignore the field gradient term. For simplicity in this discussion and to make contact with some numerical results to follow, we again make the quasistatic approximation and drop the ∂_t term. This amounts to assuming the advance of the interface is so slow that diffusion from one part of the interface to another may be treated as effectively rapid. [There are problems of consistency here which will be discussed below.] The problem has then been reduced to solving Laplace's equation with appropriate thermodynamic conditions at the interface, the same as used in the linear analysis of Section 5 above. The "Coulomb" Green function, defined by

$$\nabla^2 G(\mathbf{r}|\mathbf{r}') = -\delta(\mathbf{r} - \mathbf{r}') \tag{22}$$

plays a central rôle. If $\mathbf{r}$ lies in the α phase and $\mathbf{r}_1$ lies on the interface, one uses Green's theorem to write

$$u(\mathbf{r})/2 = \int_S [u(\mathbf{r}_1)\nabla_1 G(\mathbf{r}|\mathbf{r}_1) - G(\mathbf{r}|\mathbf{r}_1)\nabla_1 u(\mathbf{r}_1)] \cdot \hat{n} dS_1, \tag{23}$$

where the normal points into the α phase. The factor 1/2 on the left-hand side is to take account of the fact that the point $\mathbf{r}_1$ is on the boundary of the domain. (An alternative approach [25] yields equivalent results.) A similar expression is written for a position $\mathbf{r}$ in the β phase. Then letting the two bulk points approach the *same* point on the interface and noting that, for this model, u is continuous across the interface, one finds

$$u(\mathbf{r}) = \int_S G(\mathbf{r}|\mathbf{r}_1)[\nabla_1 u|_\alpha - \nabla_1 u|_\beta] \cdot \hat{n} dS_1. \tag{24}$$

Now, as discussed in the linear analysis in Section 5 above, the difference in gradients is simply related to the current crossing the interface (recall for the present discussion the field gradient has been neglected). Furthermore, the left-hand side is evaluated from the thermodynamic boundary condition Eq. (13), in which case one finds the interface 'equation of motion'

$$\Gamma K(\varsigma) + (j_0/D)\varsigma = \int_S G(\mathbf{r}|\mathbf{r}')(\Delta c)v_n' dS', \tag{25}$$

where now $\mathbf{r}$ and $\mathbf{r}'$ lie on the interface, ς is the displacement of the interface at $\mathbf{r}$, v_n' is the normal velocity at the primed point, and the integration lets the primed point range over the interface. The structure of this equation is as follows. At time t the left-hand side is known; one must solve for the unknown velocities and advance the interface to its new configuration at $t + \delta t$.

Several comments are in order. The Green function is of the form $1/r$ in three-dimensional bulk and $-\ln r$ in two dimensions, where r is the spatial separation of the arguments. One part of the interface affects every other no matter how far away. This is a consequence of the quasistatic approximation. In reality the full diffusion propagator should be employed, [25] in which case, at fixed time, the influence of one interfacial point on another is strongly cut off owing to the finite diffusion coefficient. Only at sufficiently short distances is the Coulomb propagator a reasonable approximation to the full diffusion propagator. The emphasis in the work reported here is on the dynamics of an infinite system; to study the growth processes one wants to avoid evolution to a steady state (induced by a finite channel, for example) for as long as possible. Hence it is sensible, in view of these aims and the approximations made, to introduce a cutoff into $G(\mathbf{r}|\mathbf{r}')$ on phenomenological grounds. Effectively one introduces a diffusion length into the problem. Note from the point of view of the microscopic simulations reviewed above, a field gradient of its own accord introduces a characteristic length, inversely proportional to the field gradient. Note, in addition, that in the case of a field gradient, the cutoff enters in a spatially anisotropic form.

One may investigate the interface equation (25) in the linear regime for small distortions; [25] the results are, of course, equivalent to the alternative approach to linear stability presented in the previous section. Alternatively, one may solve the interface equation numerically. Some preliminary results are given here.

One must make the interface equation discrete (replacing the continuous curve by a sequence of "nodes") taking care to handle the (integrable) singularity in the Green function at small separations. It turns out to be advantageous (and it doesn't affect the evolution of the interface) to add some tangential velocity to the points, chosen, for example, to keep the node separation constant. Ultimately, one should include an ensemble of initial conditions to investigate scaling. (Eliminating the effect of initial conditions is no problem in the microscopic simulations, which are totally stochastic, but it is not so direct here.) The numerics must be handled in a system of finite width. To avoid edge effects, one may "wrap" interactions around so that points interact directly and with images. The point of view taken here (which is new to numerical work of this sort), is that the system is infinite, but interfacial configurations are periodic over as large a "primitive cell" as practical. A sufficient number of modes must be included in the initial conditions so that one can observe coarsening, if it is present. Of course, the shorter the wavelengths included in the initial conditions, the greater the number of nodes that might be required in setting up the discrete equations. One has the linear stability analysis at hand to give an idea of which modes are unstable and which among them grow most rapidly. The linear analysis also provides an extremely important benchmark for the numerical solution, which should reproduce the dispersion relation in the linear regime.

The time consuming elements of this prescription are the solution of the matrix equations (essentially $N \times N$, where N is the number of nodes), solving a system of $O(N)$ ordinary differential equations for the motion of the interface (typically by an implicit scheme), and the evaluation of the Green function. The linear systems and ODE solution are well suited to the CRAY X-MP/48 at the Pittsburgh Supercomputing Center, where this work has been begun.

Some preliminary results are now presented. For the sake of example we show the evolution of the interface using the quasistatic Green function, defined to be periodic over the system width W, namely,

$$G \sim -\ln[1 - 2p\cos(q_0(x - x')) + p^2], \tag{26}$$

with $q_0 = 2\pi/W$ and $p = \exp(-q_0|z - z'|)$. This is closely related to the kernel for directional solidification cells. [9] A sample case is shown in Fig. 8. The growth begins as one might expect. The flat tops which ultimately develop appear to try to fill the width to the extent possible. Note in this case there is essentially no characteristic length.

A second example of the interfacial evolution is shown in Fig. 9. Here the point of view described above is taken; the system is infinite, but the pattern is imagined to repeat over width W. The Green function is chosen phenomenologically to be pure Coulomb at short distances and to decay exponentially in all directions, for example, $G \sim -\ln[1-\exp(-\kappa(|\mathbf{r}-\mathbf{r}'|))]$. Again, a given point on the interface is coupled in the interface equation to other interface points within the width W, and to their images outside the "primitive cell." For this more realistic modeling of an infinite system there is less tendency for the interface to develop flat tops. However, reentrant shapes still develop. Experience with the microscopic simulations, in which there is high mobility along the interface so that overhanging parts of the contour can get swept away by the field (bias), suggests that the

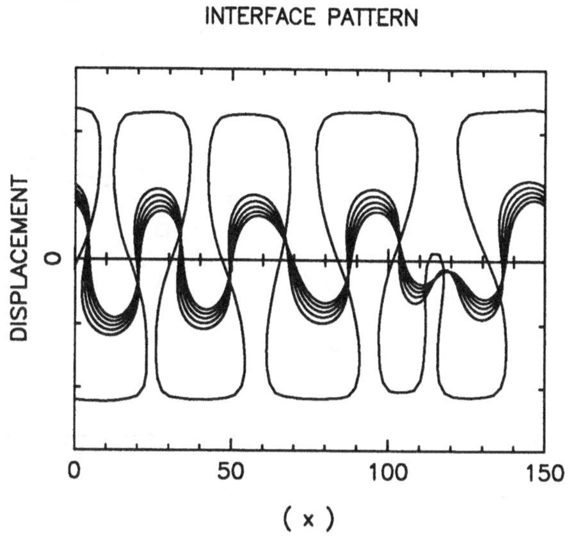

Fig. 8. Sample evolution of the interfacial structure according to the interface equation (25) using the periodic Green function.

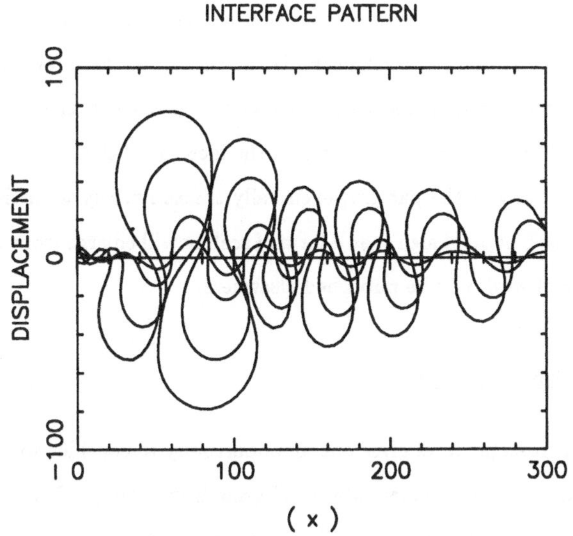

Fig. 9. Sample evolution of the interfacial pattern using the phenomenological Green function with isotropic cutoff.

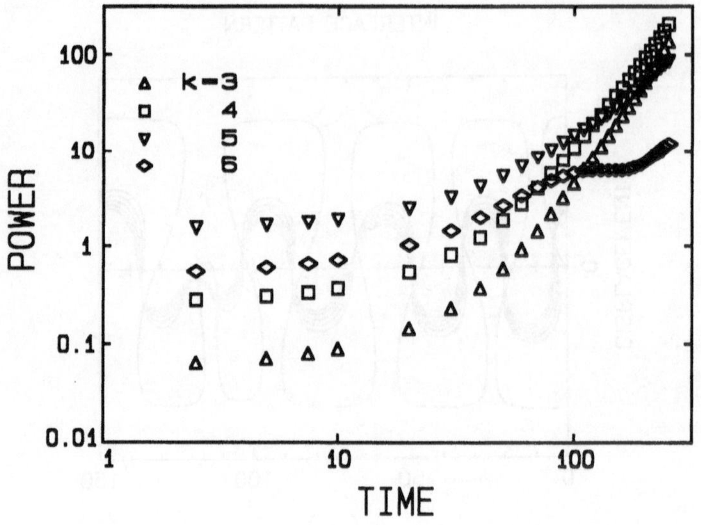

Fig. 10. Growth of several modes of the interfacial displacement as a function of length along the curve.

reentrant behavior will be reduced when the field gradient is added. The reentrant shapes preclude the simplest type of spectral analysis of the interface described in Section 3 above; however, one can consider the interface displacement as a function of the contour variable measuring length along the interface. Figure 10 shows, for the sample run in Fig. 9, the power in several growing modes in this representation. Some evidence for coarsening appears in the behavior, but these results must be considered preliminary. (The fact that the increase of the mixing zone, *i.e.*, the maximum peak-to-peak separation in the pattern, eventually tracks linearly with the length increase of the interface, is taken as a hopeful sign that a form of scaling will emerge from these solutions.) Further details of this line of work will be published elsewhere.

7 concluding remarks

Working at the microscopic level gives one an opportunity to consider somewhat more directly the processes involved in this particular interfacial growth instability. One is able to model the bulk diffusion processes which drive the instability and consider more closely relaxation processes which occur on the developing structures. The simulations don't require separate modeling of boundary conditions at the interface. The local energetics are responsible for the local relaxation processes; one need not make any assumptions about local thermodynamic equilibrium or assume values for the parameter specifying the capillarity, and so on. As noted above, as long as the structures which develop are sufficiently larger than the characteristic microscopic sizes (say, the correlation length), one may hope to make contact with the macroscopic domain. The present simulations appear to reach the borders of that domain. Additionally, the simulations at the microscopic level lead naturally to an examination of universal features of the growth instability under consideration.

The microscopic simulations are not without their own difficulties, for example, the small system sizes, the rather narrow range of parameters over which successful simulations have been carried out, and the inherent noisiness. Going over to numerical or other analysis of interface equations, as discussed in the previous section, may eliminate some of these difficulties.

It has been argued [23] that the simulations represent an instability of the Mullins - Sekerka type. The simulations indicate the presence of a scaling domain for the *dynamics* of the evolution of the flat interface. As has been noted above, the Monte Carlo simulations are the natural extension to the study of interfacial growth processes, of well-characterized techniques used to examine dynamics and nonequilibrium processes in systems with a conserved order parameter. Particularly interesting is the evidence (albeit preliminary) of *universality classes*; one may at least vary the driving flux, and the temperature within some range and still find the same scaling exponents δ and z. [From the macroscopic perspective the temperature change controls the capillarity as discussed elsewhere. [23]] Further, there is evidence for at least two diferent universality classes represented by the symmetric two-sided case and the one- sided case. Finally, an exponent relation is suggested by considering the microscopic processes involved.

These simulations have demonstrated that a prototypical interfacial instability can be studied from a microscopic level. In addition, other models closely related to the ones described here have been explored to gain insight into the breadth of universality classes. These models contain essentially the same physical processes, but differ in the details of the energetics and/or dynamics. For example, preliminary studies based on solid-on-solid energetics have been carried out. [47]

The possibility of a scaling domain having been demonstrated, further work is in progress using alternative schemes to reduce the uncertainties and allow an investigation of higher order correlations. Details of these investigations will be presented elsewhere.

ACKNOWLEDGMENT: The author would like to thank Dr. Hong Guo for his assistance with many aspects of work discussed. The author would also like to thank Dr. Jorge Viñals for his interest, for assistance in sorting out the effects of a field gradient, and for helpful suggestions on this manuscript. The author is grateful to the National Science Foundation through the Division of Materials Research for support of this work under grant No. DMR-86-13030. The author also thanks the Pittsburgh Supercomputing Center, where the calculations based on the interface equations have been carried out.

References

[1] K. A. Jackson and J. D. Hunt, Acta Metall. 13:1215 (1965).

[2] F. Heslot and A. Libchaber, Phys. Scr. T9:126 (1985).

[3] P. G. Saffman and G. I. Taylor, Proc. Roy. Soc. (London) A245:312 (1958).

[4] P. Tabeling and A. Libchaber, Phys. Rev. A33:794 (1986).

[5] S. C. Huang and M. E. Glicksman, Acta Metall. 29:701 (1981); 29:717 (1981).

[6] J. S. Langer, Rev. Mod. Phys. 52:1 (1980).

[7] See, e.g., D. Kessler, J. Koplik and H. Levine, Phys. Rev. A30:3161 (1984); E. Ben-Jacob, N. Goldenfeld, B. G. Kotliar and J. S. Langer, Phys. Rev. Lett. 53:2110 (1984).

[8] S. K. Sarkar and D. Jasnow, Phys. Rev. A35:4900 (1987).

[9] A. Karma, Phys. Rev. Lett. 57:858 (1986).

[10] T. Dombre and V. Hakim, Phys. Rev. A36:2811 (1987).

[11] M. Ben–Amar and B. Moussallam, Phys. Rev. Lett. 60:317 (1988).

[12] D. A. Kessler and H. Levine, U.C.S.D. preprint (1988).

[13] T. A. Witten and L. M. Sander, Phys. Rev. Lett. 47:1400 (1981).

[14] T. A. Witten and L. M. Sander, Phys. Rev. B27:5685 (1983).

[15] See, e.g.,articles in *Kinetics of Aggregation and Gelation*, edited by F. Family and D. P. Landau (North Holland, New York, 1984).

[16] D. Bensimon, L. P. Kadanoff, S. Liang, B. Schraiman and C. Tang, Rev. Mod. Phys. 58:977 (1986).

[17] T. Vicsek, Phys. Rev. Lett. 53:2281 (1984).

[18] L. A. Turkevich and H. Scher, Phys. Rev. Lett. 55:1026 (1985).

[19] W. W. Mullins and R. F. Sekerka, J. Appl. Phys. 34:323 (1963); 35:444(1964).

[20] L. P. Kadanoff, J. Stat. Phys. 39:267 (1985).

[21] See, e.g., J. Marro, J. L. Lebowitz and M. H. Kalos, Phys. Rev. Lett. 43:282 (1979); J. L. Lebowitz, J. Marro and M. H. Kalos, Acta Metall. 30:297 (1982); for additional background and references see J. D. Gunton, M. San Miguel and P. S. Sahni in *Phase Transitions and Critical Phenomena*, ed. by C. Domb and J. L. Lebowitz (Academic, New York, 1983)v.8.

[22] For an interesting new approach and a thorough discussion of background and references, the reader is referred to the work of Y. Oono and S. Puri, Phys. Rev. A (in press).

[23] Previous accounts have appeared in H. Guo and D. Jasnow, Phys. Rev. 34:5027 (1986).

[24] D. Jasnow, Superlattices and Microstructures 3:581 (1987).

[25] J. S. Langer and L. Turski, Acta Metall. 25:267 (1977).

[26] D.Jasnow, D. A. Nicole and T. Ohta, Phys. Rev. A23:3192 (1981).

[27] A. Onuki, K. Sekimoto and D. Jasnow, Prog. Theor. Phys. 74:685 (1985).

[28] A brief discussion of roughening with references is contained in D. Jasnow, Rept. Prog. Phys. 47:1059 (1984).

[29] K. Kawasaki, Phys. Rev. 145:224 (1966).

[30] See, e.g., A. B. Bortz, M. H. Kalos, J. L. Lebowitz and M. A. Zendejas, Phys. Rev. B10:535 (1974); A. Sadiq, Phys. Rev. b(:2299 (1974); additional references may be found in K. W. Kehr and K. Binder in *Applications of the Monte Carlo Method in Statistical Physics*, ed. K. Binder (Springer - Verlag, New York, 1984).

[31] S. Katz, J. L. Lebowitz and H. Spohn, J. Stat. Phys. 34:323 (1984); 38:725 (1985).

[32] A. Hernandez-Machado and D. Jasnow, Phys. Rev. 37:656 (1988).

[33] For a review of hyperscaling in critical phenomena see, e.g., M. E. Fisher in Lecture Notes in Physics Vol. 186, *Critical Phenomena,* (ed. F. J. W. Hahne, Springer-Verlag, N. Y., 1983).

[34] K. Leung and J. L. Cardy, J. Stat. Phys. 44:567 (1986).

[35] J. W. Cahn and J. E. Hilliard, J. Chem. Phys. 28:258 (1958).

[36] See, e. g., D. B. Abraham and P. Reed, J. Phys. A10:L121 (1977); K.K. Mon and D. Jasnow, Phys. Rev. A31:4008 (1985); J. Stat. Phys. 41:273 (1985).

[37] M. P. A. Fisher, D. S. Fisher and J. D. Weeks, Phys. Rev. Lett. 48:368 (1982).

[38] J. B. Collins and H. Levine, Phys. Rev. B31:6119 (1985).

[39] E. Ben-Jacob, N. Goldenfeld, J. S. Langer and G. Schon, Phys. Rev. Lett. 51:1930 (1983).

[40] G. Caginalp, Phys. Rev. B (in press); University of Pittsburgh preprint (1988).

[41] K. Kitahara, Y. Oono and D. Jasnow, Mod. Phys. Lett. B (in press).

[42] Y. Oono and S. Puri, Phys. Rev. Lett. 58:836 (1987).

[43] A. Hernandez-Machado, unpublished.

[44] J. S. Langer, N. Bar-On and H. D. Miller, Phys. Rev. A11:1417 (1975).

[45] L. M. Sander, P. Ramanial and E. Ben-Jacob, Phys. Rev. A32:3160 (1985)

[46] S. Sarkar, unpublished.

[47] H. Guo, unpublished.

FIELD THEORY FOR GROWTH KINETICS

Gene F. Mazenko
Department of Physics and
James Franck Institute
The University of Chicago
5640 S. Ellis
Chicago, IL 60637

I. Introduction:

I want to discuss in these lectures recent work[1] I have carried out in collaboration with Professor Oriol T. Valls of the University of Minnesota and Professor Marco Zannetti of the University of Naples. This work is toward an analytical theory for growth kinetics problems. There are numerous details discussed in the paper which I will avoid here. Instead, I will emphasize the structure of the theory.

Much of the progress made on growth kinetics problems[2] over the past ten years has been driven by the results of Monte Carlo simulations[2,3] of kinetic Ising models. Earlier studies[4] were of field theoretical models of the Langevin type. I will connect with this earlier development by focussing on the growth of order in the time dependent Ginzburg-Landau (TDGL) model for a scalar field. In particular I consider the TDGL model with gaussian noise in the case where the system is quenched from an initial symmetric disordered equilibrium state into a nonequilibrium state driven by parameters corresponding to some new final ordered equilibrium state with broken symmetry. I then consider the time evolution which characterizes the ordering of the system - the growth of the domains of the new ordered phases.

The main measure of the domain growth is the order parameter structure factor $C(\vec{q},t)$ where $\vec{q}$ is a wavevector and t is the time after the quench. Ordering is reflected in the growth in time of a Bragg peak in the structure factor. One of the major discoveries[3,5] in this field was that the peak contributions to the structure factor scale with a single characteristic length L(t) which is a measure of a typical domain size in the system. In particular, for $\vec{q}$ near the ordering wavevector,

$$C(\vec{q}\,,t) = L(t)^d F(\vec{q}\,L(t)) \tag{1.1}$$

where d is the spatial dimensionality of the system. Considerable effort has been spent trying to determine the growth law, how L depends on t, for various systems and the form of the scaling function F in (1.1). This work has been based on either direct numerical simulation,[6] renormalization group calculations,[7-9] or on simple model calculations[10] with an assumed morphological structure for the evolving domains at late times.

The earliest theoretical work on these growth kinetics problems was that of Cahn and coworkers[11] and Lifshitz and coworkers.[12] The problem of spinodal decomposition was essentially defined in the the early work of Cahn and Hilliard[11] in terms of a linearized treatment of the TDGL model for a conserved order parameter without thermal noise. This work led to the prediction of exponential growth of a particular wavenumber component of the structure factor. While this theory reflected the underlying unstable nature of the system, it did not include any appropriate stabilizing mechanism for later times and could only be viewed as a very early time approximation. Cook[13] later included the effects of thermal noise in the problem, but this did not change the basic structure of the approximation. A further substantial advance in the theory was due to Langer[14] and coworkers who took into account nonlinear feedback terms which had the effect of stabilizing the later stage growth. The resulting theory was a considerable improvement on that of Cahn-Hilliard and Cook[11,13] since it led to a peaked form for the structure factor which moved to lower values of the wavenumber as time evolved and the peak grew with a rate much slower than exponential. Both effects were in qualitative agreement with experimental observation[15] of spinodally decomposing fluid systems using light scattering techniques.

While the work of Langer, Bar-on and Miller[14] (LBM) represented a significant advance for the field, there were some important drawbacks to their theory. The approximation developed was somewhat ad hoc and difficult to treat systematically. More specifically, Binder and coworkers[3,16] pointed out conceptually important problems associated with the long-time behavior of the theory. In the original high temperature equilibrium state there is a single correlation length ξ_I associated with the system. As the system evolves toward its final broken symmetry state there are two independent lengths which characterize the system: the typical domain size L(t) and the final equilibrium correlation length ξ. The LBM theory, as described below in Sect. II, can accomodate only one length at any given time, and therefore can not describe both L(t) and ξ. The result is that the theory does not lead to the appropriate final equilibrium state. While this type of theory has been extended to more complicated physical systems,[17] they all share this same defect. This problem of developing a theory capable of treating two characteristic lengths (or masses in field theoretic language) is the fundamental unsolved problem addressed in Ref. 1.

Another line of theoretical development has been to avoid the problem of early time evolution and jump to the later stages where one imposes a certain morphological structure on the system. The work of Cahn and Allen[11] and Lifshitz and Slyozov[12] was along these lines and led to very useful predictions for the growth laws of a variety of systems. These theories focused on the evolution of a single droplet or domain and did not worry about the distribution of these domains (Lifshitz and Slyozov, for example, restricted their analysis to the case of a very dilute solution of droplets). Kawasaki and coworkers[10] have extended these techniques to the case of the late stage evolution of interacting defects. It is not clear in this method how one takes into account temperature effects, equilibration and the evolution toward this late stage. In particular it is difficult to self-consistently determine the distribution of defects. In the work of Ohta, Jasnow and Kawasaki,[10] for example, it was necessary to postulate that the distribution of interfaces satisfied a gaussian distribution.

We were encouraged to return to the problem of the time evolution over the entire unstable regime by the work[18,19] treating the N-vector model generalization of the TDGL model. In Ref. 18 this model was solved in the large-N limit and the time evolution of the system was analysed in detail over the entire time regime from the time of quench to the late time scaling regime. This system is qualitatively different from the scalar order parameter case we study here since there is a broken continuous symmetry in the final state of the N-vector model for $N > 1$. Thus one generates Nambu-Goldstone (NG) modes[20] in the final state as the system evolves. A relevant observation made in Ref. 18 is that the solution looks very similar to that found by LBM. This comes about in the large-N limit because the transverse modes dominate over the single longitudinal mode and the only length in the problem during the later stages of growth is the characteristic domain size L(t).

The method I discuss is designed to deal with the problem of separating the two length scales L(t) and ξ as the system evolves. In the very long time limit the method gives the exact equilibrium result without spurious NG modes. As described in Sect. II, the method makes systematic use of field theory techniques for classical fields as well as new techniques required for treating the ordering component of the order parameter field. The main physical motivation behind the formalism is the recognition that the order parameter field $\psi(\vec{R},t)$ can be decomposed into the sum of two fields. Early in the evolution of the system these fields are strongly coupled. However, as time passes, they assume separate roles and, for sufficiently long times, they become essentially independent. One of the two fields is then associated with the domain growth and the peak in the structure factor and the other is associated with the fluctuations of the order parameter within an ordered domain. The characteristic size associated with the "peak" field is L(t), while that for the "fluctuation" field is ξ. At long times $L(t) >> \xi$ and the two fields decouple. It seems reasonable that the fluctuating field can be treated at low temperatures as a field governed by gaussian statistics. The peak field can be visualized at long times in coordinate space as being uniform almost everywhere except near domain boundaries where the system rapidly changes its orientation from one type of domain to another. It is therefore appealing to think of these variables as being somewhat Ising-like in character.

II. Separation of Variables

A. Langevin Equation

The Langevin dynamics I consider is the standard TDGL model for a scalar field:

$$\frac{\partial \psi(\vec{R},t)}{\partial t} = -\Gamma(\vec{R}) \frac{\delta F}{\delta \psi(\vec{R},t)} + \eta(\vec{R},t) \tag{2.1}$$

where the field $\psi(\vec{R},t)$ is defined either on the continuum or on a lattice[21] characterized by a set of lattice vectors $\vec{R}$. In (2.1), $\Gamma(\vec{R})$ is a constant kinetic coefficient, Γ in the case of a nonconserved order parameter (NCOP), while for a conserved order parameter (COP) we have:

$$\Gamma(\vec{R}) = -D\,\nabla_R^2 \tag{2.2}$$

where D is a transport coefficient and ∇_R^2 is the Laplacian or its lattice version. The noise $\eta(\vec{R},t)$ appearing in (2.1) is gaussian and white and satisfies

$$\langle \eta(\vec{R},t)\eta(\vec{R}\,',t')\rangle = 2T\Gamma(\vec{R})\delta(\vec{R}-\vec{R}\,')\delta(t-t') \tag{2.3}$$

where T is the temperature [22] of the thermal bath in contact with the system. The effective free energy is assumed to be of the Ginzburg-Landau-Wilson form

$$F = \frac{1}{2}\int d^dR\,[c(\nabla\psi(\vec{R}))^2 + r\psi^2(\vec{R}) + \frac{u}{2}\psi^4(\vec{R})]\ . \tag{2.4}$$

B. Static Equilibrium Behavior.

The static equilibrium properties of this model are governed by the Boltzmann probability distribution proportional to $e^{-F[\psi]/T}$. The two independent variable parameters in the equilibrium theory are chosen to be the coupling r and the temperature T. The coefficient c and the nonlinear coupling u are kept fixed and positive. By varying r from positive to negative values the local potential $V(\psi) = \frac{r}{2}\psi^2 + \frac{u}{4}\psi^4$ changes from a single to a double well form. For $r > 0$ one has only a single disordered phase for all temperatures greater than zero. However, for $r < 0$, there will be some transition temperature $T_c = T_c(r,u,c)$, below which $(T < T_c)$ the system develops a spontaneous magnetization $\langle\psi\rangle = M(T,r,u,c) \neq 0$. I will focus on quenches to final states where $T << T_c$ and a low temperature theory can be developed since one expects fluctuations about the ordered state to be small.

The basic nature of the low-temperature theory can be seen by shifting and rescaling the field appearing in the free-energy,

$$\psi(\vec{R}) = M + \phi(\vec{R}) \tag{2.5}$$

where $<\psi(\vec{R})> = M$, and expanding in powers of $\phi \sim \sqrt{T}$. This expansion is well known,[23] and I can quote the relevant results. For the equilibrium structure factor $C(\vec{q})$, the Fourier transform of $<\psi(\vec{R})\psi(\vec{R}')>$, one finds:

$$C(\vec{q}) = M^2(2\pi)^d \ \delta(\vec{q}) + C_E(\vec{q}) \tag{2.6}$$

where the spontaneous magnetization is given by:

$$M = M_o[1 - \frac{3}{2M_o^2} <\phi^2(\vec{R})> + 0(T^2)] \tag{2.7}$$

with

$$M_o^2 = \frac{|r|}{u} \quad , \tag{2.8}$$

$C_E(\vec{q})$ is the fluctuating part of the correlation function given by

$$C_E(\vec{q}) = \frac{T}{cQ^2(\vec{q}) + r - \Sigma(\vec{q})} \quad , \tag{2.9}$$

where $Q^2(\vec{q})$ (q^2 in the continuum) is the Fourier transform of the Laplacian and, to O(T),

$$\Sigma(\vec{q}) = -\ 3uM^2 - 3uT<\phi^2> + \frac{18u^2M^2}{T}\ \Pi(\vec{q}) \tag{2.10}$$

with

$$\Pi(\vec{q}) = \int \frac{d^dk}{(2\pi)^d}\ C_E(\vec{q} - \vec{k})C_E(\vec{k}) \quad . \tag{2.11}$$

and

$$<\phi^2> = \int \frac{d^dk}{(2\pi)^d}\ C_E(\vec{q}) \quad . \tag{2.12}$$

C. Functional Integral Formulation for Quench Problems.

I consider there the case where the dynamical system associated with (2.1) is initially in equilibrium at some high temperature T_I and at time $t = t_o$ is quenched to some low temperature T. The parameters r and u may also suddenly change at $t = t_o$ from initial values r_I and u_I to final values r , u. It is now well understood from studies of dynamic correlations in equilibrium, that it is advantageous in developing a systematic perturbation theory, to recast the problem in functional integral form.[24] As first pointed out by Martin, Siggia and Rose (MSR)[25], this is most conveniently carried out by introducing a response field $\hat{\psi}(\vec{R},t)$ conjugate to $\psi(\vec{R},t)$. The transformation from averages over the noise to averages over ψ and $\hat{\psi}$ is well described in Refs. 24 and 26. We follow here the conventions developed in Ref. 27.

The additional ingredient in our development here is the quench at $t = t_o$ and the imposition of an initial condition at that time. This new feature is taken into account[28] by rewriting the Langevin equation in the form

$$\frac{\partial \psi(\vec{R},t)}{\partial t} = -\Gamma(\vec{R}) \frac{\delta F}{\delta \psi(\vec{R},t)} + \eta(\vec{R},t) + \delta(t - t_o)\psi_o(\vec{R}) \tag{2.13}$$

where $\psi(\vec{R},t)$ and $\eta(\vec{R},t)$ are zero for $t < t_o$ and the initial value of $\psi(\vec{R},t)$ is imposed as a constraint. In carrying out averages $\psi_o(\vec{R})$ is treated as an independent field with its own probability distribution $\sim e^{-F_I[\psi_o]/T_I}$.

We can gain some feeling for the functional integral approach, and develop some important ideas, by considering a linearized version of (2.13):

$$[\frac{\partial}{\partial t} + \Gamma(\vec{R})[-c\nabla_R^2 + W(t)]]\psi(\vec{R},t) = \delta(t - t_0\psi_0(\vec{R}) + \eta(\vec{R},t) \tag{2.14}$$

where $W(t)$ is a time dependent quantity which evolves to a positive value W_E as $t \to \infty$. If we multiply (2.14) by i it can be rewritten in the matrix form

$$G^{-1}(1\bar{1})\psi(\bar{1}) = -I(1\bar{1})\psi_0(\bar{1}) + i\eta(1) \tag{2.15}$$

where, for example, the index 1 stands for $(\vec{R}_1, t_1)$ and repeated barred indices, here and below, indicate integration over space and time. The "inverse" propagator is defined by

$$G^{-1}(12) = i[\frac{\partial}{\partial t_1} + i\Gamma(\vec{R}_1)(-c\nabla_{R_1}^2 + W(t_1))]\delta(12) \tag{2.16}$$

$$\delta(12) = \delta(\vec{R}_1 - \vec{R}_2)\delta(t_1 - t_2) \quad , \tag{2.17}$$

$$I(12) = -i\delta(12)\delta(t_2 - t_0) \quad . \tag{2.18}$$

The formal solution for the correlation function is given by

$$C(12) = \langle\psi(1)\psi(2)\rangle$$

$$= G(1\bar{1})G(2\bar{2})\langle[-I(\bar{1}\bar{1}')\psi_0(\bar{1}') + i\eta(\bar{1})][-I(\bar{2}\bar{2}')\psi_0(\bar{2}') + i\eta(\bar{2})]\rangle \quad , \tag{2.19}$$

where the average is over ψ_0 and η. By construction

$$\langle\psi_0(1)\eta(2)\rangle = 0 \quad , \tag{2.20}$$

$$\langle\eta(1)\eta(2)\rangle = 2\pi(12) \tag{2.21}$$

where

$$\pi(12) = \Theta(t_1 - t_0)T\Gamma(12) \tag{2.22}$$

$$\Gamma(12) = \Gamma(\vec{R}_1)\delta(12) \tag{2.23}$$

and we assume

$$<\psi_0(1)\psi_0(2)> = g_0(\vec{R}_1 - \vec{R}_2) \quad . \tag{2.24}$$

Then,

$$<\psi(1)\psi(2)> = -\, G(1\bar{1})G(2\bar{2})2\tilde{\pi}(\overline{12}) \tag{2.25}$$

where

$$2\tilde{\pi}(12) = 2\pi(12) + \delta(t_1 - t_0)\delta(t_2 - t_0)g_0(\vec{R}_1 - \vec{R}_2) \tag{2.26}$$

and the thermal noise and the initial conditions play a similar role.[29]

Consider next the evolution of the equal time correlation function

$$C(\vec{R} - \vec{R}', t) = <\psi(\vec{R}, t)\psi(\vec{R}', t)> \tag{2.27}$$

for $t > t_0$. This then satisfies, using (2.14),

$$\{\frac{\partial}{\partial t} + \Gamma(\vec{R})[-c\nabla_R^2 + W(t)] + \Gamma(\vec{R}')[-c\nabla_{R'}^2 + W(t)]\}C(\vec{R}_1 - \vec{R}_1', t)$$

$$= <\psi(\vec{R},t)\eta(\vec{R}', t)> + <\eta(\vec{R},t)\psi(\vec{R}', t)> \quad . \tag{2.28}$$

The fluctuation dissipation theorem requires

$$<\psi(\vec{R},t)\eta(\vec{R}',t)> = T\Gamma(\vec{R})\delta(\vec{R}-\vec{R}') \quad . \tag{2.29}$$

In terms of spatial Fourier transforms we obtain

$$[\frac{\partial}{\partial t} + 2\Gamma(\vec{q})[cQ^2(\vec{q}) + W(t)]]C(\vec{q},t) = 2T\Gamma(\vec{q}) \quad . \tag{2.30}$$

In the long-time limit we assume

$$\lim_{t \to \infty} W(t) = W_E > 0 \quad , \tag{2.31}$$

$$\frac{\partial C(q,t)}{\partial t} = 0 \quad , \tag{2.32}$$

and we have the solution

$$C_E(\vec{q}) = \frac{T}{cQ^2(\vec{q}) + W_E} \tag{2.33}$$

and W_E/c can be associated with the inverse equilibrium correlation length squared.

The equivalent functional integral approach says that averages over the noise can be replaced by direct functional integration over the field ψ at the expense of introducing an additional auxiliary field $\hat{\psi}$. In this formulation correlation functions such as C(12) are computed as functional integrals over a probability distribution

$$P[\psi\,,\hat{\psi}] \equiv e^{-A[\psi,\hat{\psi}]}/Z \quad , \tag{2.34}$$

where Z is a normalization factor, via,

$$C(12) = \int D[\psi]D(\hat{\psi})P[\psi,\hat{\psi}]\psi(1)\psi(2) \quad . \tag{2.35}$$

In the case of the Langevin equation (2.14), the associated "action" is given by

$$A[\psi,\hat{\psi}] = \int d1d2[\hat{\psi}(1)\pi(12)\hat{\psi}(2) + \hat{\psi}(1)G^{-1}(12)\psi(2)] \quad . \tag{2.36}$$

It is a simple exercise in Gaussian functional integrals to show that C(12) is given by (2.25). In addition we find

$$\int D(\psi)D(\hat{\psi})P[\psi,\hat{\psi}]\psi(1)\hat{\psi}(2)) = G(12) \tag{2.37}$$

and this quantity serves as the "response" function for this problem.

Returning to our original nonlinear Langevin equation (2.13), it is easy to generalize the results given above and write down the appropriate action

$$A_\psi[\hat{\psi},\psi] = \int d1d2[\hat{\psi}(1)\pi(12)\hat{\psi}(2) + \hat{\psi}(1)G_o^{-1}(12)\psi(2) + iu\,\hat{\psi}(1)\Gamma(12)\psi^3(2)] \tag{2.38}$$

where G_0^{-1} is given by (2.16) with W(t) = r, and the associated probability measure is written

$$P_\psi[\psi\,,\hat{\psi}] = e^{-A_\psi[\hat{\psi},\psi]}/Z \quad . \tag{2.39}$$

The advantage of this formalism in the case of fluctuations in equilibrium is that one has a formulation of the standard field theoretical type and one can conveniently develop perturbation theory.

D. Naive Mean Field Theory.

Starting with the action given by (2.38), it is straightforward to develop perturbation theory directly in terms of the coupling u. If we set $u = 0$, we obtain the Cahn-Hilliard theory[11] and exponential growth. A considerable improvement in the theory occurs if we work to first order in u, which is equivalent to the mean field approximation, in treating the cubic term in (2.38):

$$\psi^3 \rightarrow 3\,S(t)\psi \tag{2.40}$$

where

$$S(t) = \langle\psi^2(\vec{R},t)\rangle \quad . \tag{2.41}$$

The resulting theory is of the form given by (2.36) and (2.14) with the "mass"

$$W_{MF}(t) = r + 3uS(t) \quad . \tag{2.42}$$

The resulting equation satisfied by the Fourier transform $C(\vec{q},t)$ of the equal time correlation function (2.27) is given by (2.30) with W replaced by W_{MF} and

$$S(t) = \int \frac{d^d q}{(2\pi)^d} C(\vec{q},t) \quad . \tag{2.43}$$

This equation of motion is identical in structure to Eq. (2.31) found in Ref. 22 in the large-N limit except that the 3 multiplying $uS(t)$ in (2.42) is replaced by 1. These equations are very similar in structure to those found by LBM.[30] The important aspect of the solution of these equations for our purposes here is that

$$\lim_{t \to \infty} W_{MF} = 0 \tag{2.44}$$

and in the long time limit

$$C(\vec{q}) = M^2 (2\pi)^d \delta(\vec{q}) + \frac{T}{cQ^2(\vec{q})} \quad . \tag{2.45}$$

Inserting (2.45) into (2.43) and using (2.44), one obtains

$$M^2 = -\frac{r}{3u} - \int \frac{d^d q}{(2\pi)^d} \frac{T}{cQ^2(\vec{q})} \quad . \tag{2.46}$$

Comparing with (2.9), we see that (2.45) gives a very poor approximation for the final state. It gives an unphysical NG mode ($\sim T/q^2$) in the fluctuation spectrum. It also gives an incorrect value for the zero temperature magnetization (compare (2.46) with the correct result (2.7)) and an incorrect first order temperature correction.

As mentioned in the Introduction, the problem with theories of this type is that they allow for only a single length or mass. In (2.14) we can only identify a single mass term W and a length L via

$$W_{MF}(t)/c = r + 3uS(t) = -L^{-2}(t) \tag{2.47}$$

and the dynamics drives $W \to 0$ and $L \to \infty$ as $t \to \infty$. To overcome these fundamental defects we must construct a theory which naturally allows for two masses and the two corresponding lengths.

E. The Order Variable $\sigma(\vec{R},t)$.

In order to make progress we now implement the idea brought forward in the Introduction of decomposing $\psi(\vec{R},t)$ as the sum of two fields, one of which becomes associated, as time evolves, with the growth of domains, and the other with fluctuations within a domain. In other words we must introduce the appropriate dynamical generalization of the shift (2.5) used in the analysis of the equilibrium state, with the requirement of keeping the symmetry unbroken at all finite times.

The first step is to enlarge the function space by introducing, in addition to the order parameter $\psi(\vec{R},t)$, a new independent stochastic field $\sigma(\vec{R},t)$ with its own normalized distribution $P_\sigma[\sigma]$, which, for the moment, is not specified, except for the condition $<\sigma(\vec{R},t)> = 0$. The joint distribution for the pair (ψ,σ) then takes the product form

$$P[\hat{\psi},\psi,\sigma] = P_\psi[\hat{\psi},\psi]P_\sigma[\sigma] \tag{2.48}$$

and it is clear that at this stage the field σ does not enter into the order parameter correlation functions.

Next, we introduce the field $\phi(\vec{R},t)$ through the translation

$$\psi(\vec{R},t) = \phi(\vec{R},t) + \sigma(\vec{R},t) \tag{2.49}$$

and, switching to the pair (ϕ,σ), we obtain the joint distribution

$$P[\hat{\psi}\ ,\phi,\sigma] = P_\psi[\hat{\psi},\phi+\sigma]P_\sigma[\sigma]\quad . \tag{2.50}$$

The scheme becomes nontrivial when we use $\sigma(\vec{R},t)$, whose dynamics is governed by $P_\sigma[\sigma]$, to model the growth of order. The field $\phi(\vec{R},t)$ is left to describe fluctuations about order, which become less and less important as time proceeds. Eventually, in the asymptotic regime, the field $\phi(\vec{R},t)$ is expected to become gaussian and to decouple from $\sigma(\vec{R},t)$. It is clear that the careful construction of the appropriate form for $P_\sigma[\sigma]$ is crucial to the structure of our theory.

According to the physical picture described above, $\sigma(\vec{R},t)$ should be a two-valued variable, which keeps the same magnitude inside a given domain and changes its sign across an interface. We formalize this by taking:

$$\sigma(\vec{R},t) = \sqrt{S(t)}\ \mu(\vec{R},t) \tag{2.51}$$

where $S(t)$ is some time-dependent quantity to be discussed below, and $\mu(\vec{R},t) = \pm 1$. We can then rewrite the action in terms of ϕ and σ as

$$A_\psi[\hat{\psi},\phi,\sigma] = \int d1d2\{\hat{\psi}(1)\bar{\pi}(12)\hat{\psi}(2) + \hat{\psi}(1)[G_f^{-1}(12)\phi(2) + G_p^{-1}(12)\sigma(2)+iu\Gamma(12)(3\sigma(2)\phi^2(2) + \phi^3(2))]\}\quad , \tag{2.52}$$

where the new propagators, after a spatial Fourier transform, are given by:

$$G_p^{-1}(\vec{q},t_1,t_2) = i[\frac{\partial}{\partial t_1} + \Gamma(\vec{q})(cQ^2(\vec{q}) + W_p(t))]\delta(t_1 - t_2) \tag{2.53}$$

$$G_f^{-1}(\vec{q},t_1,t_2) = i[\frac{\partial}{\partial t_1} + \Gamma(\vec{q})(cQ^2(\vec{q}) + W_f(t))]\delta(t_1 - t_2) \tag{2.54}$$

and the mass terms are now different and given by

$$W_p(t) = r + uS(t) \tag{2.55}$$

$$W_f(t) = r + 3uS(t) \tag{2.56}$$

where the subscripts p and f will be used to denote *peak* and *fluctuating* contributions.

In specifying $P_\sigma[\sigma]$, we must, of course, be guided by the physics described by the original Langevin equation (2.13) and our intuition that ϕ becomes in some sense "small" at long times. Let us assume for the moment that we can ignore the field ϕ in (2.13) and replace $\psi \to \sigma$. If we multiply the resulting equation by i and recognize that the noise θ associated with σ should differ from η, we obtain immediately the equation of motion

$$G_p^{-1}(1\bar{1})\sigma(\bar{1}) = -I(1\bar{1}) + i\theta(1) \tag{2.57}$$

where the noise θ is gaussian and satisfies

$$<\theta(1)\theta(2)> = 2\Pi(12) \tag{2.58}$$

where Π will be discussed below and differs from $\pi(12)$. Comparing (2.57) and (2.15), we are led to the identification of G_p as the propagator associated with the peak contribution.

There is another way of viewing this discussion. Let us consider $P_\psi[\hat{\psi},\phi+\sigma]$ as the probability distribution for $\phi(\vec{R},t)$, parametrized by the stochastic process $\sigma(\vec{R},t)$. If the dynamics of $\sigma(\vec{R},t)$, **governed by** $P_\sigma[\sigma]$, are not properly chosen, then, in the long-time limit, ϕ will not be appropriately enslaved by $\sigma(\vec{R},t)$ and will not represent small fluctuations on an ordered background.

If this picture is to be valid, then it is **crucial** that the coefficient of the term linear in σ in (2.52) and proportional to G_p^{-1}, vanishes in the long time and distance limit. It is, of course, this vanishing of G_p^{-1} and W_p for small wavenumbers and long times which contributes to the building of a Bragg peak in the solution of (2.57) for the structure factor.

The choice of G_p^{-1} as the propagator associated with the σ-dynamics leads not only to the separation of the variables σ and ϕ at long times, but also to the correct separation of length scales. The characteristic length $L(t) = [-W_p(t)/c]^{-1/2}$ can then be associated with the domain size, and $\xi(t) = [W_f(t)/c]^{-1/2}$ associated with the correlation length of fluctuations. As order develops in the peak, we expect $L(t) \to \infty$ $(W_p = r + uS \to 0)$ and $W_f(t) \to 2|r| > 0$. In this case, with $W_f(\infty) = 2|r|$, we see, using (2.54) and (2.33), that the equal time correlation function associated with the ϕ variable will reach, as $t \to \infty$, its correct equilibrium value (2.9):

$$\lim_{t \to \infty} < |\phi(\vec{q},t)|^2 > = \frac{T}{cQ^2(\vec{q}) + 2|r|} \quad . \tag{2.59}$$

This indicates that the variable ϕ can be taken as $O(\sqrt{T})$ for sufficiently long times.

The above schematic analysis, although very appealing on physical grounds, is mathematically flawed, since (2.57) cannot hold for a discrete variable. However one can require that it be satisfied on average. If we formally solve (2.57) for σ and compute the average two-point correlation function, we obtain using (2.58),

$$<\sigma(1)\sigma(2)> = -G_p(1\bar{1})G_p(2\bar{2})2\tilde{\Pi}(\bar{1}\bar{2}) \tag{2.60}$$

where

$$2\tilde{\Pi}(12) = 2\Pi(12) + \delta(t_1 - t_o)\delta(t_2 - t_o)g_o(\vec{R}_1 - \vec{R}_2) \quad . \tag{2.61}$$

For consistency, we must remember that (2.61) must be supplemented by the crucial constraint

$$S(t) = <\sigma^2(1)> \tag{2.62}$$

which follows from (2.51).

F. Construction of $P[\sigma]$.

Our analysis in the last subsection has led us to a set of conclusions which are difficult to reconcile.

(i) The fundamental field ψ should contain an Ising-like component σ whose identification allows us to separate the two mass scales W_p and W_f and suggests that we identify the propagators G_p and G_f with the variables σ and ϕ.

(ii) This Ising-field σ, however, cannot directly satisfy (2.57) derived by setting $\psi = \sigma$ in the original Langevin equation because of its Ising-nature. The best we can do is to require that $<\sigma(1)\sigma(2)>$ satisfies (2.60) and (2.62).

These apparently conflicting conclusions can be reconciled through the proper construction of the distribution $P_\sigma[\sigma]$.

As a first step in constructing $P_\sigma[\sigma]$, we introduce[31,32]

$$Q[\sigma | m] = \prod_l \frac{1}{2}[1 + m(l)\sigma(l)/S(t_l)] \tag{2.63}$$

where $m(\vec{R},t)$ is a continuous field, and Q, which satisfies,

$$\sum_{\{\sigma\}} Q[\sigma | m] = 1 \tag{2.64}$$

$$\sum_{\{\sigma\}} \sigma(1) Q[\sigma | m] = m(1) \tag{2.65}$$

$$\sum_{\{\sigma\}} \sigma(1)\sigma(2) Q[\sigma | m] = m(1)m(2) + \delta(12)[S(t_1) - m^2(1)] \quad , \tag{2.66}$$

maps σ onto m except for space-time points which coincide, where, for example, $\sigma^2(i) = S(t_i)$.

Next we endow the field $m(\vec{R},t)$ with dynamical behavior through a probability distribution of the form

$$P_m[\hat{m},m] = e^{-A_m[\hat{m},m]} \tag{2.67}$$

where $A_m[\hat{m},m]$ is an action of the MSR type and we define[33]

$$P_\sigma[\sigma] = \int D[\hat{m}]D[m]Q[\sigma|m]P_m[\hat{m},m] \quad . \tag{2.68}$$

If we reorganize the product in Q in terms of a sum ordered by the number of σ variables we obtain

$$P_\sigma[\sigma] = \frac{1}{2^N}\,[1 + \sum_l \frac{<m(l)>}{S(t_l)}\,\sigma(l) + \frac{1}{2}\sum_{l_1 \neq l_2} \frac{<m(l_1)m(l_2)>}{S(t_{l_1})S(t_{l_2})}\,\sigma(l_1)\sigma(l_2)$$

$$+ \frac{1}{3!}\sum_{l_1 \neq l_2 \neq l_3} \frac{<m(l_1)m(l_2)m(l_3)>}{S(t_{l_1})S(t_{l_2})S(t_{l_3})}\,\sigma(l_1)\sigma(l_2)\sigma(l_3) + \cdots] \tag{2.69}$$

where the averages in the coefficients are taken with respect to P_m. We note, however, that a normalized distribution for an Ising-like field is of the general form

$$P_\sigma[\sigma] = \frac{1}{2^N}[1 + \sum_l \frac{<\sigma(l)>}{S(t_l)}\,\sigma(l) + \frac{1}{2}\sum_{l_1 \neq l_2} \frac{<\sigma(l_1)\sigma(l_2)>}{S(t_{l_1})S(t_{l_2})}\sigma(l_1)\sigma(l_2) \tag{2.70}$$

$$+ \frac{1}{3!}\sum_{l_1 \neq l_2 \neq l_3} \frac{<\sigma(l_1)\sigma(l_2)\sigma(l_3)>}{S(t_{l_1})S(t_{l_2})S(t_{l_3})}\,\sigma(l_1)\sigma(l_2)\sigma(l_3) + \cdots] \quad .$$

Comparing (2.69) and (2.70), we see that we are in the position of modelling the dynamics of the Ising-like field $\sigma(\vec{R},t)$ by means of the dynamics of the **continuous** field $m(\vec{R},t)$. In fact, one can see by inspection from (2.69) and (2.70) that the correlation functions of $\sigma(\vec{R},t)$ do coincide, except at distinct points with the correlation functions of $m(\vec{R},t)$, and the latter are specified by giving the action $A_m[\hat{m},m]$. Due to the intrinsic difference between gaussian and Ising statistics, deviations occur when two or more fields are taken at the same point. Consider, for example, the two point correlation function $<\sigma(1)\sigma(2)>$. If we multiply $\sigma(1)\sigma(2)$ by $Q[\sigma|m]P_m[\hat{m},m]P_\psi[\hat{\psi},\psi]$ and sum over all of the variables, using (2.66), we obtain the general result

$$<\sigma(1)\sigma(2)> = <m(1)m(2)> + \delta(12)[S(t_1) - <m^2(1)>] \tag{2.71}$$

and the two correlation functions will be identical everywhere,

$$C_m(12) = <\sigma(1)\sigma(2)> = <m(1)m(2)> \quad , \tag{2.72}$$

if we enforce the self-consistent constraint (2.62),

$$S(t_1) = <m^2(1)> \quad . \tag{2.73}$$

One can show in a similar manner, using (2.65), that

$$<\sigma(1)m(2)> = C_m(12) \quad . \tag{2.74}$$

Higher order correlation functions can not be constructed to coincide everywhere.

We can now develop a nontrivial, nongaussian dynamics for $\sigma(\vec{R},t)$ through a gaussian dynamics for $m(\vec{R},t)$. We simply stipulate that $m(\vec{R},t)$ obeys the equation of motion (2.57).

$$G_p^{-1}(1\overline{1})m(\overline{1}) = -I(1\overline{1})m_o(\overline{1}) + i\theta(1) \tag{2.75}$$

where m_o satisfies (2.24) with ψ_o replaced by m_o. The quadratic action associated with the field m is given by

$$A_m = \int d1d2[\hat{m}(1)\Pi(12)\hat{m}(2) + \hat{m}(1)G_p^{-1}(12)m(2)] \quad . \tag{2.76}$$

From (2.75) we can identify G_p^{-1} as the propagator for the m field, and, to the extent that the m and σ fields coincide at distinct points, we can also identify G_p^{-1} as the propagator for the σ variable and reconcile the two points listed at the beginning of this section. In particular we have constructed $P_\sigma[\sigma]$ such that (2.60) and (2.62) hold and

$$C_m(12) = -G_p(1\overline{1})G_p(2\overline{2})2\tilde{\Pi}(\overline{1}\overline{2}) \quad . \tag{2.77}$$

G. Determination of Π(12)

The last point to be specified in the determination of the σ and m variables is the noise term $\Pi(12)$. The construction of an acceptable form for Π is based on the observation that unless Π vanishes at long times one will generate the spurious NG modes discussed earlier. In addition, one expects that at short times the peak growth will be influenced by thermal noise and $\Pi \sim T$, while at long times thermal noise should be unimportant in determining the peak contribution.

The construction of the appropriate form for Π is discussed in detail in Ref. 1 and I will skip the development here. The final result is

$$2\tilde{\Pi}(12) = 2\tilde{\pi}(12) + G_p^{-1}(1\overline{1})G_p^{-1}(2\overline{2})C_N(\overline{1}\overline{2}) \tag{2.78}$$

where C_N satisfies

$$G_\infty^{-1}(1\overline{1})C_N(\overline{1}2) = -2\pi(1\overline{2})G_\infty(2\overline{2}) \tag{2.79}$$

where

$$G_\infty^{-1}(12) = i[\frac{\partial}{\partial t_1} + \Gamma(\vec{R}_1)c\nabla_{\vec{R}_1}^2]\delta(12) \quad . \tag{2.80}$$

By making use of (2.53) in (2.78), we obtain, after a spatial Fourier transform,

$$2\Pi(\vec{q},t_1,t_2) = \Gamma^2(\vec{q})[-2iW_p(t_2)G_\infty(q,t_2,t_1)T - 2iW_p(t_1)G_\infty(q,t_1,t_2)T - W_p(t_1)W_p(t_2)C_N(\vec{q},t_1,t_2)] \quad . \tag{2.81}$$

The key point for our purposes here is that $\Pi(\vec{q},t_1,t_2)$ vanishes as t_1 or $t_2 \rightarrow \infty$ since $W_p(t)$ vanishes as $t \rightarrow \infty$. The resulting peak correlation function $C_m(12)$ will, consequently, not contain any spurious NG modes.

The result of all these manipulations is the distribution governing the theory in the extended function space given by

$$P_T[\hat{\psi},\phi,\sigma,\hat{m},m] = e^{-A_T[\hat{\psi},\phi,\hat{m},m]}Q[\sigma\,|\,m] \tag{2.82}$$

where

$$A_T[\hat{\psi},\phi,\sigma,\hat{m},m] = A_m[\hat{m},m) + A_\psi[\hat{\psi},\phi,\sigma]$$

where A_m is given by (2.76) and A_ψ by (2.52).

III. Perturbation Theory

A. General Considerations.

The distribution P_T given by (2.82) is an exact formal rearrangement of the original field theory. We can now proceed to solve for the basic properties of this model in perturbation theory. To do this we start by choosing an appropriate quadratic approximation to A_T. Since A_m is already quadratic in the fields m and $\hat{m}$, we include it completely in our zeroth order action. There are two sources of nonquadratic terms in A which should be included in the interaction. The obvious term is proportional to $3\sigma(2)\phi^2(2) + \phi^3(2)$. Less obvious is the term

$$\int d1d2\hat{\psi}(1)G_p^{-1}(12)\sigma(2) \quad . \tag{3.1}$$

This term is quadratic in the fields, but, because of the Ising-nature of σ, it does not lead to a gaussian contribution in the associated probability distribution. Instead we write

$$\int d1d2\,\hat{\psi}(1)G_p^{-1}(12)\sigma(2) = \int d1d2\hat{\psi}(1)G_p^{-1}(12)m(2) + V_\sigma \tag{3.2}$$

and we treat

$$V_\sigma = \int d1d2\,\hat{\psi}(1)G_p^{-1}(12)[\sigma(2) - m(2)] \tag{3.3}$$

as part of the perturbation. We expect V_σ to be a small perturbation because $\sigma(1)$ is, on average, very nearly equal to $m(1)$. At long times the two fields coincide except near interfaces. Furthermore, $G_p^{-1}(12)$ goes to zero at long times and distances because $W_p \rightarrow 0$. One can show that this expectation is explicitly verified at first-order where the contribution arising from V_σ vanishes. Should one include (3.1) directly in the zeroth order probability distribution, terms would be generated at first order which are of the same magnitude as the zeroth order terms.

We can then write down the total action

$$A_T = A_o + V \quad , \tag{3.4}$$

where

$$A_o = A_m + \int d1d2\,\{\hat{\psi}(1)\bar{\pi}(12)\hat{\psi}(2) + \hat{\psi}(1)\,G_f^{-1}(12)\phi(2) + G_p^{-1}(12)m(2)]\} \tag{3.5}$$

is the zeroth-order contribution, and

$$V = \int d1d2\,\{\,\hat{\psi}(1)iu\Gamma(12)(3\sigma(2)\phi^2(2) + \phi^3(2))\} + V_\sigma \tag{3.6}$$

is the interaction. We develop perturbation theory by writing the total distribution

$$P_T = P_o e^{-V} \tag{3.7}$$

where

$$P_o = e^{-A_o} Q[\sigma \mid m] \tag{3.8}$$

and expanding in powers of V. A few general comments are in order:

(i). Even though the portion of P_o proportional to e^{-A_o} is gaussian, P_o is certainly not gaussian with respect to the basic field ψ due to the factor of $Q[\sigma|m]$. This will be demonstrated explicitly below when we calculate the singlet distribution function ρ.

(ii). The assumption that V can be treated as small for small T is supported by the expectation that at long times $\hat{\psi} \sim 1/\sqrt{T}$, $\phi \sim \sqrt{T}$ and $m \sim \sigma \sim 1/\sqrt{T}$. Thus, the term in V proportional to $\sigma\phi^2$ is of $0(\sqrt{T})$ and the ϕ^3 term of $0(T)$. V_σ has been discussed above.

(iii). Looking at the zeroth order action A_o, we see that there is an indirect coupling between the fields m and ϕ through the term $\hat{\psi}G_p^{-1}m$. We can achieve a further separation of the peak variable and the equilibrating *phonon* contribution if we make one further shift and define the field $\zeta(1)$ via

$$\phi(1) = \zeta(1) - a(1\bar{1})m(\bar{1}) \tag{3.9}$$

where

$$a(12) = G_f(1\bar{1})G_p^{-1}(\bar{1}2) \quad . \tag{3.10}$$

The zeroth order action then decouples into a sum of two parts:

$$A_o = A_m + A_\zeta \tag{3.11}$$

where A_m is given by (2.76), and A_ζ has the same general form as A_m:

$$A_\zeta = \int d1d2\{\hat{\psi}(1)\pi(12)\hat{\psi}(2) + \hat{\psi}(1)\, G_f^{-1}(12)\zeta(2)\} \quad . \tag{3.12}$$

In terms of these new variables the original field is given by

$$\psi(1) = \sigma(1) - a(1\bar{1})m(\bar{1}) + \zeta(1) \quad . \tag{3.13}$$

B. Zero Order Theory

Let us consider the zeroth order theory for the structure factor

$$C(12) = \langle\psi(1)\psi(2)\rangle \quad . \tag{3.14}$$

After replacing ψ using (3.13), we evaluate the zeroth order average with respect to P_o, which we denote by $< >_o$. Since the variables m and σ are decoupled from ζ at zeroth order, all such cross-correlations are zero. Using (2.72) and (2.74), it is easy to show that at zeroth order

$$C^0(12) = C_\zeta^0(12) + C_m(12) - C_a(12) - C_a(21) + C_{aa}(12) \tag{3.15}$$

where we have introduced the notation

$$C_\zeta^{\alpha} 12) = \langle\zeta(1)\zeta(2)\rangle_o \tag{3.16}$$

$$C_m(12) = \langle m(1)m(2)\rangle_o \tag{3.17}$$

$$C_a(12) = a(1\bar{1})C_m(\bar{1}2) \tag{3.18}$$

$$C_{aa}(12) = a(1\bar{1})a(2\bar{2})C_m(\overline{12}) \quad . \tag{3.19}$$

While we have worked out the general case in Ref. 1, I will restrict the discussion here to the case of quenches to T=0. The analysis simplifies significantly in this limit. I also limit the discussion to equal time correlation functions. It is easy to show that the various correlating functions appearing in the zeroth order theory can all be written in the compact form

$$D_p(\vec{q},t)C_m(\vec{q},t) = 0 \tag{3.20a}$$

$$D_f(\vec{q},t)C_\zeta^0(\vec{q},t) = 0 \tag{3.20b}$$

$$D_a(\vec{q},t)C_a(\vec{q},t) = 0 \tag{3.20c}$$

$$D_f(\vec{q},t)C_{aa}(\vec{q},t) = 0 \tag{3.20d}$$

with

$$D_x(\vec{q},t) = \frac{\partial}{\partial t} + 2\Gamma(\vec{q})[cQ^2(\vec{q}) + W_x(t)] \quad , \tag{3.21}$$

where W_p and W_f are defined by (2.55) and (2.56),

$$W_a(t) = r + 2uS(t) \quad , \tag{3.22}$$

and all correlation functions are initially[34] equal to $g_o(\vec{q})$. For the particular case of T=0, we see that $C_\zeta^0(\vec{q},t) = C_{aa}(\vec{q},t)$. We show explicitly in Sect. IV that the peak growth in the structure factor is driven by (3.20a), and that it indeed involves only $C_m(\vec{q},t)$ and its moment S(t). Starting from the initial instability due to $W_p(t_o) < 0$, the system equilibrates by growing a peak until $W_p(t) \to 0$. Hence, (3.20a) has a long time solution:

$$\lim_{t \to \infty} C_m(\vec{q},t) = M_o^2 \delta(\vec{q}) \tag{3.23}$$

with

$$M_o^2 = -r/u \quad . \tag{3.24}$$

Furthermore, since $W_f(t)$ and $W_a(t)$ are both positive for sufficiently long times, the other contributions to the structure factor eventually decay exponentially to zero and after the passing of these transients, the structure factor is given by the peak contribution only. This demonstrates that the set (3.20) produces the expected ordering and the correct equilibrium state for T=0.

The structure of the distribution P_o underlying the zeroth order theory is most effectively illustrated by studying the reduced singlet probability distribution $\rho(y,t)$, defined by

$$\rho(y,t) = \langle \delta(y - \psi(\vec{R},t)) \rangle_o \quad , \tag{3.25}$$

which gives the probability that the field $\psi(\vec{R},t)$ at the time t has the value y. Using the integral representation of the δ-function and the decomposition (3.13), we have

$$\rho(y,t) = \langle \int_{-\infty}^{+\infty} \frac{dx}{2\pi} e^{ix(y-\sigma(1)+a(1\bar{1})m(\bar{1})-\zeta(1))} \rangle_o \quad . \tag{3.26}$$

After tracing over σ, which can easily be performed, one is left with gaussian averages over ζ and m and the integration over x. The gaussian averages can be carried out using the basic result

$$\langle e^{iB(1\bar{1})\zeta(\bar{1})} \rangle_o = e^{-\frac{1}{2}B(1\bar{1})B(1\bar{2})\langle\zeta(\bar{1})\zeta(\bar{2})\rangle_o} \quad . \tag{3.27}$$

The final result is given by

$$\rho(y,t) = (8\pi b)^{-1/2} \{ [1 + \frac{S_a(t)}{b\sqrt{S(t)}} (y+\sqrt{S(t)})] e^{-(y+\sqrt{S(t)})^2/2b} + [1 - \frac{S_a(t)}{b\sqrt{S(t)}} (y - \sqrt{S(t)}] e^{-(y-\sqrt{S(t)})^2/2b} \} \tag{3.28}$$

where

$$b = S_\zeta(t) + S_{aa}(t) \tag{3.39}$$

and $S_\zeta(t)$, $S_a(t)$, $S_{aa}(t)$ are moments of $C_\zeta^0(\vec{q},t)$, $C_a(\vec{q},t)$, $C_{aa}(\vec{q},t)$ analogous to $S(t)$ as given by (2.73).

The general structure of the perturbation theory at higher orders is discussed in Ref. 1 and has no important surprises. We do recover the correct O(T) corrections to the zero temperature result.

IV. Results

I now discuss some of the results obtained from the lowest order theory. The calculations I report were carried out for an isotropic continuum, introducing a cutoff Λ_o in $\vec{q}$ space to avoid the ultraviolet divergence in the computation of S(t). Alternatively, we could perform the calculation on a lattice.

Defining the dimensionless variables

$$\vec{k} = \left(\frac{c}{|r|}\right)^{1/2} \vec{q} \tag{4.1}$$

and

$$\tau = 2\Gamma_a |r| t \quad , \tag{4.2}$$

with $\Gamma_a = \Gamma$ for NCOP and $\Gamma_a = D|r|/c$ for COP, all the differential equations satisfied by the various equal time correlation functions are of the general form

$$\left[\frac{\partial}{\partial \tau} + \gamma(\vec{k},\tau)\right] f(\vec{k}, \tau) = h(\vec{k},\tau) \quad . \tag{4.3}$$

We compute the dimensionless correlation functions ($x = m, a, aa, \zeta$),

$$\tilde{C}_x(\vec{k}, \tau) = \frac{u}{|r|} \left(\frac{|r|}{c}\right)^{d/2} C_x(\vec{q}, t) \quad , \tag{4.4}$$

and the corresponding moment

$$\tilde{S}_x(\tau) = \frac{u}{|r|} S_x(t) \quad . \tag{4.5}$$

with a dimensionless cutoff Λ, related to Λ_o as in (4.1). We have chosen $\Lambda = 1$ in all of the calculations reported here. The quantity:

$$\varepsilon = \frac{uT}{r^2} \left(\frac{|r|}{c}\right)^{d/2} \tag{4.6}$$

is the dimensionless coupling characterizing the final equilibrium state. The shared initial condition for the dimensionless correlation functions $\tilde{C}_x$ is given by

$$\tilde{g}(\vec{k}) = \frac{u}{|r|}\left(\frac{|r|}{c}\right)^{d/2} g_o(\vec{q}) = \varepsilon_I \quad . \tag{4.7}$$

assuming the initial state is site disordered. The time evolution of $f(\vec{k},\tau)$ can be generated by iteration of (4.3), keeping in mind that at each time step the computation of $\gamma(\vec{k}, t)$ involves an integration over $\vec{k}$ to obtain the moment $\tilde{S}(\tau)$.

The driving terms appearing on the right hand side of (4.3) are of two types: one is noise driven, proportional to the final quenching temperature (ε). The other is proportional to the initial value of the correlations. If both are zero, $\tilde{C}(\vec{k}, \tau)$ remains zero at all times. If either is nonzero there is growth.

Given the noise level (ε), the initial condition (ε_I) and the cut-off $\Lambda = 1$, we can proceed to determine the various correlation functions entering (3.15), which we rewrite in scaled form

$$\tilde{C}(\vec{k},\tau) = \tilde{C}_m(\vec{k}, \tau) + \tilde{C}^0_\zeta(\vec{k}, \tau) + \tilde{C}_{aa}(\vec{k}, \tau) - 2\tilde{C}_a(\vec{k}, \tau) \qquad (4.8)$$

and the corresponding dimensionless moments

$$\tilde{S}_T(\tau) = \tilde{S}(\tau) + \tilde{S}_\zeta(\tau) + \tilde{S}_{aa}(\tau) - 2\tilde{S}_a(\tau) \quad . \qquad (4.9)$$

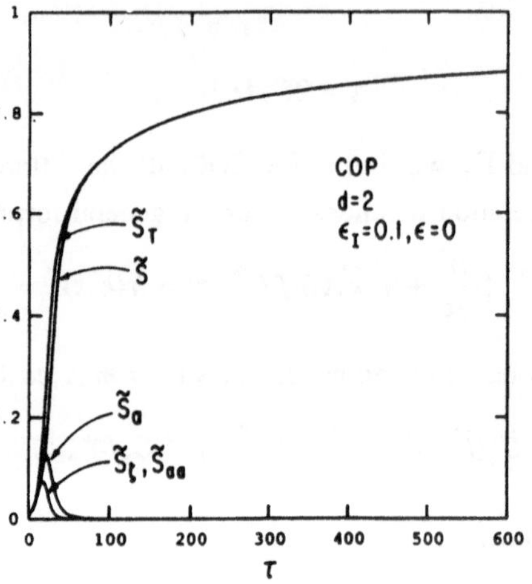

Fig. 1: Moments for COP: $\tilde{S}_{aa} = \tilde{S}_\zeta, \tilde{S}_a, \tilde{S}, \tilde{S}_T$.

In Fig. 1 we examine the moments appearing in (4.9) for a COP over a time range ($\tau \leq 600$) which, although short compared with the times we consider later on, is already comparable to the longest times ever considered in computer simulations.[35,36] The curves in Fig. 1 contain a large amount of information. The overall gross feature for very early times is that all of the moments grow rapidly ($\tilde{S}(\tau) \sim e^{\tau/4}/\sqrt{\tau}$) and the variables $\sigma(1)$, $\zeta(1)$ and $a(1\bar{1})\, m(\bar{1})$ are all initially unstable. However one finds, as $\tilde{S}$ increases, that the W(t) functions associated with the correlation functions $\tilde{C}_m$, $\tilde{C}^0_\zeta$, $\tilde{C}_a$, $\tilde{C}_{aa}$ increase from their shared initial negative value of -1 (in dimensionless units) toward zero. When $\tilde{S}(\tau) \sim 1/3$, one finds that the mass terms for $\tilde{C}_\zeta^0$ and $\tilde{C}_{\alpha\alpha}$ pass through zero and their subsequent positive value causes $\tilde{S}_\zeta$ and $\tilde{S}_{\alpha\alpha}$ to reach maximum values and begin exponential decay toward their final equilibrium values (for $\varepsilon \neq 0$):

$$\tilde{S}_\zeta(\infty) = \int \frac{d^d k}{(2\pi)^d} \frac{\varepsilon}{k^2 + 2} \tag{4.10}$$

and $\tilde{S}_{\alpha\alpha} = 0$. As time evolves and $\tilde{S}(\tau)$ approaches $\frac{1}{2}$ the same change of sign occurs in the $S_\alpha(\tau)$ mass term, causing the subsequent decay of this moment to zero. Since the area under the central peak tends to the magnetization squared for long times:

$$M^2 = \int \frac{d^d}{(2\pi)^d} C_m(\vec{q}) = S(\infty) \tag{4.11}$$

we obtain, using (4.5),

$$M_0^2 = \frac{-r}{u} \tag{4.12}$$

as expected.

Once the various moments shown in Fig. 1 are obtained one can proceed to evaluate the singlet distribution function given by (3.28). Figure 2 shows $\rho(y)$, in terms of the dimensionless time defined by (4.2) and y measured in units of M_0. The parameters characterizing the system are $\varepsilon = 0.1$ and $\varepsilon_I = 0$.

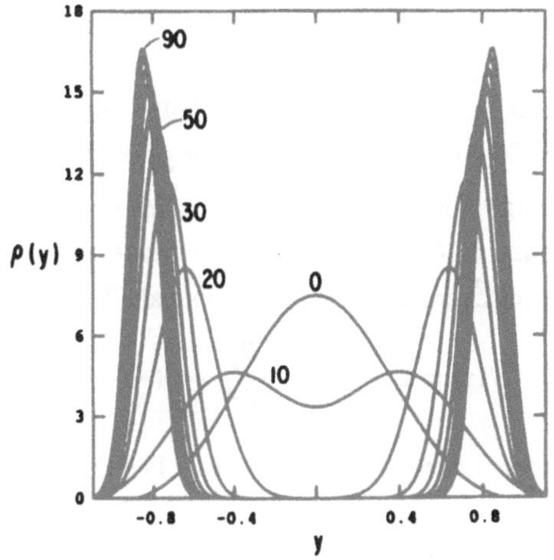

Fig. 2: Reduced Singlet Distribution Function with COP in d = 2 for a quench from $\varepsilon_I = 0$ to $\varepsilon = 0.1$. Time steps of 10 units. Normalized to 2π.

In Fig. 3, we show $\tilde{C}(\vec{k}, \tau)$ as a function of k for several values of τ in the intermediate time regime. The evolution of the wavevector $k_p(\tau)$, corresponding to the maximum value of $\tilde{C}(\vec{k}, \tau)$, toward smaller values can be clearly seen. Its behavior with time is analyzed in more detail below.

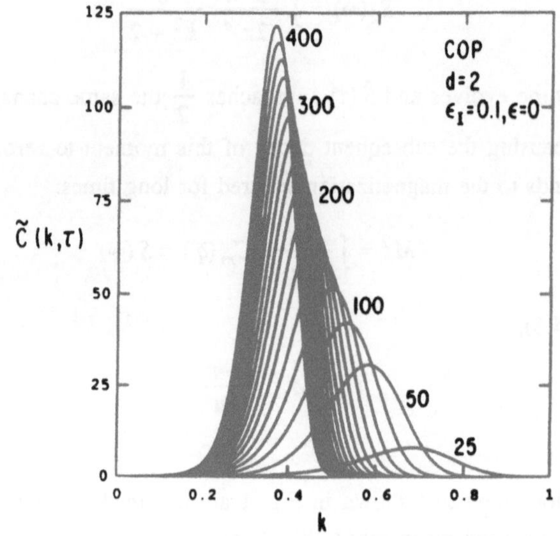

Fig. 3: Time evolution of $\tilde{C}(k,\tau)$ in time steps of 25 units for COP.

We turn, in Figs. 4 and 5, to the case where the order parameter is not conserved. The moment $\tilde{S}_T(\tau)$ and its constituents are shown in Fig. 4. Except for a large change in the time scale reflecting a much faster equilibrium, the qualitative behavior is as in the conserved case. In Fig. 5 results are shown for $\tilde{C}(\vec{k}, \tau)$ for a NCOP. Here we have the very obvious difference, compared to the COP case, that the ordering peak now grows at $k = 0$.

We turn next to the long-time behavior. We consider first the behavior of the moments at long times. In this regime $\tilde{S}_T(\tau) = \tilde{S}(\tau) + \tilde{S}_\zeta(\infty)$. Since the latter is a constant at long times, the time dependences of $\tilde{S}_T(\tau)$ and $\tilde{S}(\tau)$ are identical. We find that as $\tau \rightarrow \infty$:

$$\tilde{S}(\tau) = 1 - \frac{A_s}{\tau^{1/2}} + \frac{B_s}{\tau^{34}} + 0(\frac{1}{\tau}) \quad . \tag{4.13}$$

The parameters A_s and B_s depend weakly on ε and ε_I. For the parameters ($\varepsilon = 0.1$, $\varepsilon_I = 0$), $A_s = 2.77$ and $B_s = 0.15$).

When the order parameter is not conserved we find that $\tilde{S}(\tau)$ reaches its asymptotic value according to:

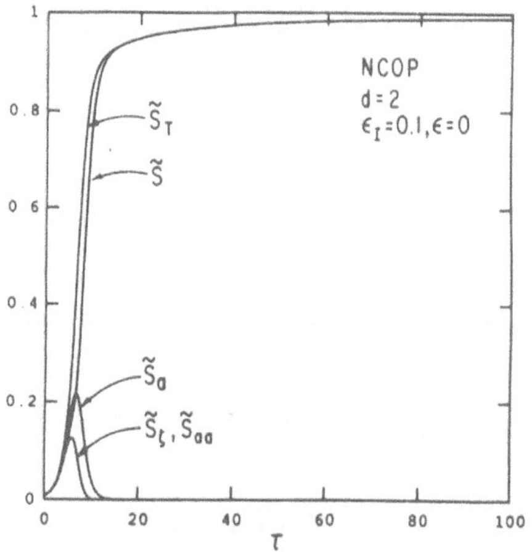

Fig. 4: Moments for NCOP: $\tilde{S}_{aa} = \tilde{S}_\zeta$, $\tilde{S}_a$, $\tilde{S}$, $\tilde{S}_T$.

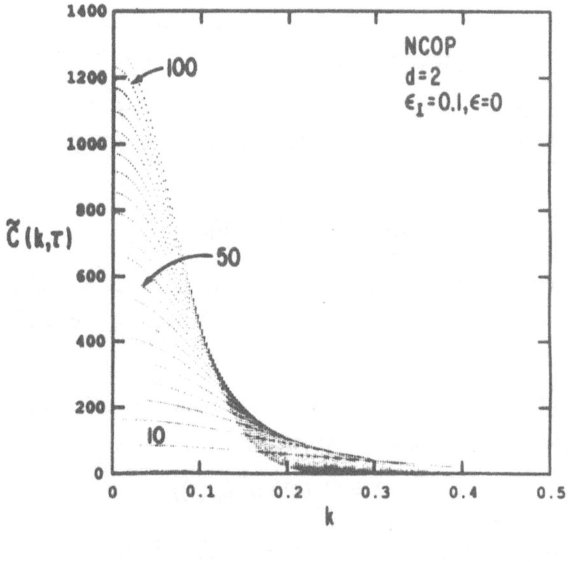

Fig. 5 Time evolution of $\tilde{C}(k,\tau)$ in time steps of 5 units for NCOP.

$$\tilde{S}(\tau) = 1 - \frac{A_s}{\tau} + \frac{B_s}{\tau^{3/4}} + 0\,(\frac{1}{\tau^2}) \tag{4.14}$$

where, for example, for two dimensions, $\varepsilon = 0.1$ and $\varepsilon_l = 0$, the values $A_s = 0.97$, $B_s = -\,0.40$ give a good fit over the time interval $20 < \tau < 300$. It is not surprising that the asymptotic limit is reached earlier and with a faster power law in the absence of a conservation law.

We now turn to the long time behavior of the structure factor itself. To characterize the domain growth, in the conserved case we use the position $k_p(\tau)$ of the peak in $\tilde{C}(\vec{k}\,,\tau)$. We also define the peak height, $\tilde{C}_p(\tau) = \tilde{C}(k_p(\tau) < \tau)$. We find that $k_p^{-1}(\tau)$ and $\tau_p(\tau)$ are given asymptotically by:

$$k_p^{-1}(\tau) = L_0 + A\,\tau^{1/4} \tag{4.15}$$

$$\tilde{C}_p(\tau) = A_p\,\tau^{1/2} + B_p\,\tau^{1/4} + C_p \tag{4.16}$$

where L_0, A, A_p, C_p depend weakly on ε and ε_l. For $\varepsilon = 0.1$, $\varepsilon_l = 0$, we find, in the range $2000 < \tau < 15000$, $L_0 = 0.025$, $A = 0.62$ and for the peak height $A_p = 6.83$, $B_p = -\,0.69$, $C_p = -\,15.4$ in the range $2000 < \tau < 15000$.

In three dimensions (4.15) remains valid while (4.16) is replaced by:

$$\tilde{C}_p(\tau) = A_p\,\tau^{3/4} + B_p\,\tau^{1/2} + C_p\,\tau^{1/4} + D_p \quad . \tag{4.17}$$

This is in agreement with the expectation that the dimensionality should not affect the growth exponent n, while $\tilde{C}_p \sim t^{nd}$ implying the scaling law $\tilde{C}_p k_p^d = A_p / A^d$.

The development of asymptotic behavior occurs much earlier when the order parameter is not conserved. We then find in this case that the width of the peak at $k = 0$ decreases asymptotically as:

$$k_W^{-1}(\tau) = A_W\;\tau^{1/2} + B_W \tag{4.18}$$

which is in agreement with the Lifshitz-Cahn-Allen law, as expected. For the case $\varepsilon = 0.1$, $\varepsilon_l = 0$, the fit to (4.18) with $A_W = 1.21$ and $B_W = -\,0.13$, can not be distinguished from the solution for k_W.

In the long time regime we have also verified that scaling is satisfied. In the asymptotic region the structure factor for a COP can be written as

$$\tilde{C}(\vec{k}\,,\tau) = \tilde{C}_p(\tau) F\,(\vec{k}/k_p(\tau)) \tag{4.19}$$

with $F(1) = 1$. The scaling function $F(x)$ is plotted in Fig. 6 for $d = 2$ and $d = 3$ (narrower peak) in the case of a quench with $\varepsilon_l = 0$, $\varepsilon = 0.1$.

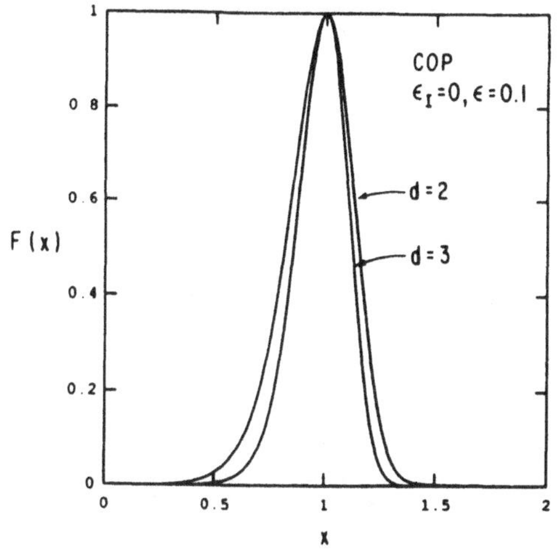

Fig. 6: Scaling function $F(x)$ vs. x for COP in two and three dimensions.

We have also obtained the shape function for the nonconserved case where we define $F(x)$ by:

$$\tilde{C}(\vec{k}, \tau) = \tilde{C}(0,\tau) F(\vec{k}/k_W(\tau)) \quad . \tag{4.20}$$

In this case the shape function is gaussian for x not too large, as it is demonstrated by plotting (Fig. 7) log F(x) vs. x^2. In Fig. 7 the shape functions for a quench in $d = 2$ and $d = 3$ (with $\varepsilon_I = 0$, $\varepsilon = 0.1$) are compared and, as in the COP case, a narrower peak corresponds to the higher dimensionality as it should be expected. We have found in all cases (COP and NCOP) that F(x) decreases as $|x|^{-4}$ for large values of $|x|$ in both two and three dimensions. This is in agreement with Porod's law ($x^{-(d+1)}$) in three dimensions, but not in two. It is possible to show that the $|x|^4$ result follows analytically from substituting (4.13) into (3.20) and carefully examining the long time limit. Numerical methods, these results are compatible with the asymptotic behavior obtained here. A detailed comparison between the two methods will be made in future work.

V. Conclusions

I have discussed here a theory which is capable of describing the entire time evolution of a system undergoing a non-equilibrium process from an initial disordered state to a final ordered state. This theory was developed in the context of a field theoretic method for Langevin dynamics and involves a systematic low temperature perturbation expansion. The

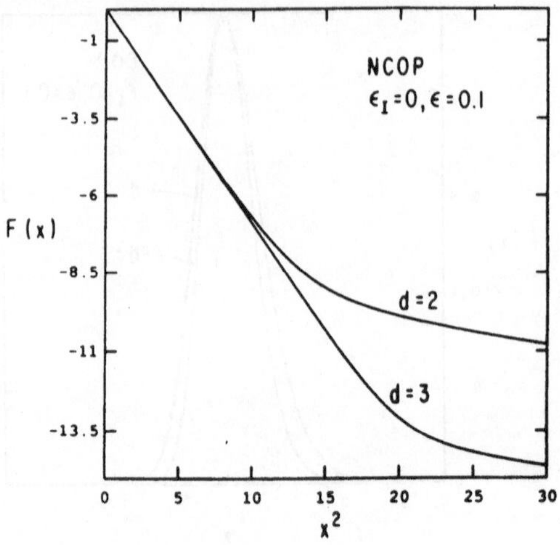

Fig. 7: Scaling function $F(x)$ vs. x^2 for NCOP in two and three dimensions.

lowest order theory has been analyzed in detail producing all the expected features of the global time evolution: a reduced singlet probability distribution evolving from an initial gaussian to a final bimodal distribution, scaling behavior for the structure factor, the separation of the domain size L(t) from the correlation length $\xi(t)$ generated by the dynamics, and the equilibration to a final ordered state without Nambu-Goldstone modes. The growth law at long time $L(t) \sim t^n$ has also been computed to lowest order, obtaining n=1/4 for COP and n=1/2 for NCOP. We find that these models are of the class I type introduced in Ref. (9), with no activated processes at very low temperatures. The stability of these results has been checked when first order corrections are taken into account and it has been found that no change occurs in the growth law, while the final equilibrium quantities, like the magnetization and the correlation length, acquire the appropriate first order corrections in the temperature.

The result n = 1/4 for a conserved order parameter differs from the Lifshitz-Slyozov result (n = 1/3). There are conflicting reports in the literature[36] as to whether simulation results are in better agreement with 1/3 or 1/4. This is a very difficult question to resolve from purely numerical methods, since the small difference between 0.25 and 0.33 is easily buried in numerical uncertainly and the high cost of the numerical calculations makes it prohibitively expensive to reach the indisputably asymptotic time region. The method presented here, being purely theoretical, is obvioiusly quite free of these limitations. The perturbation theory is systematic and well-defined, and we can show formally that higher-order terms in perturbation theory decay to zero at long times. Consequently we expect that our low-order results should not be affected by higher-order terms for low enough temperatures and long enough times. If, indeed the long-time growth law for the COP case is n = 1/3, then the source of this cross-over from 1/4 ro 1/3 must come from nonperturbative solutions to the field equations and not picked up by our peak variable σ.

The method discussed here is not limited to purely uniform boundary conditions. It is rather straightforward to use the basic equations developed for the case of nonuniform initial conditions. One can study, for example, the evolution of hot drops in a fluid[36] or the evolution of an initially hot system quenched in the presence of an ordered boundary. These problems contain the germ of the physics which must be introduced to study problems such as dendritic growth, domain wall motion and nucleation theory. We believe that the extension of the methods discussed here to more complicated and physically relevant models is not unduly difficult.

Acknowledgements:

This work was supported by NSF grant No. DMR 84-12901 at the University of Chicago.

References

1. G.F. Mazenko, O.T. Valls and M. Zannetti, to be published, Phys. Rev. B.
2. An excellent review of the subject is contained in: J.D. Gunton, M. San Miguel and P.S. Sahni, in *Phase Transitions and Critical Phenomena*, ed. by C. Domb and J.L. Lebowitz (Academic, London, 1983), Vol. 8. More recent reviews include: K. Binder, Physica **140A**, 35 (1986), and *Mechanism for the Decay of Unstable and Metastable Phases: Spinodal Decomposition, Nucleation and Late-Stage Coarsening*, preprint (1987). M. Grant, *Kinetics of First-Order Phase Transitions*, preprint (1987). J.D. Gunton, *Recent Theoretical Developments in the Kinetics of First Order Phase Transitions*, preprint (1987). H. Furukawa, Adv. Phys. **34**, 703 (1985).
3. For example, the existance of scaling behavior was first confirmed by J. Marro, J.L. Lebowitz and M.H. Kalos, Phys. Rev. Lett. **43**, 282 (1979).
4. For reviews of the early theory see: K. Binder, *Spinodal Decomposition* Lecture Notes in Physics, Vol. **132** (1980), ed. L. Garrido (Springer-Verlag, Berlin). J.D. Gunton and M. Droz, *Introduction to the Dynamics of Metastable and Unstables Systems*, Lecture Notes in Physics, Vol. **183**, (1983), ed. J. Zittartz (Springer-Verlag, Berlin).
5. K. Binder and D. Stauffer, Phys. Rev. Lett. **33**, 1006 (1974).
6. M.K. Pahni, J.L. Lebowitz, M.H. Kalos and O. Penrose, Phys. Rev. Lett. **45**, 366 (1980); P.S. Sahni, G. Dee, J.D. Gunton, M.K. Phani, J.L. Lebowitz, and M.H. Kalos, Phys. Rev. B **24**, 410 (1981); J.L. Lebowitz, J. Marro and M.H. Kalos, Acta. Metall. **30**, 297 (1982). P. Fratzl, J.L. Lebowitz, J. Marro and M.H. Kalos, Acta. Metall. **31**, 1849 (1983); K. Kaski, M.C. Yalabik, J.D. Gunton, and P.S. Sahni, Phys. Rev. B **28**, 5263 (1983).
7. G.F. Mazenko, O.T. Valls and F.C. Zhang, Phys. Rev. B **31**, 4453 (1985).
8. J. Vinals, M. Grant, M. San Miguel, J.D. Gunton and E.T. Gawlinski, Phys. Rev. Lett. **54**, 1264 (1985); S. Kumar, J. Vinals and J.D. Gunton, Phys. Rev. B **34**, 1908 (1986); J. Vinals and J.D. Gunton, Phys. Rev. B **33**, 7795 (1986).
9. Z.W. Lai, G.F. Mazenko and O.T. Valls, to be published, Phys. Rev. B.

10. K. Kawasaki and T. Ohta, Progr. Theor. Phys. **68**, 129 (1982); T. Ohta, D. Jasnow and K. Kawasaki, Phys. Rev. Lett. **49**, 1223 (1983); K. Kawasaki and T. Ohta, Physica (Utrecht) **118A**, 185 (1983); M. Tokuyama and K. Kawasaki, Physica (Utrecht) **123A**, 386 (1984).

11. J.W. Cahn and J.E. Hilliard, J. Chem. Phys. **28**, 258 (1958); **31**, 668 (1959). J.W. Cahn, Trans. Metall. Soc. AIME **242**, 166 (1968); J.W. Cahn, J. Chem. Phys. **42**, 93 (1965); S.M. Allen and J.W. Cahn, Acta. Metall. **27**, 1085 (1979); J. Phys. (Paris) Colloq. C **7**, 54 (1977).

12. I.M. Lifshitz and V.V. Slyozov, J. Phys. Chem. Solids, **19**, 35 (1961); I.M. Lifshitz, Zh. Teor. Fiz. **42**, 1354 (1962). [Sov. Phy. JETP **115**, 939 (1962)].

13. H.E. Cook, Acta. Metall. **18**, 297 (1970).

14. J.S. Langer, M. Bar-on, and H.D. Miller, Phys. Rev. A **11**, 1417 (1975).

15. N.C. Wong and C.M. Knobler, J. Chem. Phys. **69**, 725 (1978); Y.C. Chou and W.I. Goldburg, Phys. Rev. A **23**, 858 (1981); **20**, 2105 (1979).

16. K. Binder, C. Billotet, and P. Mirold, Z. Phys. B **30**, 183 (1978).

17. K. Kawasaki and T. Ohta, Prog. Teor. Phys. **59**, 362 (1978).

18. G.F. Mazenko and M. Zannetti, Phys. Rev. Lett. **53**, 2106 (1984); Phys. Rev. B **32**, 4565 (1985).

19. J.K. Bhattacherjee, P. Meakin and D.L. Scalapino, Phys. Rev. A **30**, 1026 (1984). F. de Pasquale and P. Tartaglia, Phys. Rev. B **33**, 2081 (1986).

20. D. Forster, *Hydrodynamic Fluctuations, Broken Symmetry and Correlation Functions* (Benjamin, N.Y. 1975).

21. While the theory is written in continuum notation, the discretized theory on a lattice with spacing a is obtained by the replacements

$$\int d^dR \rightarrow \frac{1}{a^d}\sum_{\vec{R}}$$

where the sums run over lattice vectors and

$$(\nabla\psi(\vec{R}))^2 = \frac{1}{2a^2}\sum_{\vec{\delta}}[\psi(\vec{R}) - \psi(\vec{R}+\vec{\delta})]sup\,2$$

$$\nabla^2\psi(\vec{R}))^2 = \frac{1}{a^2}\sum_{\vec{\delta}}[\psi(\vec{R}+\vec{\delta}) - \psi(\vec{R})]$$

where the sum runs over nearest neighbors lattice vectors.

22. Units with $k_B = 1$ are assumed.

23. D.J. Amit and M. Zannetti, J. Stat. Phys. **7**, 31 (1973).

24. C. de Dominicis, J. Phys. C **1**, 247 (1976); H.K. Janssen, Z. Phys. B **24**, 113 (1976); R. Bausch, H.J. Janssen and H. Wagner, Z. Phys. B **23**, 377 (1976).

25. P.C. Martin, E.D. Siggia and H.A. Rose, Phys. Rev. A **8**, 423 (1973).

26. C. de Dominicis and L. Peliti, Phys. Rev. B **18**, 353 (1978).

27. S.P. Das and G.F. Mazenko, Phys. Rev. A **34**, 2265 (1986).

28. U. Deker, Phys. Rev. A. **19**, 846 (1979). R. Phythian, J. Phys. A. **10**, 777 (1977). R.V. Jensen, J. Stat. Phys. **25**, 183 (1981).

29. In the subsequent development we assume that the initial conditions can be included into an effective noise as indicated by (2.26). This is applicable only if the initial free-energy is quadratic in ψ_0 or if F is quadratic in ψ.

30. In the LBM theory the structure factor obeys an equation of the same form as (2.24), except for the replacement of $r + uS(t)$ by a quantity $A(t)$ which was evaluated in a rather elaborate analysis of the one-body distribution. Nevertheless $A(t) \rightarrow 0$ as $t \rightarrow \infty$ as in (2.26) and the subsequent criticisms of the late time behavior remain.

31. An object similar to Q is introduced in H. Sommers, Phys. Rev. Lett. **58**, 1268 (1987).

32. At this stage, in order to carefully define things, we must assume that both space and time are defined on a discrete mesh. In the end we take the continuum time limit.

33. Note that $Q[\sigma | m]$ is not positive definite (since, for example, with $\sigma(l) = -\sqrt{S(t_1)}$, $1 - m(l)/\sqrt{S(t_1)}$ is negative for large positive $m(l)$ and $P_\sigma[\sigma]$ can not literally be identified as a probability distribution. However, as it will be clear from the computations of Sect. IV, this is not a serious drawback.

34. In, for example, (2.15) the initial condition is governed by the δ-function term on the right hand side. This arises because the field is defined with a multiplying step function which generates the δ-function when acted upon by the time derivative in the equation of motion. It is convenient and equivalent to treat the coefficient of the step function directly, drop the δ-function term and treat the resulting differential equation as an initial value problem.

35. R. Petschek and H. Metiu, J. Chem. Phys. **79**, 3443 (1985); O.T. Valls and G.F. Mazenko, Phys. Rev. B **34**, 7941 (1986); A. Milchev, K. Binder and D.W. Heermann, Z. Phys. B **63**, 521 (1986); Y. Oono and S. Puri, Phys. Rev. Lett. **58**, 836 (1987); G.F. Mazenko and O.T. Valls, Phys. Rev. Lett. **59**, 680 (1987); T.M. Rogers, K.R. Elder and R.C. Desai, *A Numerical Study of the Late Stages of Spinodal Decomposition*, preprint (1987).

36. O.T. Valls, Phys. Rev. B **37**, in press.

PARTICIPANTS

Prof. ABRAHAM, N.B., Dept. of Physics, Bryn Mawr College, Bryn Mawr, PA 19010 USA

Mr. BAFALUY, F.J., UAB, Dpto. de Física, 08193 Bellaterra, Barcelona, Spain

Mr. BESTEHORN, M., Inst. f. Theor. Phys. und Synergetik, 7000 Stuttgart-80, Germany

Prof. BLANCO, R., Dpto. de Física Moderna, Universidad de Cantabria, 39005 Santander, Spain

Mr. BONET, J., Fac. de Informática, Universidad Politécnica, 08082 Barcelona, Spain

Mr. BRITO, R., Depto. de Física Aplicada I, Universidad Complutense, 28040 Madrid , Spain

Dr. van den BROECK, C., Limburgs Universitair Centrum, B 3610 Diepenbeek, Belgium

Mr. CASADEMUNT, Dpto. Estructura i Constituents de la Matèria, Universitat de Barcelona, 08028, Spain

Dr. CARNEIRO, G. Dpto. de Física, UAB, Bellaterra, Barcelona, Spain

Prof. CASADO, J.M., Facultad de Física, Universidad de Sevilla, 41080 Spain

Mr. COLET, P., Dpto. de Física, Universidad de las Islas Baleares, 07071 Palma de Mallorca, Spain

Prof. COULLET, P., Lab. de la Mat. Condensée, Observatoire de Nice, 06003 France

Prof. CROSS, M.C., Caltech, Division of Physics, Pasadena, CA 91125, USA

Mr. CUESTA, J.A., Dpto. de Física Aplicada I, Universidad Complutense, 28040 Madrid, Spain

Prof. DERRIDA, B., SPhT, CEN-Saclay, F 91191 Gif-sur-Yvette, France

Mr. DIAZ, A., Dpto. de Física, UAB, 08193 Bellaterra, Barcelona, Spain

Mr. DIAZ-BORREGO, F., Dpto. de Física Teórica, Universidad de Sevilla, 41080 Spain

Prof. DICKMAN, R., Herbert H. Lehman College, C.U.N.Y., Bronx, NY 10468 USA

Prof. DRUGOWICH de FELICIO, J.R., Inst . de Física e Química de São Carlos, USP 13560 SP, Brasil

Prof. DUFTY, J.W., Dpt. of Physics, University of Florida, Gainesville, FL 32611, USA

Prof. EICKE, H.F., Inst. for Physical Chemistry, University of Basel, CH 4056 Basel, Switzerland

Prof. ELIZALDE, E., Facultad de Física, Universidad de Barcelona, 08028 Spain

Mr. ESPAÑOL, J., Dpto. de Física, UAB, 08193 Bellaterra, Barcelona, Spain

Prof. FERNANDEZ TEJERO, C., Dpto. de Física Aplicada I, Universidad Complutense, 28040 Madrid, Spain

Dr. FOGEDBY, H.C., Inst. of Physics, University of Aarhus, 8000 Aarhus C, Denmark

Prof. GARRIDO, L., Facultad de Física, Universidad de Barcelona, 08028 Spain

Dr. GARRIDO GALERA, P.L., Dpto. de Física Aplicada, Universidad de Granada, 18071, Spain

Dr. AGUADO GOMEZ, M., Dpto. de Física, Universidad de las Islas Baleares, Palma de Mallorca, Spain

Dr. GOBRON, T., Lab. de Physique de la Matière Condensée, Ecole Polytechnique, 91128 Palaiseau, France

Prof. GONZALEZ MIRANDA, J.M., Facultad de Física, Universidad de Barcelona, 08028, Spain

Prof. GOULART ROSA Jr., S., IFQSC, C.P. 369, 13560 São Carlos SP, Brasil

Prof. GUYON, E., Lab. d'Hydrodynamique et Mécanique Physique ESPCI, 10 Rue Vauquelin, 75231 Paris, France

Mr. HERNANDEZ GARCIA, E., Dpto. de Física, Universidad de las Islas Baleares, Palma de Mallorca, Spain

Prof. HERNANDEZ MACHADO, A., Facultad de Física, Universidad de Barcelona 08028 Barcelona, Spain

Prof. HIGUERA, F.J., ETSI Aeronáuticos, Universidad Politécnica de Madrid, 28040 Spain

Dr. ISOMÄKI, H., Main Building, Helsinki University of Technology, SF 02150 Espoo 15, Finland

Dr. JIMENEZ AQUINO, J.I., Dpto. de Física, Universidad de Barcelona, 08028 Spain

Prof. JASNOW, D., Dept. of Physics and Astronomy, University of Pittsburgh PA 15260, USA

Dr. JAUSLIN, H.R., Dept. Physique Théorique, Université de Genève, CH 1211 Genève, Switzerland

Prof. JOU, D., Dpto. de Física, UAB, Bellaterra 08193, Barcelona, Spain

Prof. LAGE, E.J.S., Lab. de Física, U.P. 4000 Porto, Portugal

Mrs. LEMOS, M.C., Dpto. de Termología, Facultad de Física, Universidad de Sevilla, 41080, Spain

Dr. LE BERRE, A., Lab. Photophysique Moléc. du CNRS, Université Paris-Sud, 91405 Orsay, France

Prof. LOPEZ de HARO, M., Instituto de Investigación en Materiales U.N.A.M., Ap.Postal 70-360, México 04510 D.F., México

Prof. LOPEZ BONILLA, L., Facultad de Física, Universidad de Sevilla 41080 Spain

Prof. LINDENBERG, K., Dept. of Chemistry, University of Calif. at San Diego, La Jolla, CA 92093, USA

Prof. LOPEZ LACOMBA, A., Dpto. de Física Aplicada, Universidad de Granada, 18071 , Spain

Prof. LUGIATO, L., Politecnico di Torino, Dipto. di Física 10129 Torino, Italy

Prof. LÜCKE, M., FB 11 der Universität des Saarlandes, 6000 Saarbrücken Germany

Dr. LUIS GONZALEZ, D., Dpto. de Física Fundamental y Experimental, Universidad de La Laguna, Tenerife, Spain

Prof. MANDEL, P., Université Libre de Bruxelles, Campus Plaine, CP 231, 1050 Bruxelles, Belgium

Prof. MARRO, J., Facultad de Ciencias, Universidad de Granada, 18071 Spain

Dr. MARTANO, P., Dept. of Physics, University of Lancaster, LA1 4YB, Great Britain

Dr. MARTINEZ MARDONES, J., Dpto. de Física, UAB, Bellaterra 08193, Barcelona, Spain

Prof. MAZENKO, G.F., The James Franck Institute, The University of Chicago, IL 60637, USA

Dr. MELROSE, J.R., Dept. of Physics, The Blackett Laboratory, Imperial College , London, Great Britain

Prof. MIKHAILOV, A.S., Dept. of Physics, Moscow State University, Moscow, USSR

Prof. NAUDTS, J., Dept. of Physics, University of Antwerpen UIA, B2610 Antwerpen, Belgium

Dr. NUNES da SILVA, J.M., Lab. de Física, Univeristy of Porto, 4000 Porto, Portugal

Prof. de OLIVEIRA, M.J., Instituto de Física, Universidade de São Paulo, CP 20516, 01498 São Paulo, Brasil

Mr. ORTS, J., Dpto. E.C.M., Universidad de Barcelona, 08028 Spain

Prof. de PASQUALE, F., Dept. of Physics, University of L'Aquila, 67100 Italy

Dr. PAVON, D., Dpto. de Física, UAB, 08193 Bellaterra, Barcelona, Spain

Mr. PEREZ CRUZ, J., Dpto. de Física, Fac. de Química, Universidad de La Laguna, Tenerife, Spain

Prof. PEREZ GARCIA, C., Dpto. de Física Estadística, UAB, 08193 Bellaterra, Barcelona, Spain

Dr. PEREZ MADRID, A., Dpto. de Física Estadística, UAB, 08193 Bellaterra Barcelona, Spain

Prof. PESQUERA, L., Dpto. de Física Moderna, Universidad de Santander, 39005 Spain

Prof. PLATTEN, J., Dept. of Thermodynamics, State University of Mons, B 7000 Mons, Belgium

Dr. RAMIREZ de la PISCINA, L., Dpto. de Física, Universidad de Barcelona 08028 Barcelona, Spain

Dr. RESSAYRE, E., Lab. Photophysique Moléculaire du CNRS, Université Paris-Sud, 91405 Orsay, France

Dr. RENZ, W., Institut für Festkörperforschung (IFF), KFA, Pf 1913, 5170 Jülich, Germany

Prof. RISKEN, Abt. für Theoretische Physik, Universität Ulm, Pf 4066, 7900 Ulm, Germany

Mr. RODRIGUEZ PARRONDO, J.M., Dpto. de Física Aplicada I, Universidad Complutense de Madrid, 28040, Spain

Dr. RODRIGUEZ DIAZ, M.A., Dpto. de Física Moderna, Universidad de Santander, 39005 Spain

Prof. RODRIGUEZ, J.R., Dpto. de Física de la Materia Condensada, Universidad de Santiago, 15703 Santiago de Compostela, Spain

Dr. RODRIGUEZ, R.F., Dpto. de Física, Universidad de las Islas Baleares, Palma de Mallorca, Spain

Mr. ROSELL, J., Facultad de Física, UAB, Bellaterra 08193, Barcelona, Spain

Prof. RUBÍ, J.M., Dpto. de Física Fundamental, Universidad de Barcelona, 08028 Spain

Prof. SAGUÉS, F., Facultad de Física, Universidad de Barcelona, 08028, Spain

Dr. SALAN, J., Dpto. E.C.M., Universidad de Barcelona, 08028, Spain

Dr. SALUEÑA, C., Universidad Politécnica de Barcelona, 08028, Spain

Prof. SANCHO, J.M., Facultad de Física, Universidad de Barcelona, 08028, Spain

Mr. SANCHEZ SANCHEZ, A., Facultad de Física, Universidad Complutense de Madrid, 28040 Spain

Prof. SAN MIGUEL, M., Facultad de Física, Universidad de las Islas Baleares, Palma de Mallorca, Spain

Dr. SANTOS, M.A., O.P., Dpto de Física, Universidad de Porto, 4000 Porto, Portugal

Prof. SANTOS, A., Dpto. de Física Teórica, Universidad de Sevilla, 41080 Spain

Prof. SERTORIO, L., Dept. of Theoretical Physics, University of Torino, 10125 Torino, Italy

Dr. SCHILLING, R., Inst. Physik der Universität Basel, CH 4056 Basel, Switzerland

Dr. TALLET, A., Lab. Photophysique Moléculaire du CNRS, Université Paris-Sud 91405 Orsay, France

Mrs. TORRENT, M.C., Facultad de Física, Universidad de Barcelona 08028 Spain

Prof. VAN DAEL, W., Molecular Physics, Catholic University of Leuwen, 3030 Leuwen, Belgium

Mr. E. VIVES, Dpt. Estructura i Constituents de la Matèria, Universitat de Barcelona, 08028 Spain

Dr. WIO, H., Facultad de Física, Universidad de las Islas Baleares, Palma de Mallorca, Spain

Mrs. ZEGHLACHE, H., Université Libre de Bruxelles, CP 231, 1050 Bruxelles Belgium

Lecture Notes in Mathematics

Vol. 1236: Stochastic Partial Differential Equations and Applications. Proceedings, 1985. Edited by G. Da Prato and L. Tubaro. V, 257 pages. 1987.

Vol. 1237: Rational Approximation and its Applications in Mathematics and Physics. Proceedings, 1985. Edited by J. Gilewicz, M. Pindor and W. Siemaszko. XII, 350 pages. 1987.

Vol. 1250: Stochastic Processes – Mathematics and Physics II. Proceedings 1985. Edited by S. Albeverio, Ph. Blanchard and L. Streit. VI, 359 pages. 1987.

Vol. 1251: Differential Geometric Methods in Mathematical Physics. Proceedings, 1985. Edited by P.L. García and A. Pérez-Rendón. VII, 300 pages. 1987.

Vol. 1255: Differential Geometry and Differential Equations. Proceedings, 1985. Edited by C. Gu, M. Berger and R.L. Bryant. XII, 243 pages. 1987.

Vol. 1256: Pseudo-Differential Operators. Proceedings, 1986. Edited by H.O. Cordes, B. Gramsch and H. Widom. X, 479 pages. 1987.

Vol. 1258: J. Weidmann, Spectral Theory of Ordinary Differential Operators. VI, 303 pages. 1987.

Vol. 1260: N.H. Pavel, Nonlinear Evolution Operators and Semigroups. VI, 285 pages. 1987.

Vol. 1263: V.L. Hansen (Ed.), Differential Geometry. Proceedings, 1985. XI, 288 pages. 1987.

Vol. 1265: W. Van Assche, Asymptotics for Orthogonal Polynomials. VI, 201 pages. 1987.

Vol. 1267: J. Lindenstrauss, V.D. Milman (Eds.), Geometrical Aspects of Functional Analysis. Seminar. VII, 212 pages. 1987.

Vol. 1269: M. Shiota, Nash Manifolds. VI, 223 pages. 1987.

Vol. 1270: C. Carasso, P.-A. Raviart, D. Serre (Eds.), Nonlinear Hyperbolic Problems. Proceedings, 1986. XV, 341 pages. 1987.

Vol. 1272: M.S. Livšic, L.L. Waksman, Commuting Nonselfadjoint Operators in Hilbert Space. III, 115 pages. 1987.

Vol. 1273: G.-M. Greuel, G. Trautmann (Eds.), Singularities, Representation of Algebras, and Vector Bundles. Proceedings, 1985. XIV, 383 pages. 1987.

Vol. 1275: C.A. Berenstein (Ed.), Complex Analysis I. Proceedings, 1985–86. XV, 331 pages. 1987.

Vol. 1276: C.A. Berenstein (Ed.), Complex Analysis II. Proceedings, 1985–86. IX, 320 pages. 1987.

Vol. 1277: C.A. Berenstein (Ed.), Complex Analysis III. Proceedings, 1985–86. X, 350 pages. 1987.

Vol. 1283: S. Mardešić, J. Segal (Eds.), Geometric Topology and Shape Theory. Proceedings, 1986. V, 261 pages. 1987.

Vol. 1285: I.W. Knowles, Y. Saitō (Eds.), Differential Equations and Mathematical Physics. Proceedings, 1986. XVI, 499 pages. 1987.

Vol. 1287: E.B. Saff (Ed.), Approximation Theory, Tampa. Proceedings, 1985–1986. V, 228 pages. 1987.

Vol. 1288: Yu. L. Rodin, Generalized Analytic Functions on Riemann Surfaces. V, 128 pages, 1987.

Vol. 1294: M. Queffélec, Substitution Dynamical Systems – Spectral Analysis. XIII, 240 pages. 1987.

Vol. 1299: S. Watanabe, Yu.V. Prokhorov (Eds.), Probability Theory and Mathematical Statistics. Proceedings, 1986. VIII, 589 pages. 1988.

Vol. 1300: G.B. Seligman, Constructions of Lie Algebras and their Modules. VI, 190 pages. 1988.

Vol. 1302: M. Cwikel, J. Peetre, Y. Sagher, H. Wallin (Eds.), Function Spaces and Applications. Proceedings, 1986. VI, 445 pages. 1988.

Vol. 1303: L. Accardi, W. von Waldenfels (Eds.), Quantum Probability and Applications III. Proceedings, 1987. VI, 373 pages. 1988.

Lecture Notes in Physics

Vol. 297: A. Lawrence (Ed.), Comets to Cosmology. X, 415 pages. 1988.

Vol. 298: M. Georgiev, F' Centers in Alkali Halides. XI, 282 pages. 1988.

Vol. 299: J.D. Buckmaster, T. Takeno (Eds.), Mathematical Modeling in Combustion Science. Proceedings, 1987. VI, 168 pages. 1988.

Vol. 300: B.-G. Englert, Semiclassical Theory of Atoms. VII, 401 pages. 1988.

Vol. 301: G. Ferenczi, F. Beleznay (Eds.), New Developments in Semiconductor Physics. Proceedings, 1987. VI, 302 pages. 1988.

Vol. 302: F. Gieres, Geometry of Supersymmetric Gauge Theories. VIII, 189 pages. 1988.

Vol. 303: P. Breitenlohner, D. Maison, K. Sibold (Eds.), Renormalization of Quantum Field Theories with Non-linear Field Transformations. Proceedings, 1987. VI, 239 pages. 1988.

Vol. 304: R. Prud'homme, Fluides hétérogènes et réactifs: écoulements et transferts. VIII, 239 pages. 1988.

Vol. 305: K. Nomoto (Ed.), Atmospheric Diagnostics of Stellar Evolution: Chemical Peculiarity, Mass Loss, and Explosion. Proceedings, 1987. XIV, 468 pages. 1988.

Vol. 306: L. Blitz, F.J. Lockman (Eds.), The Outer Galaxy. Proceedings, 1987. IX, 291 pages. 1988.

Vol. 307: H.R. Miller, P.J. Wiita (Eds.), Active Galactic Nuclei. Proceedings, 1987, XI, 438 pages. 1988.

Vol. 308: H. Bacry, Localizability and Space in Quantum Physics. VII, 81 pages. 1988.

Vol. 309: P.E. Wagner, G. Vali (Eds.), Atmospheric Aerosols and Nucleation. Proceedings, 1988. XVIII, 729 pages. 1988.

Vol. 310: W.C. Seitter, H.W. Duerbeck, M. Tacke (Eds.), Large-Scale Structures in the Universe – Observational and Analytical Methods. Proceedings, 1987. II, 335 pages. 1988.

Vol. 311: P.J.M. Bongaarts, R. Martini (Eds.), Complex Differential Geometry and Supermanifolds in Strings and Fields. Proceedings, 1987. V, 252 pages. 1988.

Vol. 312: J.S. Feldman, Th.R. Hurd, L. Rosen, "QED: A Proof of Renormalizability." VII, 176 pages. 1988.

Vol. 313: H.-D. Doebner, T.D. Palev, J.D. Hennig (Eds.), Group Theoretical Methods in Physics. Proceedings, 1987. XI, 599 pages. 1988.

Vol. 314: L. Peliti, A. Vulpiani (Eds.), Measures of Complexity. Proceedings, 1987. VII, 150 pages. 1988.

Vol. 315: R.L. Dickman, R.L. Snell, J.S. Young (Eds.), Molecular Clouds in the Milky Way and External Galaxies. Proceedings, 1987. XVI, 475 pages. 1988.

Vol. 316: W. Kundt (Ed.), Supernova Shells and Their Birth Events. Proceedings, 1988. VIII, 253 pages. 1988.

Vol. 317: C. Signorini, S. Skorka, P. Spolaore, A. Vitturi (Eds.), Heavy Ion Interactions Around the Coulomb Barrier. Proceedings, 1988. X, 329 pages. 1988.

Vol. 319: L. Garrido (Ed.), Far from Equilibrium Phase Transitions. Proceedings, 1988. VIII, 340 pages. 1988.